I0044385

Renewable Energy: Sustainable Energy for Next Generation

Renewable Energy: Sustainable Energy for Next Generation

Edited by Kurt Marcel

SYRAWOOD
PUBLISHING HOUSE

New York

Published by Syrawood Publishing House,
750 Third Avenue, 9th Floor,
New York, NY 10017, USA
www.syrawoodpublishinghouse.com

Renewable Energy: Sustainable Energy for Next Generation
Edited by Kurt Marcel

© 2019 Syrawood Publishing House

International Standard Book Number: 978-1-68286-717-4 (Hardback)

This book contains information obtained from authentic and highly regarded sources. Copyright for all individual chapters remain with the respective authors as indicated. All chapters are published with permission under the Creative Commons Attribution License or equivalent. A wide variety of references are listed. Permission and sources are indicated; for detailed attributions, please refer to the permissions page and list of contributors. Reasonable efforts have been made to publish reliable data and information, but the authors, editors and publisher cannot assume any responsibility for the validity of all materials or the consequences of their use.

Trademark Notice: Registered trademark of products or corporate names are used only for explanation and identification without intent to infringe.

Cataloging-in-Publication Data

Renewable energy : sustainable energy for next generation / edited by Kurt Marcel.
    p. cm.
Includes bibliographical references and index.
ISBN 978-1-68286-717-4
1. Renewable energy sources. 2. Renewable natural resources. I. Marcel, Kurt.
TJ808 .R46 2019
333.794--dc23

TABLE OF CONTENTS

Permissions

List of Contributors

Index

PREFACE

This book was inspired by the evolution of our times; to answer the curiosity of inquisitive minds. Many developments have occurred across the globe in the recent past which has transformed the progress in the field.

Renewable energy refers to the energy that is naturally replenished with time. Some of the sources of renewable energy include sunlight, wind, tides, rain, etc. Adoption of renewable energy has resulted in increased energy security, climate change mitigation and economic benefits. The availability of energy is a major driver of human development. Renewable energy helps in facilitating this even in rural and remote areas of the world. Research is being done to develop newer and better energy sources like algal fuel or algal-derived biomass and artificial photosynthesis for energy storage using nanotechnology. This book presents the important theories and concepts that are central to the development of the technologies of renewable energy production. It further elucidates the cutting-edge technological innovations of recent years that have revolutionized this field. Students and researchers will benefit alike from an in-depth study of this book.

This book was developed from a mere concept to drafts to chapters and finally compiled together as a complete text to benefit the readers across all nations. To ensure the quality of the content we instilled two significant steps in our procedure. The first was to appoint an editorial team that would verify the data and statistics provided in the book and also select the most appropriate and valuable contributions from the plentiful contributions we received from authors worldwide. The next step was to appoint an expert of the topic as the Editor-in-Chief, who would head the project and finally make the necessary amendments and modifications to make the text reader-friendly. I was then commissioned to examine all the material to present the topics in the most comprehensible and productive format.

I would like to take this opportunity to thank all the contributing authors who were supportive enough to contribute their time and knowledge to this project. I also wish to convey my regards to my family who have been extremely supportive during the entire project.

Editor

Renewable energy auctions in South Africa outshine feed-in tariffs

Anton Eberhard[1] & Tomas Kåberger[2]

[1]Graduate School of Business, University of Cape Town, Private Bag X3, Rondebosch 7701, South Africa
[2]Energy and Environment, Chalmers University of Technology, SE-412 96 Gothenburg, Sweden

Keywords
Auctions, competitive tenders, prices, renewable energy, South Africa

Correspondence
Anton Eberhard, Graduate School of Business, University of Cape Town, Private Bag X3, Rondebosch 7701, South Africa.
E-mail: eberhard@gsb.uct.ac.za

Funding Information
No funding information provided.

Abstract

South Africa's Renewable Energy Independent Power Producer Procurement Program has run four competitive tenders/auctions since 2011, which have seen US$19 billion in private investment, and electricity prices of wind power falling by 46% and solar PV electricity prices by 71%, in nominal terms. Competitive tenders were introduced after an unsuccessful attempt to implement feed- in tariffs. The tenders incorporated standard, nonnegotiable contract documents, including 20- year Power Purchase Agreements and an Implementation Agree-ment whereby the Government of South Africa back- stops IPP payments by the national utility, Eskom. All of these projects have reached financial close to date and some are already delivering power to the grid. The financing success has been due in part to the requirements for commercial banks to undertake a thorough due diligence of projects prior to bids being offered. The details of the policy package described may be useful for other policy makers in countries developing policies for renewable energy deployment.

Introduction

Costs of renewable energy technologies have fallen during recent years. The cost reductions are the result of many different factors, some related to technologies, others to finance, national institutional development, and increased competition. These developments would not have been possible without generous supporting policies in some pioneering countries, as suggested by Wene [1].

Developments in different countries may vary significantly depending on policy, legislation and regulatory frameworks, procurement practices, and market conditions. There are significant opportunities for learning lessons from countries that are achieving price reductions [2]. South Africa offers an interesting example of a renewable energy auctioning system that is attracting significant investment at highly competitive prices. The data and analysis in this article draw, in part, from an earlier study by Eberhard et al. [3] and have been updated with the latest data extracted from the South African Department of Energy's Independent Power Producer Office.

The latest grid-connected renewable energy auctions in South Africa have seen prices fall to among the lowest in the world with solar PV prices as low as USc 6.4/kWh and the cheapest wind at USc 4.7/kWh.[1] Over four bid rounds, between 2012 and 2015, wind energy has fallen by 46% and solar PV by 71% (in nominal, local currency terms) (Fig. 1).

South Africa occupies a central position in the global debate regarding the most effective policy instruments to accelerate and sustain private investment in renewable energy.

To date, a total of 92 projects have been contracted and private sector investment totaling US$19 billion has been committed for projects totaling 6327 megawatt (MW). There have also been notable economic development commitments in the form of local manufacture, employment creation, black economic empowerment, and community development [4]. Important lessons can be learned for both South Africa and other emerging markets contemplating investments in renewables and other critical infrastructure investments [3].

South African Renewable Energy Programme		
Bid round	Wind ZARc/kWh	PV ZARc/kWh
Round 1	114	276
Round 2	90	165
Round 3	74	99
Round 4	62	79
Round 4		
Cheapest ZARc/kWh	56	77
Average USc/kWh	5.2	6.6
Cheapest USc/kWh	4.7	6.4

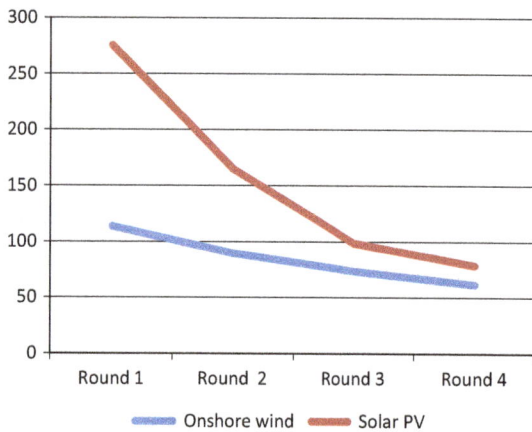

Figure 1. Average nominal bid prices in South Africa's renewable energy IPP program (ZARc/kWh). Source: Authors' compilation based on data provided by South Africa's DOE IPP Office.

From REFIT to REIPPPP

A REFIT policy was approved in 2009 by the national energy regulator of South Africa, NERSA. Tariffs were designed to cover generation costs plus a real, after tax return on equity of 17% that would be fully indexed for inflation. Initial published feed-in tariffs were generally regarded as generous by developers – 15.6 USc/kWh for wind, 26 USc/kWh for solar PV, and 49 USc/kWh for concentrated solar (troughs, with 6 h storage).[2] But considerable uncertainty about the nature of the procurement and licensing process remained. And the national utility, Eskom was less than enthusiastic in fully supporting the REFIT program by concluding power purchase agreements and interconnection agreements.

In March 2011, NERSA introduced a new level of uncertainty with a surprise release of a consultation paper calling for lower feed-in tariffs, arguing that a number of parameters, such as exchange rates and the cost of debt had changed. The new tariffs were 25% lower for wind, 13% lower for concentrated solar, and 41% lower for photovoltaic. Moreover, the capital component of the tariffs would no longer be fully indexed for inflation.

Importantly, in its revised financial assumptions, NERSA did not change the required real return for equity investors of 17% [3].

More policy and regulatory uncertainty was to come. Already concerned that NERSA's FITs were still too high, the Department of Energy and National Treasury commissioned the legal opinion that concluded that the feed-in tariffs amounted to noncompetitive procurement, and were therefore prohibited by the government's public finance and procurement regulations. The Department of Energy and National Treasury then took the lead on a reconsideration of the government's approach. The fundamental goal of achieving large-scale renewable energy projects with private developers and financiers remained the same. However, the structure of the transactions, including the feed-in tariffs, was to change significantly.

A series of informal consultations were held with developers, lawyers, and financial institutions throughout the first half of 2011. These meetings proved to be extremely important in terms of allaying market concerns resulting from the earlier REFIT process and providing informal feedback from the private sector on design, legal, and technology issues.

In August 2011, the Department of Energy (DOE) announced that a competitive bidding process for renewable energy would be launched, known as the Renewable Energy Independent Power Procurement Program. Subsequently, NERSA officially terminated the REFITs. Not a single megawatt of power had been signed in the 2 years since the launch of the REFIT program as a practical procurement process was never implemented, and the required contracts were never negotiated or signed. The abandonment of feed-in tariffs was met with dismay by a number of renewable energy project developers that had secured sites and initiated resource measurements and environmental impact assessments. But, it was these early developers who would later benefit from the first round of competitive bidding under REIPPPP.

Competitive tenders

In August 2011, a Request for Proposals was issued, and the next month a compulsory bidder's conference was held to address questions on bid requirements, documentation, power purchase agreements, etc. Some 300 organizations attended this conference. The REIPPPP program initially envisioned the procurement of 3625 MW of power over a maximum of five tender rounds. Another 100 MW was reserved for small projects below 5 MW that were procured in a separate small projects IPP program. Caps were set on the total capacity to be procured for individual technologies, the largest allocations were for wind and solar photovoltaics, with smaller amounts for concentrated solar, biomass, biogas, landfill gas, and hydro. The rationale for these caps was to limit the supply to be bid out, and therefore increased the level of competition among the different technologies and potential bidders.

The tenders for different technologies were held simultaneously. Interested parties could bid for more than one project and more than one technology. Projects had to be larger than 1 MW, and an upper limit was set on bids for different technologies, for example, 75 MW for a photovoltaic project, 100 MW for a concentrated solar project, and 140 MW for a wind project. Caps were also set on the price for each technology (at levels not dissimilar to NERSA's 2009 REFITs). Bids were due within 3 months of the release of the RFP, and financial close was to take place within 6 months after the announcement of preferred bidders.

The RPF was divided into three sections detailing: 1) general requirements, 2) qualification criteria, and 3) evaluation criteria. The documents also included a standard Power Purchase Agreement (PPA), an Implementation Agreement (IA), and a Direct Agreements (DA). The PPA was to be signed by the IPP and the Eskom, the off-taker. The PPAs specified that the transactions should be denominated in South Africa Rand and that contracts would have 20 year tenures from Commercial Operation Date (COD). The IAs were to be signed by the IPPs and the DOE and effectively provided a sovereign guarantee of payment to the IPPs, by requiring the DOE to make good on these payments in the event of an Eskom default. The IA also placed obligations on the IPP to deliver economic development targets. The DAs provided step-in rights for lenders in the event of default. The PPA, the IA and the DA were nonnegotiable contracts and were developed after an extensive review of global best practices and consultations with numerous public and private sector actors. Despite some bidder reservations regarding the lack of flexibility to negotiate the terms of the various agreements, the overall thoroughness and quality of the standard documents seemed to satisfy most of the bidders participating in the three rounds.

Bids were required to contain information on the project structure, legal qualifications, land, environmental, financial, technical, and economic development qualifications.

An important element of the design of the procurement process was to maximize the likelihood that winning bidders would able to execute the projects. Bidders had to submit bank letters indicating that the financing was locked-in highly unusual and basically a way to outsource due diligence to the banks. Effectively this meant that lenders took on a higher share of project development risk and this arrangement dealt with the biggest problem with auctions – the "low-balling" that results in deals not closing.

Further, the developers were expected to identify the sites and pay for early development costs at their own risk. A registration fee of US$1875 was due at the outset of the program. Bid bonds or guarantees had to be posted, equivalent to US$12,500 per megawatt of nameplate capacity of the proposed facilities, and the amount was doubled once preferred bidder status was announced. The guarantees are to be released once the projects come on line or if the bidder was unsuccessful after the RFP evaluation stage.

Project selection was based on a 70/30 split between price and economic development considerations. REIPPPP was able to adjust the normal government 90/10 split favoring price considerations in the procurement selection process. An exemption was obtained from the Public Preferential Procurement Framework Act in order to maximize economic development objectives.

The DOE IPP unit used a group of international and local experts to assess the bids. Many of these advisors had been involved in the initial design process. Given the scale of the investments, the competition anticipated, and the reputational risk identified, security, and confidentiality surrounding the evaluation process was extremely tight with 24-h voice and CCTV monitoring of the venue. Approximately 130–150 local and international advisors were used to develop the RFP and evaluate the bids in the first round, at a total cost of approximately US$10 million.

The bid evaluation involved a two-step process. First, bidders had to satisfy certain minimum threshold requirements in six areas: environment, land, commercial and legal, economic development, financial, and technical. For example, the environmental review examined approvals, while the land review looked at tenure, lease registration, and proof of land use applications. Commercial considerations included the project structure and the bidders' acceptance of the Power Purchase Agreement. The financial review included standard templates used for data collection that were linked to a financial model used by the evaluators. The technical specifications were set for each of the technologies. For example, wind developers were required to provide 12 months of wind data for the designated site and an independently verified generation forecast. The economic development requirements, in particular, were complex and generated some confusion among bidders.

Bids that satisfied the threshold requirements then proceeded to the second step of evaluation, where bid prices counted for 70% of the total score, with the remaining 30% of the score given to a composite score covering job creation, local content, ownership, management control, preferential procurement, enterprise development, and socioeconomic development. Bidders were asked to provide two prices: one fully indexed for inflation and the other partially indexed, with the bidders initially allowed to determine the proportion that would be indexed. In subsequent rounds, floors and caps were instituted for the proportion that could be indexed. The bids were evaluated using a standard financial model.

In the first round, 53 bids for 2128 MW of power-generating capacity were received. Ultimately 28 preferred bidders were selected offering 1416 MW for a total investment of nearly US$6 billion. Successful bidders realized that not enough projects were ready to meet the bid qualification criteria and that all qualifying bids were thus likely to be awarded contracts. Bid prices in the first round were thus close to the price caps set in the tender documents. Major contractual agreements were signed on November 5, 2012, with most projects reaching full financial close shortly thereafter. Construction on all of these projects has commenced with the first project coming on line in November 2013.

A second round of bidding was announced in November 2011. The total amount of power to be acquired was reduced, and other changes were made to tighten the procurement process and increase competition. Seventy-nine bids for 3233 MW were received in March 2012, and 19 bids were ultimately selected. Prices were more competitive, and bidders also offered better local content terms. Implementation, power purchase, and direct agreements were signed for all 19 projects in May 2013.

A third round of bidding commenced in May 2013, and again, the total capacity offered was restricted. In August 2013, 93 bids were received totaling 6023 MW. Seventeen preferred bidders were notified in October 2013 totaling 1456 MW. Prices fell further in round three. Local content again increased, and financial closure was expected in July 2014, but has been delayed a number of times because of uncertainties around Eskom transmission connections. A fourth round of bidding commenced in August 2014; 77 bids were received with 64 being compliant and 13 preferred bidders were announced in April 2015, totaling 1121 MW. Prices were so competitive that a further 13 projects were awarded totaling 1084.

Over the four bidding rounds, US$19 billion has been invested in 92 projects totaling 6327 MW.

Increased competition was no doubt the main driver for prices falling over the bidding rounds. But, there were other factors as well. International prices for renewable energy equipment have declined over the past few years due to a glut in manufacturing capacity, as well as ongoing innovation and economies of scale. REIPPPP was well positioned to capitalize on these global factors. Transaction costs were also lower in subsequent rounds, as many of the project sponsors and lenders became familiar with the REIPPPP tender specifications and process.

Now RE prices are reaching grid parity and there is the potential for other countries to explore how they can learn from the SA REIPPPP through lowering transaction costs and designing competitive tenders appropriate to local markets.

Conclusion

Over the past 4 years, South Africa's Renewable Energy Independent Power Producer Procurement Program has delivered remarkable investment and price outcomes which offer lessons for other countries on the potential benefits of competitive tenders or auctions.

Notes

[1] Prices fully indexed with inflation. ZAR/USD exchange deteriorated from 8 to 12 over period.
[2] These values are calculated at the exchange rate at the time of ZAR8/USD.

Conflict of interest

None declared.

References

1. Wene, C.-O., (2000). Experience Curves for Energy Technology Policy. OECD/IEA Paris. http://www.oecd-ilibrary.org/energy/experience-curves-for-energy-technology-policy_9789264182165-en

2. IRENA. 2015. Renewable energy auctions: a guide to design. International Renewable Energy Agency, Vienna.

3. Eberhard, A., J. Kolker, and J. Leigland. 2014. South Africa's renewable energy IPP procurement programme: success factors and lessons. Public Private Infrastructure Advisory Facility, World Bank, Washington, DC.

4. Baker, L., and H. L. Wlokas. 2015. South Africa's renewable energy procurement: a new frontier? Energy Research Centre. University of Cape Town, Cape Town, South Africa.

Study of thermal behavior of deoiled karanja seed cake biomass: thermogravimetric analysis and pyrolysis kinetics

Radhakumari Muktham[1,2], Andrew S. Ball[2], Suresh K. Bhargava[2] & Satyavathi Bankupalli[1]

[1]Chemical Engineering Division, CSIR – Indian Institute of Chemical Technology, Hyderabad 500007, India
[2]School of Applied Sciences, Royal Melbourne Institute of Technology, Melbourne 3083, Australia

Keywords
Biochar, karanja cake, kinetics, pyrolysis, TGA

Correspondence
Satyavathi Bankupalli, Chemical Engineering Division, CSIR – Indian Institute of Chemical Technology, Hyderabad, 500007, India.
E-mail: satya@iict.res.in

Funding Information
This study was supported by the CSIR-IICT.

Abstract

Karanja is a medium sized evergreen tree which has minor economic importance in India. The nonedible seed kernel contains 27–30% oil that is used for biodiesel production, leaving the remaining nonedible seed cake as a waste product. The aim of the present work was to obtain kinetic parameters in relation to technological parameters in nonedible seed cake biomass pyrolysis conversion process to bio-oil and biochar. Effects of heating rate on karanja seed cake slow pyrolysis behavior and kinetic parameters were investigated at heating rates of 5, 10, and 20°C/min using thermogravimetric analysis (TGA). Thermogravimetric experiments showed the onset and offset temperatures of the devolatilization step shifted toward the high-temperature range, and the activation energy values increased with increasing heating rate. In the present study, isoconversional method was applied for the pyrolysis of karanja seed cake biomass by TGA and the activation energies (118–124 kJ/mol) and the pre-exponential factors obtained using progressive conversion. Proximate–ultimate analyses, energy value, surface structure, and Fourier transform infrared spectra of the biomass processed under conditions were reported. The pyrolysis resulted in upgradation of the energy value of seed cake biomass from 18.1 to 24.5 MJ/kg; importantly with high carbon and low oxygen contents. The approach represents a novel method for the upgrading of karanja seed cake that has significant commercial potential.

Introduction

With the growing concerns about fossil fuel exhaustion and environmental problems such as global warming, efforts are underway to decrease CO_2 emission rates and to develop alternative energies which substitute for limited fossil fuels. Biodegradable nontoxic biomass having low emission profiles compared to petroleum-based fuels (petrol and diesel) is now widely recognizing as a potential energy source.

Biomass pyrolysis is a thermal decomposition of biomass taking place in the absence of oxidizing agent with products biochar, bio-oil, and gases such as carbon dioxide, carbon monoxide, hydrogen, and methane. Pyrolysis is a promising technology for the production of bio-oil and biochar. Bio-oil, also known as pyrolytic oil, can be upgraded to light hydrocarbons which contain low levels of aromatics with the absence of sulfur compared to petroleum-based fuels. Biochar, on the other hand, is an ecofriendly carbon-rich product from the pyrolysis of renewable feedstocks like nonedible seed cakes after the extraction of oil and has been widely used in agriculture as a soil amendment and for improving soil fertility due to its carbon sequestration ability [1]. Biochar is also reported to promote nitrogen fixation thereby decreasing the emission of N_2O [2] and other greenhouse gases from agricultural soils [3, 4]. In addition, biochar has been used as an adsorbent in the removal of arsenic [5], cadmium [6], and chromium from aqueous solutions.

Pyrolysis process occurs with high heat flux to the biomass with a corresponding high heating rate of the biomass particle and heat transfer to the biomass in a very short time period. Applying heat to biomass at high rate will result into smaller fragments (volatiles) due to the cleavage of cellulose, hemicellulose, protein, and lignin, and other constituents of biomass, unstable above 400°C due to the presence of oxygen in the fragments. Hence, very short residence times are required in the thermal treatment step followed by immediate quenching to impede secondary chemical change of unstable volatiles. Pyrolysis with a short residence time and optimum temperature of exposure with immediate cooling of the vapors can be achieved by taking into account the appropriate size of the reactor and heating rate in terms of heat flux. Heat transfer rate to biomass particles and residence time of vapors in the reactor have a strong influence on the nature and distribution of pyrolysis reaction products. Therefore, the first and foremost consideration for designing biomass pyrolysis reactor (residence time, size, and material of construction) is to draw the clear picture of thermal behavior of the feedstock biomass with the application of heat at different heating rate which corresponds to heat transfer rate to biomass particle in order to estimate the amount of energy to be supplied for pyrolysis.

Various biomass feedstocks show different thermal behavior profiles mainly due to the variation in the composition of biomass constituents. For instance, Parthasarathy and Sheeba [7] have investigated the thermal behavior of bagasse, coir pith, groundnut shell, and casuarinas leaves, which exhibited different degradation temperatures of individual biomass components (hemicellulose, cellulose, and lignin).

Vhathvarothai et al. [8] have studied the copyrolysis of cypress wood chips, macadamia nut shells biomass, and coal using thermogravimetric technique and Kissinger's corrected kinetic equation for kinetic parameters determination and concluding that the activation energy of coal is greater than that of both types of biomass.

Pyrolysis kinetics of raw and hydrothermally carbonized loblolly pine lignocellulosic biomass, performed in a thermogravimetric analyzer, was carried out by Yan et al. [9]. The study revealed that pyrolysis of hydrothermally carbonized biomass progressed less aggressively than that of raw biomass which also showed that after an initial significant decomposition, the pyrolysis reactions continued at much slower rates with temperature.

The current research interest is to create a platform for deoiled karanja seed cake biomass in biorefinery by producing energy and value-added products. The study is aimed at thermal behavior of the biomass under pyrolysis conditions which can be used in the design and development of an effective pyrolysis process. Recent research [10] has indicated the potential of bio-oil and biochar obtained from pyrolysis of the seed cake biomass

in areas such as energy production and the production of activated carbon.

Pongamia pinnata or Pongamia glabra known as the karanja tree in India is an evergreen tree of south and Southeast Asia. The Indian Government has initiated karanja tree plantation project in the year 2003 since then 20 million trees have been planted resulting into an annual production of 800 metric tons of karanja seeds in the year 2008 and was increased to 2,00,000 metric tons by 2011 [11]. National oilseeds and vegetable oils Development Board, Ministry of Agriculture, Government of India (2008) has showed the statistics of karanja seed kernals production per one hectare in the range 1958–4405.5 kg kernals/hectare. Nonedible seed oils (30% of entire seed weight) have been successfully processed for biodiesel production (transesterification), the remaining 70% of seed material and deoiled seed cake can be processed by a variety of treatment methods (hydrolysis and fermentation, pyrolysis, and gasification) for energy and chemicals production, which gives an added economic value to the seed cake biorefinery [12].

The aim of the present work was to study the thermal behavior and pyrolysis kinetics of the nonedible karanja seed cake with different temperature programs using thermogravimetric technique. Characteristic properties including proximate–ultimate analyses, surface structure, surface active functional groups, and higher heating value (HHV) of the raw and thermally treated biomass in a fixed bed pyrolysis reactor are evaluated. The present work is the first report on the use of the nonisothermal methods including model-fitting and model-free methods to calculate the kinetic parameters for the pyrolysis of karanja seed cake.

Material and Methods

Materials

Deoiled karanja seed cake used in the present work was obtained from a local producer in Andhra Pradesh, India. The biomass was processed to an average particle size of <1 mm and used as such for thermogravimetric analysis (TGA).

Thermogravimetric analysis

TGA of karanja seed cake was performed in a Mettler Toledo TGA/SDTA 851^e analyzer (Mettler Toledo, Switzerland). The analysis of the samples was performed at three different heating rates, 5, 10, and 20°C/min, using nitrogen with a flow rate of 30 mL/min to a final temperature of 800°C with a sample mass of ~11 ± 0.1 mg. The loss in the weight of the sample was recorded as a function of temperature.

Kinetic studies

Karanja pyrolysis kinetics was studied using the nonisothermal data obtained from different temperature programs at constant heating rate. International Confederation for Thermal Analysis and Calorimetry (ICTAC) kinetics committee recommendations for performing kinetic computations on thermal analysis data were followed to evaluate all the reliable kinetic parameters. Mathematical analysis to determine the kinetic triplet, activation energy (E), frequency factor (A), and reaction model $f(\alpha)$, was performed by the method of Coats and Redfern (model-fitting approach), and model-free isoconversional Kissinger–Akahira–Sunose (KAS) and Ozawa–Flynn–Wall (OFW) methods [13].

The conversion of biomass in the pyrolysis process is represented by the following reaction scheme:

$$\text{Biomass} \rightarrow \text{Biochar} + \text{Volatile}$$

and the kinetic expression with temperature dependence of the kinetic rate constant, given by Arrhenius equation as:

$$\frac{d\alpha}{dt} = k(T)f(\alpha) \qquad (1)$$

$$k(T) = A \exp\left(\frac{-E}{RT}\right) \qquad (2)$$

where $f(\alpha)$ is the reaction model, and it depends on the reaction mechanism, $k(T)$ is the rate constant, A is frequency factor, E the activation energy, T the absolute temperature, t the time, α the degree of conversion, and R the universal gas constant (8.314 J mol^{-1} K^{-1}). The constant heating rate is defined as:

$$\text{Heating rate, } \beta = \frac{dT}{dt} = \text{Constant} \qquad (3)$$

$$dt = \frac{dT}{\beta} \qquad (4)$$

Substituting $k(T)$ from equation (2) and dt from equation (4) in equation (1) and after rearrangement, the integration of equation (1) is given as:

$$\int_0^\alpha \frac{d\alpha}{f(\alpha)} \int_0^T A \exp\left(\frac{-E}{RT}\right) \frac{dT}{\beta} = g(\alpha) \qquad (5)$$

Model-fitting approach

The solution of equation (5) based on the approximation given in Coats and Redfern [14] gives rise to the equation form expressed as Coats–Redfern equation, the most frequently used expression to evaluate nonisothermal data to compute kinetic parameters is:

$$\ln\left(\frac{g(\alpha)}{T^2}\right) = \ln\left(\frac{AR}{\beta E}\right)\left(1 - \frac{2RT}{E}\right) - \frac{E}{RT} \qquad (6)$$

Conversion (α) in the pyrolysis of karanja seed cake is calculated from TGA data using the equation given below:

$$\text{Conversion, } \alpha = \frac{m_i - m_\alpha}{m_i - m_f} \qquad (7)$$

where m_i, m_α, and m_f are initial, instantaneous, and final mass of the samples, respectively.

The data obtained from TGA were analyzed by different kinetic models representing chemical reaction, nucleation and nuclei growth, surface reaction between both the phases, and diffusion models [13, 15]. The TGA data in the pyrolysis temperature range were analyzed by applying 14 kinetic models given in Table 1 using Coats–Redfern equation. The kinetic model which resulted in a

Table 1. Kinetic models used in the solid state reactions.

Model	Mechanism	$f(\alpha)$	$g(\alpha)$
Chemical reaction			
1	First order	$1 - \alpha$	$[-\ln(1-\alpha)]$
2	Second order	$(1-\alpha)^2$	$(1-\alpha)^{-1} - 1$
3	Third order	$(1-\alpha)^3$	$[(1-\alpha)^{-2} - 1]/2$
4	nth order	$(1-\alpha)^n$	$[(1-\alpha)^{1-n} - 1]/n - 1$
Random nucleation and nuclei growth			
5	Two-dimensional	$2(1-\alpha)[-\ln(1-\alpha)]^{1/2}$	$[-\ln(1-\alpha)]^{1/2}$
6	Three-dimensional	$3(1-\alpha)[-\ln(1-\alpha)]^{2/3}$	$[-\ln(1-\alpha)]^{1/3}$
Limiting surface reaction between both phases			
7	One dimension	1	α
8	Two dimensions	$2(1-\alpha)^{1/2}$	$1 - (1-\alpha)^{1/2}$
9	Three dimensions	$3(1-\alpha)^{2/3}$	$1 - (1-\alpha)^{1/3}$
Diffusion			
10	One-way transport	$1/2\alpha$	α^2
11	Two-way transport	$[-\ln(1-\alpha)]^{-1}$	$\alpha + (1-\alpha)\ln(1-\alpha)$
12	Three-way transport	$(2/3)(1-\alpha)^{2/3}/[1-(1-\alpha)^{1/3}]$	$[1-(1-\alpha)^{1/3}]^2$
13	Ginstling–Brounshtein equation	$(2/3)(1-\alpha)^{1/3}/[1-(1-\alpha)^{1/3}]$	$1 - 2\alpha/3 - (1-\alpha)^{2/3}$
14	Zhuravlev equation	$(2/3)(1-\alpha)^{5/3}/[1-(1-\alpha)^{1/3}]$	$[(1-\alpha)^{-1/3} - 1]^2$

straight line fit with high regression coefficient for ln $[g(\alpha)/T^2]$ versus $1/T$ data was concluded to be the best reaction model representing the experimental data.

Model-free approach (isoconversional methods)

A number of isoconversional methods are available for evaluating the biomass pyrolysis kinetics. These methods mainly differ in the approximations of the temperature integral in equation (5) and many of them give rise to linear equations. In equation (5) after introducing the following terms:

$$\text{Let } \frac{E}{RT} = p$$

$$dp = \frac{-E}{R}T^{-2}dT$$

$$\text{As } T \to 0; p \to \infty \text{ and } T \to T; p \to p$$

and by substituting $E/RT = p$ on rearrangement the equation is transformed to:

$$g(\alpha) = \int_0^\alpha \frac{d\alpha}{f(\alpha)} = \frac{A}{\beta}\frac{E}{R}\int_p^\infty p^{-2}\exp(-p)dp \quad (8)$$

KAS method takes the following approximation for the integral in right-side term of equation (8):

$$\int_p^\infty p^{-2}\exp(-p)dp = p^{-2}\exp(-p) \quad (9)$$

The solution of equation (9) based on the approximation and suitable modification given as equation (10) was used to determine the activation energy and frequency factor.

$$\ln\frac{\beta}{T^2} = \ln\left(\frac{AE}{Rg(\alpha)}\right) - \frac{E}{RT} \quad (10)$$

OFW method applies Doyle's approximation [16] for the integral term on the right side of equation (8), which becomes

$$\log\left[\int_p^\infty p^{-2}\exp(-p)dp\right] = -2.315 + 0.457p \quad (11)$$

By substituting equation (11) in equation (8) and on rearrangement equation (12) was obtained and was used for calculating the Arrhenius parameters:

$$\log(\beta) = \log\left[\frac{AE}{Rg(\alpha)}\right] - 2.315 - 0.457\frac{E}{RT} \quad (12)$$

From KAS method, the plot of $\ln(\beta/T^2)$ versus $1/T$ gives the slope $-E/R$. From OFW method, activation energy was calculated from the slope $(-0.457E/RT)$ of the straight line obtained by plotting $\log\beta$ and $1/T$ data.

Frequency factor from the model-free isoconversional methods was determined from the intercept of the plots by making use of the best kinetic model representing the experimental data, $f(\alpha)$, evaluated from Coats–Redfern method.

Results and Discussion

TGA of karanja seed cake

The thermal behavior of karanja seed cake in the temperature range $26 \pm 2°C$ to $800°C$ was studied at three different heating rates. Reduction in the mass of the sample $(11 \pm 0.1 \text{ mg})$ with temperature at a heating rate $20°C/$ min is shown in Figure 1. The complete pyrolysis reaction proceeds with an initial dehydration step followed by decomposition of protein, hemicellulose, cellulose, and lignin [17]. In the initial dehydration step, free/bound moisture and volatile extractives of biomass are carried away by the inert gas flow. The dehydration step occurs to a temperature of 200°C. Most of cellulose and hemicellulose decomposes above 200°C [18] and a vigorous mass loss was observed in the temperature range 250–450°C. Lignin decomposition takes place through a wide temperature range (160–900°C) [19] because of its recalcitrant structure, and in the present study a very slow mass loss after a temperature of 450°C was attributed to the lignin mass loss.

In the initial dehydration step, a 1.35% weight loss was observed as moisture was removed. In the decomposition of biomass constituents, considered as an active pyrolysis zone, a 73.8% weight loss was observed with the remaining 24.85% representing residual matter. During the pyrolysis, the intermolecular associations and weaker chemical bonds are destroyed. The side aliphatic chains may be broken and some small gaseous molecules are produced at lower temperature. At higher temperatures of decomposition, chemical bonds in lignin are broken and the parent molecular skeletons are destroyed [20, 21].

Pyrolysis kinetics

Biomass pyrolysis includes complex heat and mass transfer phenomena, therefore only about 11 mg of karanja seed cake biomass samples were used for TGA to minimize/overcome the endothermic and exothermic effects on the sample due to the furnace temperature. The thermogravimetric data were evaluated using 14 different reaction models given in Table 1. The best correlation was obtained with the third-order reaction model $[f(\alpha) = (1 - \alpha)^3]$ at three different heating rates and the regression plots according to Coats–Redfern method are shown in Figure 2A.

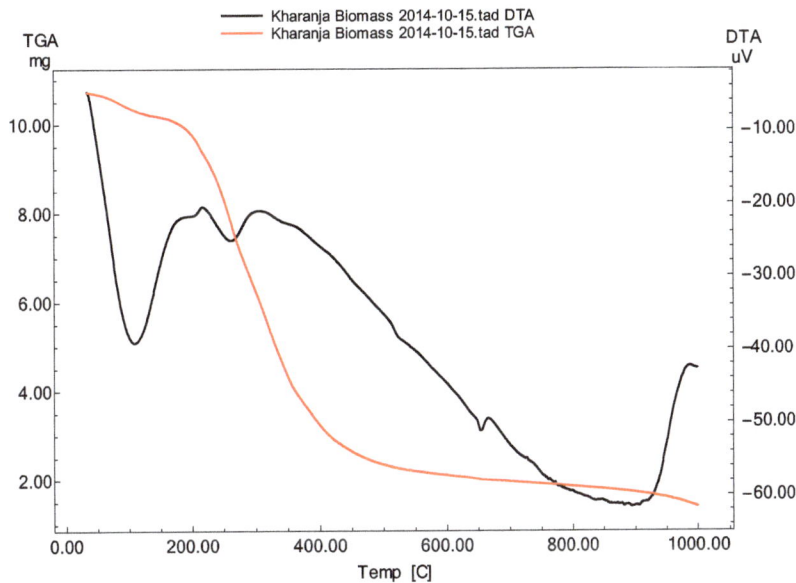

Figure 1. Thermogram of karanja seed cake at 20°C/min heating rate.

The calculation results including the values for E and A by Coats–Redfern method are listed in Table 2. The activation energy data thus obtained ranged from 82.95 to 98.21 kJ/mol. An increase in reaction kinetic parameters with the heating rate employed was observed in the present study. The obtained kinetic triplet was used to simulate the karanja seed cake pyrolysis process. A good agreement was observed between the model and experimental data as shown in Figure 2B.

The data obtained from TGA were also analyzed for isoconversional kinetics according to the KAS and the OFW methods, and standard methods for isoconversional kinetics evaluation. The regression lines obtained using KAS and OFW methods are depicted in Figure 3A and B, respectively.

Activation energies, calculated from the KAS and OFW methods, are presented in Table 3. It is evident from the table that the values obtained employing both methods are similar. Regression coefficients of the plots increased with progressive conversions. Activation energies from KAS method ranged from 122.8 kJ/mol to 106 kJ/mol for conversions ranging from $\alpha = 0.2$ to $\alpha = 0.9$. The activation energies from the OFW method were in the range 126–110 kJ/mol for the same conversions.

Activation energies decreased with an increase in conversion, as high conversions were observed at high

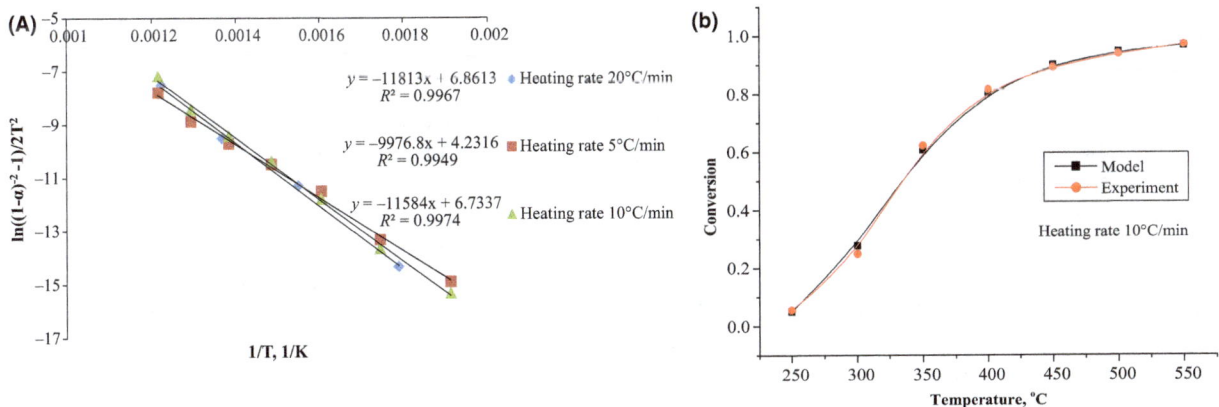

Figure 2. Coats–Redfern method. (A) Third-order reaction model fitted to thermogravimetric (TG) data at different heating rates. (B) Comparison of data from experiment (10°C/min heating rate) and model.

Table 2. Kinetic triplet for karanja seed cake pyrolysis reaction from Coats–Redfern method.

Heating rate, °C/min	Reaction model, $f(\alpha)$	Activation energy (E), kJ mol^{-1}	Frequency factor (A), min^{-1}	Correlation coefficient, R^2
5	$(1 - \alpha)^3$	82.95	3.43×10^6	0.9949
10	$(1 - \alpha)^3$	96.31	9.73×10^7	0.9974
20	$(1 - \alpha)^3$	98.21	2.25×10^8	0.9967

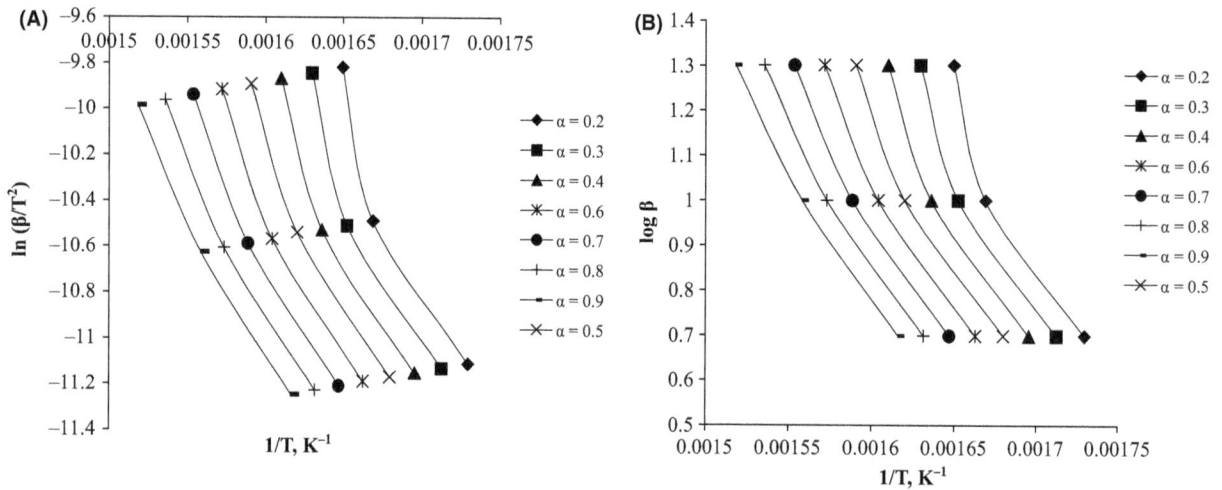

Figure 3. Regression lines for the determination of activation energy by (A) Kissinger–Akahira–Sunose (KAS) and (B) Ozawa–Flynn–Wall (OFW) methods.

Table 3. Arrhenius parameters and regression factors at different conversions from KAS and OFW methods.

Conversion, α	KAS method		OFW method		
	E_a, kJ/mol	R^2	E_a, kJ/mol	R^2	A, min^{-1}
0.2	122.81	0.907	126.03	0.919	1.75×10^{10}
0.3	120.41	0.927	123.87	0.937	1.54×10^{10}
0.4	117.92	0.943	121.61	0.951	1.28×10^{10}
0.5	114.26	0.958	117.95	0.962	8.11×10^9
0.6	112.97	0.967	117.10	0.972	9.23×10^9
0.7	108.24	0.975	114.92	0.979	9.08×10^9
0.8	108.24	0.982	112.83	0.985	1.15×10^{10}
0.9	106.00	0.987	110.81	0.989	2.54×10^{10}

KAS, Kissinger–Akahira–Sunose method; OFW, Ozawa–Flynn–Wall method.

temperatures, due to the requirement of comparatively lower activation energies at elevated temperatures. More reactions were triggered at higher temperatures, leading to a sharp rise in reaction rates with more unstable intermediates and lower activation energies. The reported values of pyrolysis reaction activation energies for hazelnut husk are in the range 128–131 kJ/mol [22], 129 kJ/mol for corn cob [23], and 62–206 kJ/mol for pine wood waste [24]; the activation energy of pyrolysis of karanja seed cake are in the range of the hazelnut husk and corn cob, but there was considerable deviation from the reported activation energy of pine wood biomass.

The activation energy of pyrolysis is highly dependent on the extent of conversion and the type and composition of biomass feedstock as seen in the activation energy for pine wood waste which is a woody biomass, rich in lignin, and very different from that of seed cake biomass and nut husk and other biomass types. KAS and OFW methods gave reliable activation energies that were used for the calculation of the pre-exponential factor in order to get a complete picture of the kinetic parameters with progressive conversions, the pre-exponential factors were determined by making use of the reaction model obtained from Coats-Redfern (CRF) method. Calculated

Study of thermal behavior of deoiled karanja seed cake biomass: thermogravimetric analysis...

11

pre-exponential factor values from OFW method are shown in Table 3. Therefore, the kinetic expression obtained for pyrolysis of karanja seed cake was:

$$\frac{d\alpha}{dt} = Ae^{-\frac{E}{RT}}(1-\alpha)^3 \qquad (13)$$

where the Arrhenius parameters obtained from model-free isoconversional methods were in good agreement with the values determined using model-fitting method.

Obtaining of the pyrolysis kinetic parameters based on TGA experimental data is extremely useful in the design and control of pyrolysis process especially large scale operation with desired composition and yield of products. Moreover, pyrolysis kinetic data link technological parameters, importantly biomass hold up time at preset temperature program that determines pyrolysis products profile.

Characterization of thermally treated biomass

Thermal treatment of karanja seed cake was performed in a fixed bed reactor of 1 cm inner diameter and 35 cm length under slow pyrolysis conditions with a continuous supply of nitrogen gas at 0.1 liter per minute (LPM) flow at different temperatures including 200, 250, 300, 350, 400, 450, and 500°C. The pyrolysis transformed the karanja seed cake to biochar (thermally treated biomass retained in the reactor), bio-oil (condensed liquid product), and the gases. The solid and the liquid obtained were weighed and the yields of the same were expressed gravimetrically. Biochar was characterized by proximate analysis carried out according to ASTM standard methods [25] using Mettler Toledo TGA/SDTA 851^e analyzer (Switzerland), ultimate/elemental analysis using a CHNS Analyzer–ELEMENTARVario microcube model, surface structure was obtained by scanning electron microscopy (SEM) using Hitachi S-3000N scanning electron microscope, Japan and FTIR (Fourier transform infrared spectroscopy) using a Perkin Elmer System 100 Fourier transform infrared spectrometer (Perkin Elmer Ltd., Seer Green, Beaconsfield, Bucks HP9 2FX, United Kingdom) [26].

Pyrolysis temperature showed a profound influence on the product distribution in the reaction. Figure 4 shows the pyrolysis products, biochar, and total volatiles and yields obtained in relation to temperature. Pyrolysis at higher temperature (≥500°C) resulted in increased carbon conversion into volatiles [27, 28] and reduced biochar.

Karanja seed cake, processed in the present work, composed of 25.2% carbohydrates, 17.4% lignin, 13.4% protein, and 4% ash with 40% water and ethanol soluble matter, details of which were presented in our earlier work [29]. Biochar obtained from the thermal treatment of karanja seed

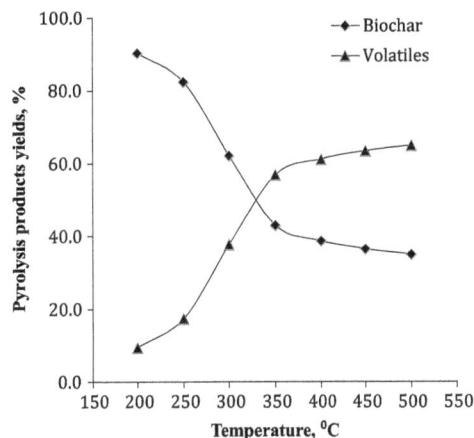

Figure 4. Effect of temperature on biochar and total volatiles yield from pyrolysis.

cake was analyzed for its composition in terms of proximate and ultimate analysis and the data were presented in Table 4 along with evaluated HHV, carbon carryover, H/C and O/C ratios. From the elemental analysis of feed karanja seed cake and biochar at different temperatures it was observed that the carbon content increased, whereas the hydrogen and oxygen contents decreased with increase in temperature. H/C and O/C values give a picture of the degree of carbonization, aromaticity, and maturity of organic materials. A decrease in oxygen (O/C) with increase in temperature was due to the increased decomposition of oxygen containing functional groups. The percent biochar obtained from the seed cake by experimentation was 56.82%, 53.29%, 52.55%, and 51.52% at 350, 400, 450, and 500°C, respectively, indicating a decrease in the carbon carryover from the seed cake to biochar with temperature. Therefore, the value addition of deoiled karanja seed cake by thermal treatment is a promising pretreatment method to obtain biochar for multiple applications.

Scanning electron microscopy

As porosity play a vital role in determining the efficiency of biosorbents in the adsorption of contaminants in effluent treatment, with maximum adsorption occurring when pores are large enough to admit contaminant molecules, SEM of karanja cake and biochar was carried out to assess the surface structure of the samples. The SEM pictures obtained are presented in Figure 5A and B. As is evident from the pictures, the char formed (Fig. 5B) has a highly porous structure compared to the original seed cake biomass (Fig. 5A). This can be attributed to the fact that thermal treatment improves the porous structure of biochar due to the loss of hydrogen, oxygen (major quantity), and carbon (minor quantity) atoms in the form of volatiles from the parent biomass material leaving the skeletal structure.

Table 4. Composition and energy value of biochar obtained from pyrolysis of karanja seed cake at different temperatures.

Element, %wt.	Karanja cake	Biochar from pyrolysis at			
		350°C	400°C	450°C	500°C
Proximate analysis					
Moisture	06.7				03.20
Ash	04.00				15.42
Volatile matter	71.44				14.13
Fixed carbon	17.86				67.25
Ultimate analysis					
N	04.30	05.84	05.91	06.07	05.95
C	44.20	58.39	60.80	63.64	65.06
H	06.50	05.21	04.45	04.61	04.30
S	–	–	–	–	–
O	45.00	30.56	28.84	25.68	24.69
HHV, MJ/kg	18.11	22.9	23.14	24.23	24.5
H/C	1.76	1.07	0.91	0.83	0.80
O/C	0.68	0.35	0.31	0.27	0.26
Carbon carryover[1], %	–	56.82	53.29	52.55	51.52

[1](Carbon content in biochar × Biochar yield)/(Carbon content in karanja seed cake).

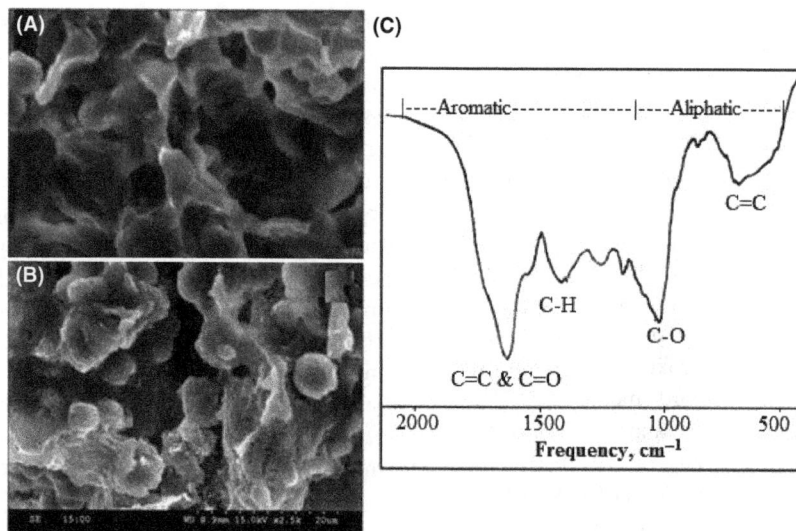

Figure 5. Surface structure (from scanning electron microscopy [SEM]) of (A) karanja seed cake biomass and (B) biochar from karanja seed cake, and (C) Fourier transform infrared spectra of biochar obtained at 500°C.

Fourier transform infrared spectroscopy

Fourier transform infrared spectra of biochar was collected using the KBr pelletization method on a Perkin Elmer System 100 Fourier transform infrared spectrometer (FTIR) with a cesium iodide (CsI) micro-focus accessory and triglycine sulfate (TGS) detector by scanning from 4000 to 400 cm^{-1}, with a resolution of 4 cm^{-1}. The biochar spectra as shown in Figure 5C depicts the vibrations in the frequency range 700–1200 cm^{-1} and 1200–2000 cm^{-1} confirming the aliphatic and aromatic nature of the sample. The peaks at frequencies of 698 (aromatic ring C-H), 1020 (C-O stretching vibration), 1420 (C-H stretching vibration), and 1626 cm^{-1} (C=C and C=O stretching vibrations) show the carbon structure of biochar. This spectrum is comparable to reference spectra obtained by Rutherford et al. [26] who studied changes in composition and porosity occurring during the thermal degradation of wood and wood components. The result further emphasizes the fact that thermal treatment of biomass converts aliphatic carbon to aromatic on prolonged treatment [30].

From the characterization of the seed cake biomass and biochar based on the elemental composition, surface structure, and FTIR, it is evident that the biochar is rich in carbon and has porous structure and can be effectively used for energy generation and as biosorbent for industrial effluent treatment and electrode material in activated carbon electrodes manufacture.

Conclusions

This study is the first report on nonedible seed cake biomass pyrolysis kinetics in particular karanja seed cake biomass, and the work presented is extremely helpful to understand the usefulness of value addition to karanja seed cake through thermal treatment. The TGA pyrolysis reactions at constant heating rate revealed that a temperature of 450–500°C was sufficient for carbonization of the biomass. The KAS, OFW, and CRF methods gave reliable activation energies and pre-exponential factors with third-order reaction model for the seed cake pyrolysis reaction. Characterization results showed that the thermally treated biomass has high porous structure and rich in carbon, suitable for various applications as biosorbent and precursor for carbon materials. The kinetic data thus obtained are extremely useful in design of pyrolysis system for different seed cake biomass.

Conflict of Interest

None declared.

References

1. Song, X. D., X. Y. Xue, D. Z. Chen, P. J. He, and X. H. Dai. 2014. Application of biochar from sewage sludge to plant cultivation: influence of pyrolysis temperature and biochar to soil ratio on yield and heavy metal accumulation. Chemosphere 109:213–220.

2. Suddick, E. C., and J. Six. 2013. An estimation of annual nitrous oxide emissions and soil quality following the amendment of high temperature walnut shell biochar and compost to a small scale vegetable crop rotation. Sci. Total Environ. 465:298–307.

3. Watanabe, A., I. KosukeIkeya, N. Kanazaki, S. Makabe, Y. Sugiura, and A. Shibata. 2014. Five crop season's records of greenhouse gas fluxes from upland fields with repetitive applications of biochar and cattle manure. J. Environ. Manage. 144:168–175.

4. Creamer, E. A., B. Gao, and M. Zhang. 2014. Carbon dioxide capture using biochar produced from sugarcane bagasse and hickory wood. Chem. Eng. J. 249:174–179.

5. Jin, H., S. Carapeda, Z. Chang, J. Gao, Y. Xu, and J. Zhang. 2014. Biocharpyrolytically produced from municipal solid wastes for aqueous As(V) removal: adsorption property and its improvement with KOH activation. Bioresour. Technol. 169:622–629.

6. Xu, D., Y. Zhao, K. Sun, B. Gao, Z. Wang, J. Jin, et al. 2014. Cadmium adsorption on plant- and manure-derived biochar and biochar amended sandy soils: impact of bulk and surface properties. Chemosphere 111:320–326.

7. Parthasarathy, P., and K. N. Sheeba. 2014. Determination of kinetic parameters of biomass samples using thermogravimetric analysis. Environ. Prog. Sustain. Energy 33:256–266.

8. Vhathvarothai, N., J. Ness, and Q. J. Yu. 2014. An investigation of thermal behaviour of biomass and coal during copyrolysis using thermogravimetric analysis. Int. J. Energy Res. 38:1145–1154.

9. Yan, W., S. Islam, C. J. Coronella, and V. R. Vásquez. 2012. Pyrolysis kinetics of raw/hydrothermally carbonized lignocellulosic biomass. Environ. Prog. Sustain. Energy 31:200–204.

10. Rebitanim, N. Z., W. A. W. Ab Karim Ghani, N. A. Rebitanim, and M. A. Mohd Salleh. 2013. Potential applications of wastes from energy generation particularly biochar in Malaysia. Renew. Sustain. Energy Rev. 21:694–702.

11. Gaurav, D., J. Siddarth, and P. S. Mahendra. 2011. Pongamia as a source of biodiesel in India. Smart Grid Renew. Energy 2:184–189.

12. Ramachandran, S., K. S. Sudheer, L. Christian, C. R. Soccol, and A. Pandey. 2007. Oil cakes and their biotechnological applications – a review. Bioresour. Technol. 98:2000–2009.

13. Vyazovkin, S., and C. A. Wight. 1999. Model-free and model-fitting approaches to kinetic analysis of isothermal and nonisothermal data. Thermochim. Acta 340–341:53–68.

14. Coats, A. W., and J. P. Redfern. 1964. Kinetic parameters from thermogravimetric data. Nature 201:68.

15. Vlaev, L. T., I. G. Markovska, L. A. Lyubchev. 2003. Non-isothermal kinetics of pyrolysis of rice husk. Thermochim. Acta 406:1–7.

16. Doyle, C. D. 1965. Series approximations to the equations of thermogravimetric data. Nature 207:290–291.

17. Kok, M. V., and E. Ozgur. 2013. Thermal analysis and kinetics of biomass samples. Fuel Process. Technol. 106:739–743.

18. Carrier, M., A. L. Serani, D. Denux, J. M. Lasnier, F. H. Pichavant, F. Cansell, et al. 2011. Thermogravimetric analysis as a new method to determine the lignocellulosic composition of biomass. Biomass Bioenergy 35:298–307.

19. Burhenne, L., J. Messmer, T. Aicher, and M. P. Laborie. 2013. The effect of the biomass components lignin, cellulose and hemicellulose on TGA and fixed bed pyrolysis. J. Anal. Appl. Pyrol. 101:177–184.

20. Raman, P., W. P. Walawender, L. T. Fan, and J. A. Howell. 1981. Thermogravimetric analysis of biomass, devolatilization studies on feedlot manure. Ind. Eng. Chem. Process Des. Dev. 20:630–636.

21. Singh, R. K., and K. P. Shadangi. 2011. Liquid fuel from castor seeds by pyrolysis. Fuel 90:2538–2544.

22. Ceylan, S., and Y. Topcu. 2014. Pyrolysis kinetics of hazelnut husk using thermogravimetric analysis. Bioresour. Technol. 156:182–188.

23. Gai, C., Y. Dong, and T. Zhang. 2013. The kinetic analysis of the pyrolysis of agricultural residue under non-isothermal conditions. Bioresour. Technol. 127:298–305.

24. Amutio, M., and G. Lopez. 2012. Kinetic study of lignocellulosic biomass oxidative pyrolysis. Fuel 95:305–311.

25. Cantrell, K. B., Martin, J. H., and Ro, K. S. 2010. Application of thermogravimetric analysis for the proximate analysis of livestock wastes. J. ASTM Int. 7: 1–13

26. Rutherford, D. W., L. Robert Wershaw, and L. G. Cox. 2004. Changes in composition and porosity occurring during the thermal degradation of wood and wood components. U.S. Geol. Sci. Invest. Rep. 5292:79.

27. Antony, R. S., D. S. Smart Robin Son, B. C. Pillai, and C. Lee Robert Lindon. 2011. Parametric studies on pyrolysis of pungam oil cake in electrically heated fluidized bed research reactor. Res. J. Chem. Sci. 1:70–80.

28. Singh, R. C., R. Kataki, and T. Bhaskar. 2014. Characterization of liquid and solid product from pyrolysis of pongamia glabra deoiled cake. Bioresour. Technol. 165:336–342.

29. Radhakumari, M., A. S. Ball, K. Suresh Bhargava, and B. Satyavathi. 2014. Optimization of glucose formation in karanja biomass hydrolysis using Taguchi robust method. Bioresour. Technol. 166:534–540.

30. Rutherford, D. W., L. RobertWershaw, E. ColleenRostad, and N. CharleneKelly. 2012. Effect of formation conditions on biochars: compositional and structural properties of cellulose, lignin, and pine biochars. Biomass and Bioenergy 46:693–701.

3

The progressive routes for carbon capture and sequestration

Sonil Nanda[1], Sivamohan N. Reddy[2], Sushanta K. Mitra[3] & Janusz A. Kozinski[1]

[1]Department of Earth and Space Science and Engineering, Lassonde School of Engineering, York University, Toronto, Ontario, Canada
[2]Department of Chemical Engineering, Indian Institute of Technology Roorkee, Roorkee, Uttarakhand, India
[3]Department of Mechanical Engineering, Lassonde School of Engineering, York University, Toronto, Ontario, Canada

Keywords
Adsorption, biofuels, carbon dioxide, carbon sequestration, flue gas, oceanic carbon storage

Correspondence
Janusz A. Kozinski, Lassonde School of Engineering, York University, Toronto, Ontario M3J 1P3, Canada.
E-mail: janusz.kozinski@lassonde.yorku.ca

Funding Information
The authors thank Natural Sciences and Engineering Research Council of Canada (NSERC) for the financial support toward this renewable energy research.

Abstract

The global warming is directly related to the increased greenhouse gas emissions from both natural and anthropogenic origins. There has been a drastic rise in the concentration of CO_2 and other greenhouse gases since the industrial revolution primarily due to the intensifying consumption of fossil fuels. With the need to reduce carbon emissions and mitigate global warming certain strategies relating to carbon capturing and sequestration are indispensable. This paper comprehensively describes several physicochemical, biological and geological routes for carbon capture and sequestration. The trend of the increase in greenhouse gases over the years is illustrated along with the global statistics for fossil fuels usage and biofuels production. The physicochemical carbon capturing technologies discussed include absorption, adsorption, membrane separation and cryogenic distillation. The algal and bacterial systems, dedicated energy crops and coalbed methanogenesis have been vividly explained as the biological routes for carbon sequestration. The geological carbon sequestering route centers on biochar application and oceanic carbon storage. A systematic survey has been made on the origin and impact of greenhouse gases along with the potential for sequestration based on some fast-track and long-term sequestration technologies.

Introduction

Today, climate change and global warming are two of the hot topics for discussion at the global environmental panorama. Climate change is a long-lasting and irrevocable shift in the weather conditions recognized by the variations in atmospheric temperature, precipitation, air quality, the wind, and other indicators. The climate change has led to the experiencing of several extreme weather events worldwide, which are mostly attributed to anthropogenic global warming. Some of these unusual and unseasonal weather events include, but are not limited to, heat and cold waves, melting of ice cover, a rise in the sea level, drought, floods, violent storms, and tropical cyclones. The Earth's climate is seasonal and naturally variable on all timescales. Since the yesteryears, the increased concentrations of GHG (greenhouse gases) have led to the induced greenhouse effect resulting in the warming of the planetary surface.

According to IPCC, climate change can be defined as "any change in the climate over time, whether due to natural variability or as a result of human activity." Likewise, the UNFCCC (United Nations Framework Convention on Climate Change) defines climate change as "change in the climate that is attributed directly or indirectly to human activity thereby altering the composition of the global atmosphere and natural climate variability observed over comparable periods of time" [1]. Since the time of industrial revolution, there have been dramatic changes in the global agriculture, material manufacturing, transportation and infrastructure. The rapid urbanization has led to an increase in the emissions of the popular GHGs viz. CO_2 (carbon dioxide), CH_4 (methane) and N_2O (nitrous oxide). In addition to CO_2, CH_4, and N_2O, the GHGs also include SF_6 (sulfur hexafluoride), O_3 (ozone), water vapor, hydrofluorocarbon, and perfluorocarbon groups of gases.

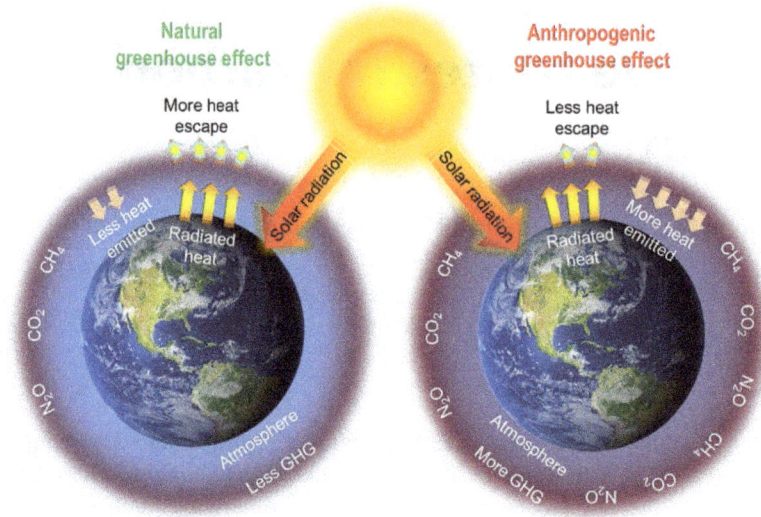

Figure 1. Graphical illustration of natural and anthropogenic greenhouse effects.

A major proportion of the GHGs in the atmosphere is due to anthropogenic reasons. The increased consumption of fossil fuels, use of chlorofluorocarbons in refrigerants, solvents, foam blowing agents and spray propellants are accountable not only for increased GHGs but also in the depletion of ozone layer. The industrial practices such as processing of minerals, metals, chemicals, solvents together with the production and utilization of halocarbons, and SF_6 also contribute to this effect. The GHGs aid in the greenhouse effect by trapping the outgoing infrared radiation from the Earth's surface and adding the heat to the net energy input of lower atmosphere [2]. Figure 1 is a graphical illustration of the natural and anthropogenic greenhouse effects. The higher levels of CO_2, CH_4 and N_2O cause depletion of the ozone layer along with the thickening of the layer of GHGs that trap the heat within the atmosphere making the Earth's surface warmer. The anthropogenic greenhouse effect has been recognized to adversely impact the global climate and functioning of the oceanic and terrestrial ecosystems by altering the temperature and rainfall patterns.

The attributions of CO_2, CH_4, and N_2O towards global warming can be considered to be 60%, 15%, and 5%, respectively [3]. While the concentration of CO_2 and CH_4 is increasing at the rate of 0.4–3% per year, N_2O is rising by 0.2% annually [4]. The Intergovernmental Panel on Climate Change (IPCC) has estimated a global rise in Earth's temperatures by 0.6°C over the last century. However, it can be predicted that the temperature will rise from 1.4°C to 5.8°C in the next two centuries if the anthropogenic emissions of GHGs keep enduring. The month of April 2015 was by far the warmest month on

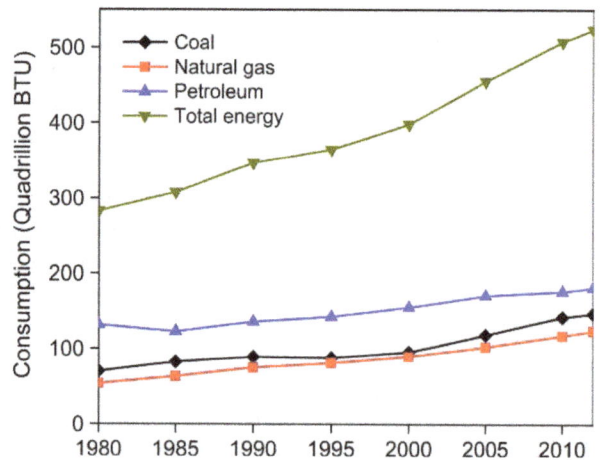

Figure 2. Worldwide consumption of fossil fuels from 1980 to 2012 (Data source: [8]).

record since 1880s. The global average land surface temperature in April 2015 was 1.11°C above the twentieth century average [5].

The rapid global industrialization has led to the unprecedented consumption of fossil fuels including coal, petroleum and natural gas that releases surplus amounts of CO_2 into the atmosphere. The world human population is expected to grow from 7.3 billion today [6] to 9.2 billion by 2050 [7]. The per capita consumption of energy also increases with the gradual increase in population. The total energy use in 2012 was 524 Quadrillion Btu (quad); however, it is expected to rise by 60% by 2030 [8]. The use of petroleum and other liquid fossil fuels was 85.7 million barrels per day in 2008 with projections for an increase up to 112.2 million barrels per day by 2035 [9]. Figure 2 shows the trend of fossil fuels consumption since 1980. The use of petroleum increased from 156 quads in 2000 to 181 quads in 2012 [8]. Similarly, the demands for coal and dry natural gas also rose to 147 and 124 quads, respectively in 2012. It entreats for a smarter and sustainable way to manage the energy demand for the growing world population. In such a scenario, energy efficiency and conservation along with decarbonizing our energy sources are essential [10].

Climate change is already having significant impacts on the ecosystems, communities and global economy irrespective of any particular geographical region. The increased GHG emissions also threaten the human health due to freshwater shortages, smog, acid rain and other ecological disturbances. The government, policy makers and individuals should act in synergy for mitigating GHG emissions and subsequently lowering their impacts, risks, and associated vulnerabilities. Although the climate change is irreversible, yet it can be alleviated by curbing the GHG emissions (especially CO_2 and CH_4) with strategies to capture and sequester the carbon.

Many papers have reported the potential of CCS (carbon capture and sequestration) to mitigate global warming. Among these many reports, the information about the routes of carbon capturing is scattered, which makes it difficult to evaluate the efficiency of one technology over the other. This paper attempts for a comprehensive review of the underlying principles and current trends in the field of CCS to deter the global warming caused by GHGs. The review is focused on careful integration of technologies for sequestering CO_2 through physicochemical, biological and geological processes. However, it is important to note that the CCS technologies discussed in this paper are either under developmental stage or at a precommercial scale. The CCS technologies have not yet been fully integrated in full-scale commercial operation due to restrained CO_2 capturing efficiency, high costs and lack of regulatory framework.

Current Trend of Greenhouse Gas Emissions

The current top ten CO_2-emitting countries are China, USA, India, Russia, Japan, Africa, Germany, South Korea, Iran, and Saudi Arabia (Fig. 3). China and India rank as the first and third largest CO_2-emitting countries because of their escalating demands for fossil fuels, which are increasing at the rates of 3.5% and 3.9% per year, respectively [11]. As some developed nations such as USA, Russia, Japan, and Germany are in the list of high CO_2 emitters, the Kyoto Protocol places a heavier burden on the principle of "common but differentiated

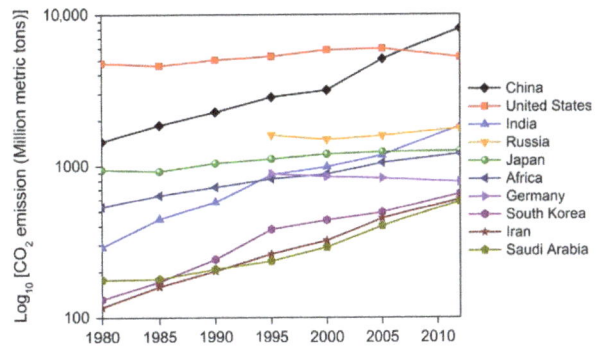

Figure 3. Top ten CO_2-emitting countries in 2012 (Data source: [8]).

responsibilities." The Kyoto Protocol is an international pact linked to UNFCCC that sets international GHG emission reduction targets. According to the Kyoto Protocol (an international environmental treaty proposed by UNFCCC on December 11, 1997 in Kyoto, Japan), it is mandatory for industrialized nations to reduce their anthropogenic GHG emissions by 5.2% from their 1990 levels within the commitment period of 2008–2012.

The "Doha Amendment to the Kyoto Protocol" was adopted in Doha, Qatar, in 2012. During its first commitment period (2008–2012), 37 industrialized countries including the European Union committed to reducing GHG emissions by 5% from their 1990 levels. In the second commitment period (2013–2020), the parties agreed to reduce the GHG emissions by 18% below their 1990 levels by 2020. The arrangement of parties in the second commitment period is different from the first, with Canada, Japan, and Russia withdrawing their commitments from the Protocol in 2011. The Canadian government withdrew from the Kyoto Protocol after ratification in December 2011. Although, Canada was committed to restraint its GHG emissions to 6% below 1990 levels, it showed 17% higher emissions in 2012. The penalties of $13.6 billion for not achieving the targets led to its withdrawal from the treaty [12]. The GHG emissions are debatable for the continual increase as two of the largest CO_2 emitters, namely, Russia and Japan also walked out of the Kyoto agreement.

Recently (November–December 2015) in Paris, France, the United Nations Climate Change Conference, that is, COP21 or CMP11 was held as the 21st annual session of the Conference of the Parties (COP) to the 1992 UNFCCC and the 11th session of Meeting of the Parties to the 1997 Kyoto Protocol. In this global meeting, 200 nations agreed to cut their carbon emissions to set a goal of curbing global warming to <2°C compared to the preindustrial levels [13]. The GHG emissions have increased by 80% since 1970 and 30% since 1990, totaling 49 gigatonnes (Gt) of CO_2 equivalent in 2010 [14]. According

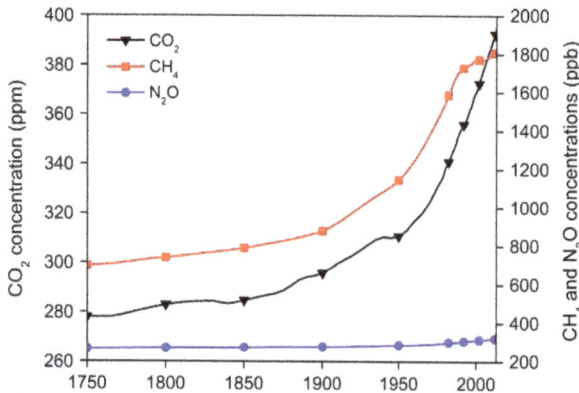

Figure 4. Worldwide greenhouse gas emissions from 1750 to 2013 (Data source: [16]).

to COP21, the parties will purse efforts to reduce the global GHG emissions by 40–70% by 2050 compared to 2010 levels and drop to zero emissions by 2100.

During the past one and half century, there has been a consistent increase in CO_2 concentration by 33% with the rate expected to rise by 4% per year [15]. Figure 4 illustrates the trend of GHG emissions since 1750. The current atmospheric concentration of CO_2 (404 ppm) compared to its 1750 levels (278 ppm) shows a dramatic increase in 265 years [16]. The CO_2 emission from fossil fuels combustion in 2014 was 36 Gt [17]. CO_2 accounted for 77% of the total anthropogenic GHG emissions in 2004 as its annual emissions have grown between 1970 and 2004 by 80%, that is, from 24 to 38 Gt, respectively [18]. CH_4 and N_2O are 20 and 300 times as potent as CO_2, respectively in retaining the atmospheric heat. The levels of CH_4 increased from 700 ppb in 1750 to 1814 ppb in 2013 indicating a 61% rise. Similarly, N_2O concentration grew by 17%, that is, from 270 ppb in 1750 to 326 ppb in 2013.

Despite the anthropogenic sources, the sources of natural GHG emissions are also of parallel consideration. Soil can be considered both as a chief source of atmospheric CO_2 and also a storehouse for carbon storage. Soil contributes a fraction of the total emission of CO_2 through soil respiration (from the plant root and heterotrophic respiration) [19]. Carbon released through soil respiration accounts for around 10% of the total atmospheric carbon pool [20]. Agricultural methods of manure management, field burning of agricultural residues, and rice cultivation are also classified as the sources of GHG emissions [21]. About 80% of CH_4 is produced biologically from rice cultivation, wetlands and sediments, landfills, enteric microbial fermentation and organic waste composting under conditions of low redox potential [22]. About 35% of the annual N_2O emission is from agriculture and change in land use [23]. N_2O is biologically generated through

denitrification by heterotrophic microorganisms in oxygen deficient environments, nitrification by autotrophic and heterotrophic nitrifying microorganisms, and dissimilation of nitrate to ammonium by heterotrophic microorganisms in aerobic conditions [24].

The natural catastrophic events such as volcanic eruptions, forest fires and hydrothermal vents also release significant amount of GHGs into the atmosphere. The CO_2 released from volcanoes can be from the erupting magma and degassing of unerupted magma (i.e., recycled sub-ducted crustal materials and decarbonation of shallow crustal materials). They can restore the lost CO_2 from the atmosphere and oceans through weathering of silicates, carbonate deposition and burial of organic carbon. The CO_2 typically makes up to 10 mol% of the total volcanic gas emissions [25]. A hydrothermal vent is a fissure in Earth's surface from which geothermally heated water and hot gas bubbles explode. The dissolution of the gases causes local increases in water density and acidity resulting in sequestration of CO_2 [26]. The emissions from forest fires also have a direct impact on the regional and global carbon cycles by increasing CO_2 levels and affecting carbon sequestration by forests. However, compared to the land use changes (3.4 Gt CO_2 per year) and vehicular emissions (3 Gt CO_2 per year), volcanoes emit less CO_2 (up to 0.44 Gt) [27].

Climate change is predicted to influence biodiversity and agriculture to a large extent. The changes in climate system is unambiguous as evident from: (1) the increase in global average temperatures of soil, air, and water bodies; (2) widespread melting of snow, ice and glacier; (3) shorter freezing seasons of lake and river ice; (4) decrease in permafrost extent; and (5) rise in the average sea level [18]. The climate change and unseasonal weather conditions have also led to habitat shifting and alterations for several ecological species, for example, desertification, coral bleaching, tundra thawing, etc. The rising temperatures cause the meltdown of mountain ice caps, snow, glacier and permafrost. This results in a shift or loss of many species from their natural habitat. The behaviors of adaptation, endangering and extinction are the anticipated in several plant and animal species. The bird migration, egg-laying, leaf-unfolding, as well as the poleward and upward shift in plant and animal species are most common examples [18]. Community shift between fishes, coral reefs, aquatic plants and animals, plankton, algae, and marine bacteria are also associated with rising water temperatures, changes in ice cover, salinity, and oxygen availability.

The emission trading is an economically sensitive strategy at the national and international level for reducing the concentrations of GHGs, particularly CO_2. Carbon credits, carbon trading and carbon markets are some of the CO_2 emission trading approach. A carbon credit is a tradable permit or certificate representing the right to emit one ton of CO_2 or the mass of another GHG with one ton CO_2 equivalent. Carbon trading allows the industries (that cannot practicably reduce CO_2 emissions) to buy credits from those industries that have already reduced their emissions more than the committed level [28]. However, each carbon credit is worth one metric ton of CO_2. The carbon trading market is an economic approach as the cost of emission reduction is higher than the cost of credits. Alternatively, the industries can invest in reforestation projects for removing CO_2 from the atmosphere biologically via photosynthesis and carbon fixation [29]. Although this approach is typically called reduction rather than credit, the process is monitored over time and units are measured in tons of CO_2. Upon the reduction in other GHGs, carbon equivalents can be earned and traded.

Carbon Capturing and Storage Technologies

Carbon capture and storage (or sequestration), often abbreviated as CCS, is the process of capturing CO_2 from large-scale emitters such as fossil fuel refineries, power plants and product manufacturing industries, and transporting it to a storage site thus preventing its re-entry into the atmosphere. The cost of CO_2 separation and compression (to 11 MPa) is estimated to be $30–50 per ton CO_2, whereas transportation (for every 100 km) and sequestration cost is about $1–3 per ton CO_2 [30]. There are several routes available for CO_2 removal from the atmosphere or industrial flue gas stream. These routes can be categorized into physicochemical, biological and geological as shown in Figure 5.

The physicochemical route involves the application of absorption, adsorption, gas separation membranes and

cryogenic distillation. This route is most popular for capturing CO_2 from industrial flue gas streams. The CO_2 can be isolated from the absorbent material for compression and transportation. The biological route includes algal systems and energy crops that can utilize CO_2 from the atmosphere for photosynthesis and fix carbon in the form of carbohydrates (in lignocellulosic plants) and polysaccharides (in algae). The carbohydrates and polysaccharides can be converted to hydrocarbon fuels via thermochemical (e.g., pyrolysis, torrefaction, liquefaction and gasification) and biochemical (e.g., fermentation and anaerobic digestion) conversion technologies [31]. Coalbed methanogenesis is also included in biological route owing to the involvement of methanogenic bacteria that lead to the biodegradation of coalbed and generating biogenic CH_4 instead to CO_2 that would otherwise be released upon coal combustion. The geological route involves storing carbon in the soil in the form of biochar or through underwater storage.

Carbon dioxide capture can be done in three ways, such as, postcombustion, precombustion, and oxy-fuel combustion [32–34]. The fossil fuels (e.g., coal and natural gas) are subjected to pyrolysis, gasification or reforming reactions in precombustion method to produce clean gas fuels such as H_2 that on combustion produces water [35]. The carbon present in the fuel gets converted to CO_2, and the effluent gas from reforming or gasification majorly comprises of H_2 and CO. The complete conversion of carbon content into CO_2 leads to its higher concentration with high pressure of the CO_2-rich effluent stream that implies lower operating costs for the separation process. High-pressure steam or pure oxygen is a prerequisite for reforming and partial oxidation to generate H_2 from fossil fuels.

The postcombustion process can capture CO_2 on-site from the flue gases emitted as a result of fossil fuel combustion. As the CO_2 concentrations in the flue gas are usually ≤15 vol%, indirect mode of operation (introducing a new phase to capture the required component) is employed to separate and sequester CO_2 [36]. The postcombustion process includes absorption, adsorption, membrane separation, and cryogenic distillation used in the capturing CO_2 from the flue gases. Each of these technologies has been vividly discussed in the following section. Table 1 summarizes different aspects of the pre-, post- and oxy-fuel combustion processes.

The oxy-fuel combustion process is the combustion of fuels with pure oxygen to generate huge quantities of energy at high temperatures [37]. The demand of pure O_2 for the combustion requires its separation from air, which is quite expensive and requires cryogenic distillation. Water and CO_2 are the typical products of the oxy-fuel combustion process. The exhaust gases obtained at

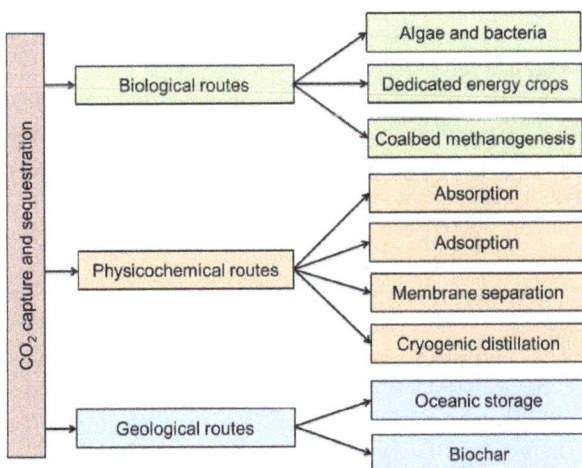

Figure 5. Routes of carbon capture and sequestration.

Table 1. Summary of CO_2 capturing technologies.

Parameter	Precombustion	Postcombustion	Oxy-fuel combustion
CO_2 concentration	15–40 vol%	4–14 vol%	75–80 vol%
Acid gases	Sulfur compounds need to be removed.	Contains NO_x, SO_x, COS and H_2S.	NO_x absent but gas desulfurization is required.
Combustion medium	Steam/air is required for gasification to generate CO_2.	Air is used.	High purity oxygen for combustion.
Equipment size	Medium size equipment.	Large size equipment required with high investment.	Low size equipment.
Temperature and pressure	Low temperature and high pressure (depends on the process employed).	Flue gas need to be cooled and pressure depends on CO_2 capture process.	Cryogenic temperature for separation of O_2. High temperatures are obtained for oxy-fuel combustion hence the flue gas is recycled to reduce the temperature.
Potential	Integrated gasification combined cycle and turbines which can effectively use H_2-rich syngas.	Can be applied to the existing coal combustion plants.	Novel cycles have been employed in synergy with integrated gasification combined cycle.
Pros	Low energy penalty than postcombustion processes. Regeneration can be achieved by altering pressure and temperature.	Small concentrations of CO_2 can be captured. Retrofit technology.	Efficiency of CO_2 capture reaches 100%. No presence of harmful NO_x.
Cons	Drying of syngas and its treatment prior to CO_2 capture. High investment and capital costs.	High operating and regeneration costs. High solvent losses.	High capital and operating cost. Annexing to the existing plants is difficult.
State of the art	Integrated gasification combined cycle and ammonia production plants are running currently.	Amine scrubbing plants (reaction with monoethanolamine) are in practice. Power plants currently employ this technology	Efficient CO_2 separation.

high temperatures are cooled and further recycled back to reduce the temperature of the flue gases. As a result, the exit stream contains higher concentrations of CO_2 for capture and storage. The economy of the process depends on separating O_2 from N_2, which is very expensive [38]. The typical techniques employed to separate CO_2 from the flue gas includes absorption, adsorption, membrane separations, and cryogenic distillation [34, 35].

Chemical looping combustion, first developed by Richter and Knoche [39], is similar to oxy-fuel combustion where O_2 for combustion is supplied by using a metal oxide. By using a metal oxide, the exit flue gas is free of N_2 resulting in high concentrations of CO_2 making it viable for easy capturing. Chemical looping combustion is usually operated in two fluidized-bed reactors where the fuel comes in contact with metal oxide carrier (that supplies O_2 for combustion). Further, the metal oxide is transferred to a regenerative bed where it is reoxidized. The supply of O_2 from the metal oxide carrier depends on its oxidation and reduction cycles. The potential metal oxide carriers are copper, manganese, iron and nickel [40]. However, some technical issues encountered during this process are deactivation of metals, inefficient metal oxide–fuel contact and sulfur poisoning.

Physicochemical Routes for CO_2 Capture and Separation

Absorption

Sorption is a composite process involving absorption and adsorption. Sorption studies can evaluate the surface areas and pore structures of solid materials providing information on the sorption abilities of adsorbates by porous or nonporous solids [41]. Absorption is a gas-liquid operation where the gas components are absorbed into the new liquid phase. The solute is recovered from the solution either by reversing the process conditions or by other mass-transfer operation. The solubility requirements along with constraints such as nonvolatility, low viscosity, nontoxicity, nonflammability, and chemical stability restrict the choice of absorbent solvent to capture gases from effluents.

The capturing of gas molecules from the exhaust gases can be achieved by physical or chemical means. In the physical process, higher concentrations of flue gases at high pressures to capture CO_2 in a suitable solvent by dissolution process. The gas molecules dissolve into the chosen solvent at the given operating conditions to capture harmful gas effluents including CO_2, NO_x, SO_x, and H_2S. Henry's law comes into picture for the absorption of gases into the liquid solvents (partial pressure α concentration of the component in the liquid phase). High pressures and low temperatures are recommended for greater dissolution of gases in liquid solvents while the solvents are regenerated by altering the process temperature and pressure. An additional advantage of physical absorption process is that the solvents are capable of absorbing H_2S, carbonyl sulfide (COS) and hydrocarbons along with CO_2 [42].

The physical absorption of CO_2 by the solvent mixture of dimethyl ethers and polyethylene glycol solution is known as the Selexol process [43]. The typical formula for Selexol solvent is $CH_3(CH_2CH_2O)_{(3-9)}CH_3$ with pressures of 1.5–14 MPa and temperatures of 1–25°C [42, 44]. In the Rectisol process, methanol is employed as the absorbing solvent at −10 to −70°C and 3–8.1 MPa pressure [45]. The solubility of CO_2 in methanol is five times that of water at ambient temperature, although it can be enhanced by 8–15 times at temperatures lower than −0.15°C [46]. In addition to the methanol and Selexol solvents, other solvents used to capture CO_2 are N-methyl-2-pyrrolidine, morpholine, and propylene carbonate. Purisol process involves N-methyl-2-pyrrolidine as the solvent at −20 to −40°C and 1 MPa pressure. Similarly, Fluor process uses propylene carbonate as the solvent at 3–7 MPa and temperatures below 25°C [42]. Since the physical absorption processes require low temperatures (i.e. <25°C) and high pressures, the exit flue gas should be cooled and pressurized to the desired operating conditions. The solubility of CO_2 is usually high in Fluor process and appropriate for the gas streams where the partial pressure of CO_2 is greater than 0.4 MPa [42, 47].

Chemical absorption is one of the traditional techniques in which CO_2 from flue gases reacts with the solvent in an absorption column. The exhaust gas from power plants is obtained at high temperatures and cooled to temperatures below 50°C before feeding it to an absorption column. The cooled gas stream is allowed to pass through the absorption column in which nearly 85–90 vol% of CO_2 is absorbed. The CO_2-rich stream is fed to a regenerator or stripping column. The absorption of CO_2 operates at moderate conditions, that is, 0.1 MPa and 40–50°C while the regeneration of absorbent solvents takes place at 0.2 MPa pressure and high temperature (100–120°C) [47]. The solvents react with CO_2 present in the effluent gas and form a stable product in the absorber that is further passed to a stripping tower where the solvent can be regenerated by altering the temperature and pressure. The CO_2 lean stream solvent is recycled back to the absorption unit, and the energy intensity for recovering the solvent is high.

Table 2 summarizes the current worldwide status of research and development on several physicochemical CCS technologies. It is noteworthy that the technologies mentioned in this table are not fully implemented on a commercial scale; hence limited practical data is accessible on the fate of captured CO_2. The first choice of a feasible solvent is aqueous alkanolamines. The widely used alkanolamine to capture CO_2 from the gaseous stream is monoethanolamine. The reactivity of amines with CO_2 decreases with the increase in alkanol group (e.g., $RNH_2 < R_2NH_2 < R_3NH_2$) [43]. The reactions between amines and CO_2 can be considered similar to the reactions between a weak acid and a weak base. The zwitterions are formed as a result of the reactions between primary and secondary amines reaction. The zwitterions further form carbamates that can be reversed at rising temperatures. The carbamates thus formed are unstable in the presence of tertiary amine solvent, and bicarbonate ions become the main product. The carbamates undergo hydrolysis to form free amine molecules and bicarbonates from which the captured CO_2 is recovered in the regenerator.

The application of these solvents has constraints to capture CO_2 from the flue gas streams. The acidic gases such as SO_x and NO_x pose serious risks to these amine solutions due to their capability to form heat-stable salts that can be corrosive to the reactor in the power plants. The inert gases such as nitrogen and argon do not account for any safety or operational issues during the pipeline transport process or final storage. If not removed, 1 mol% of nitrogen can increase the energy requirement as well as CAPEX (capital expenditure) and OPEX (operational expenditure) in the transport chain with by 1% [48]. The presence of inert gases tends to reduce the volume of actual CO_2 in the transport pipeline and require the same compression energy. In addition, NO_x tends to react with some amine-based solvents and increase the overall reclaiming cost [49].

Solvent losses usually occur at high temperatures due thermal degradation and carbamate polymerization. The absorption usually happens at low temperatures while the regeneration of solvent involves high temperatures, and the solvent loss occurs in the stripping or regeneration tower. Furthermore, in the presence of oxygen, these amines degrade into undesired compounds demanding oxygen-free gas stream. To overcome these issues associated with amine solutions, mixed aqueous amines are employed to enhance their chemical characteristics to capture CO_2.

The economics of the process depends on solvent regeneration and compression of the captured CO_2 for transporting. The energy required for solvent regeneration is 3.2–4.2 GJ/ton of CO_2 and comprises of nearly 60%

Table 2. Current status of different physicochemical CO_2 capturing technologies.

CO_2 capture process	Separation technique	Status of research and development
Physical absorption	Rectisol, Selexol, etc. Mostly integrated gasification combined cycle.	1. Summit Power Group, LLC, Seattle, USA (Texas Clean Energy Project) 2. Don valley, Yorkshire, UK (Don Valley Power Project) 3. Nuon Power, Buggenum, The Netherlands (Integrated gasification combined cycle plant) 4. Elcogas, Puertollano, Spain (Integrated gasification combined cycle plant)
Chemical absorption	Amine, chilled ammonia, and amino acid salt solvent.	1. SaskPower, Saskatchewan, Canada (Boundary Dam Carbon Capture Project) 2. TransAlta Corporation, Alberta, Canada (Project Pioneer Keephills 3 Power Plant) 3. American Electric Power, Ohio, USA (Mountaineer Power Plant) 4. Dow Chemicals, West Virginia, USA (South Charleston plant) 5. PGE Bełchatów Power Station, Łódź Voivodeship, Poland 6. Hazelwood Power Station, Victoria, Australia (Hazelwood Carbon Capture Project) 7. Abu Dhabi Future Energy Company Masdar, Abu Dhabi, UAE (Siemens PostCap™ technology)
Adsorption	Pressure swing adsorption and Pressure-temperature swing adsorption	1. Under developmental stage.
Cryogenics	Cryogenic distillation	1. Air Products and Chemicals, Inc., Pennsylvania, USA
Membrane separation	Polymeric, inorganic and mixed membranes	1. Schwarze Pumpe power station, Spremberg, Germany (Oxy-fuel technology) 2. CS Energy: Callide Power Plant A, Queensland, Australia (Callide Oxy-fuel Project) 3. OxyCoal, UK (Oxy-fuel technology)
Chemical looping combustion	FeO, CuO, MnO, and NiO	1. Less large-scale demonstration plants.

of energy consumption [50]. Additionally, to capture a ton of CO_2, nearly 1.4 kg of monoethanolamine is required. In a study by Knudsen et al. [51], different process upgrades were employed to a 1 ton/h of CO_2 capture test facility operating on flue gas slipstream from a coal-fired power plant. Nearly 25% saving in regeneration energy was achieved with monoethanolamine (3.7 GJ/ton CO_2), process improvements and novel solvents. The improved process configurations and better solvents can reduce the overwhelming power demand of CO_2 removal by amine scrubbing from 0.37–0.51 MWh/ton CO_2 to 0.19–0.28 MWh/ton CO_2, equivalent to 20–30% of a power plant supply [52].

Recently, a novel solvent from amine family namely piperazine (cyclic diamine-$C_4H_{10}N_2$) has been found to enhance the performance both in capturing of CO_2 and maintaining its stability at higher temperatures and aerobic environment with minor solvent losses [53]. The piperazine solvent has a limited solubility in water; hence high temperatures are usually employed for absorption of CO_2. Aqueous ammonia solution has been found to be an alternative to the conventional amine solutions to capture CO_2 due to its inexpensiveness and low reaction heat. Due to the high volatility of ammonia, there is a high probability for its escape into the gas stream creating a technical impediment. The temperature employed in chilled ammonia process ranges from 0 to 20°C where the formed reaction products precipitate in the absorber.

An advantage of ammonia solution is its ability to capture the acid gases such as SO_x, NO_x and Hg. The stripping tower operates at 50–200°C in the pressure range of 0.2–13.7 MPa [54].

Amino acid salts are also found to be beneficial as CO_2-absorbing solvents because they are nonvolatile (eliminating solvent loss via gas phase), nontoxic, nonexplosive, odorless, and biodegradable [55]. Recently, Siemens in partnership with Abu Dhabi Future Energy Company (Masdar) in UAE developed a proprietary postcombustion CCS technology called PostCap™. The Siemens PostCap™ technology involves selective absorption of CO_2 from flue gas using amino acid salt solvent and subsequent desorption to gain high purity CO_2. The technology was successfully validated with more than 9000 operational hours in a CO_2-capturing pilot plant adapted to a coal-fired and gas-fired power plant. The PostCap™ technology, capturing 1.8 million tonnes of CO_2 per year, is planned for use in enhanced oil recovery in the local depleted fields [55]. By capturing and transporting the CO_2 emitted from a steel-making process, the enhanced oil recovery project is targeted to benefit from approximately 5 million tonnes of CO_2 captured per year [49].

Another postcombustion process of CO_2 capture is the combination of carbonation and calcination. In carbonation, CaO reacts with CO_2 to form $CaCO_3$, whereas in the regenerator $CaCO_3$ decomposes to CaO and CO_2. After regeneration, CO_2 is stored while CaO is recycled saving the operating cost. While carbonation takes place at 600–700°C, calcination occurs at temperatures greater than 900°C [56]. The carbonation and calcination occur simultaneously in absorption and regeneration fluidized-bed towers. Although it has been proved to be economical than other CO_2 capture processes (i.e., amine-based process), a primary concern for this calcium looping is the decay in the activity of the absorbent.

Recently, ionic liquids have also gained interest in CO_2 capture. Ionic liquids are salts in the liquid state made of ions and short-lived ion pairs. The thermophysical properties such as low vapor pressure, high polarity, and thermal stability with nontoxic nature make ionic liquids as promising solvents in postcombustion CO_2 capture [57]. Depending on the ionic liquid, CO_2 can be both physically and chemical absorbed. The anionic ionic liquids exhibit higher absorbing capacity over the cationic ionic liquids. Amine-based ionic liquids can be synthesized to react with CO_2 to form complexes. The absorption capacity can be enhanced by loading the synthesized ionic liquids onto stable supports (e.g., silica gel) for chemical absorption [58].

Adsorption

The shortcomings of absorption process such as lower gas-liquid contact area, low CO_2 loading, and absorbent losses have shifted the attention toward adsorption. Adsorption is both a pre- and postcombustion CO_2 capture technology that involves a fluid-solid interface for effective capturing of CO_2. It is a gas-liquid operation, in which the molecules from the gas stream are adsorbed on the surface of the solid adsorbent. The component that is adsorbed is termed as the adsorbate while the solid that adsorbs is known as the adsorbent. The fluid molecules adhere to the surface of adsorbent either by physical forces or chemical reactions. While physical adsorption is called physisorption, the chemical adsorption is known as chemisorption [41]. Since adsorption is a surface phenomenon, the properties of solid adsorbents such as surface area, polarity, and porosity along with surface-reactive species play key roles in determining an ideal adsorbent to capture CO_2 from flue gases. The criteria for selecting effective adsorbents are: (1) low cost; (2) high availability; (3) fast reaction kinetics (i.e., adsorption and desorption); (4) low heat capacity; (5) high CO_2 loading and selectivity; and (6) chemical and thermal stability both during adsorption and regeneration. As the adsorption process is an exothermic process, it favors low temperatures, whereas desorption process (i.e., regeneration) is favored at high temperatures.

Physical adsorption is based on the interactions between the molecules of CO_2 and the solid adsorbent surface. The gas solute molecules of CO_2 preferentially interact

and adhere to the adsorbent surface if the intermolecular forces between the adsorbent and gas molecules dominate the existing interactive forces between the gases. The heat of adsorption for physical adsorption process ranges from −20 to −50 kJ/mol [59]. The typical operating conditions are with partial pressures of CO_2 close to 0.1 MPa at normal temperatures (i.e., 25°C) due to the exothermic nature of the adsorption process.

Carbonaceous materials, microporous and mesoporous zeolites, chemically surface-modified polymeric materials along with metal-organic frameworks are found to be the potential adsorbents [47, 59]. Carbonaceous materials such as activated carbon, carbon nanotubes, and carbon molecular sieves can capture CO_2 by physical adsorption. Although with high thermal stability, activated carbons have low adsorption capacities. For improved adsorption characteristics by the carbon-based materials, a high surface area is desired along with chemical modification of adsorbent surface with basic groups to enhance the acidic nature CO_2. Activated carbons exhibit fast kinetics rendering less time for adsorption and desorption cycles. The major drawback of activated carbons is lower selectivity toward CO_2. Microporous carbon molecular sieves with narrow pore size distribution and high pore volume increase both selectivity and loading capacity. Carbon nanotubes (both single- and multi-walled) are growing in demand due to their selectivity toward CO_2 compared to other adsorbents.

Both natural and synthetic zeolites have been applied to capture CO_2 from flue gas streams. Among natural zeolites, mordenite, ferrierite, clinoptilolite, and chabazite exhibit an improved performance in the separation of CO_2 from N_2. The crystalline structure of zeolites results in narrow pore size distribution (i.e., 0.5–1.2 nm), which acts as molecular sieves for separation of gas mixtures [60]. The performance of natural zeolites with the high concentration of sodium and large surface area exhibit greater adsorption capacities with enhanced adsorption kinetics [61]. Conventional zeolites are synthesized by altering the silica-to-aluminum ratio and substitution of other metals (e.g., alkali and alkaline earth metals) which lead to rendering charge inside the pores [47, 59]. The structure of adsorbents can be modified by incorporating the reactive basic sites with amine or alkali/alkaline earth metals and enhance the loading capacities. The strong interaction between the basic groups with acidic CO_2 not only improves the adsorption but also increases the selectivity of CO_2 over other gases.

The adsorption capacities of zeolites depend on the pore size and characteristic ions within the pores. The charge induced by the cations in zeolite enhances the selective adsorption of oppositely charged molecules such as CO_2 (quadrupole moment 14.3×10^{-40} C.m^2)[62]. The adsorption kinetics on zeolites has been found to be quick in attaining the equilibrium loadings within a few minutes. The poor selectivity of zeolites toward CO_2 has directed to explore the incorporation of various cations into the adsorbent structures. The performance of synthetic adsorbents depends on the optimization of operating conditions (i.e., low temperature and high partial pressures), basicity (i.e., type of cations that are incorporated into the pore structure of zeolites), pore size, and pore volume.

Another class of materials named as metal-organic frameworks has received considerable attention to capture CO_2 due to high surface area and flexibility in altering the pore structures and surface properties [63]. The solid networks with the metal ion or cluster vertices coupled with organic spacers are usually known are metal-organic frameworks. The choice of organic linkers in the cluster of metal networks provides an opportunity to tune pore structure and its shape to alter CO_2 selectivity, kinetics and carbon loading capacities. Although metal-organic frameworks demonstrate high loading capacities of pure CO_2, yet in the presence of other gases the CO_2 adsorption is reduced dramatically. A considerable amount of research is being focused for improving the selectivity, stability, and recycling of metal-organic frameworks for multiple cycles. The physical adsorbents have poor selectivity and small capacities relatively at low pressures.

Unlike physical adsorption, chemisorption involves the reaction of CO_2 with the reactive groups on the surface of adsorbent, and the heat of adsorption is in the range of −60 to −90 kJ/mol [59]. Chemisorbents are classified into two classes, primarily amine-based adsorbents and alkali metal-based adsorbents [47, 59]. Amine-based adsorbents are synthesized with the help of impregnation and grafting methods. In impregnated adsorbents, weak interactions exist between the support and amine, whereas grafted amine exhibit strong covalent binding. Based on the interactions, the impregnated adsorbents demonstrate low thermal stability compared to the grafted amine-based adsorbents. Among alkali metal-based adsorbents, the carbonates of calcium, potassium, sodium and lithium have demonstrated better performance toward CO_2 capture. Although lithium-based adsorbents show high adsorption capacities, their high diffusional resistance hampers the commercial application [64, 65]. Sodium and potassium-based carbonated chemisorbents have showed high CO_2 adsorption capacities of 9.43 and 7.23 mmol/g, respectively [59]. The moderate adsorption and desorption cycle temperatures (60–200°C) make them potential adsorbents for carbon capture in the flue gases in the postcombustion processes. Microporous materials are competent with other adsorbents due to their high surface area. The polyphenylene material (PPN-4) showed adsorption capacities close to saturation due to its high surface area of 6460 m^2/g [66].

The regeneration of the applied material after the capture of CO_2 is a critical challenge to overcome the energy constraints imposed on the overall process economics. Different regeneration processes such as pressure swing, temperature swing, vacuum swing, hybrid (temperature and pressure), and electric swing adsorption are in practice [56]. The pressure swing adsorption operates at high pressures while desorption occurs at atmospheric pressures. In contrast to pressure swing adsorption, vacuum swing adsorption involves the adsorption and desorption processes at atmospheric pressure and vacuum, respectively [56]. High-temperature flue gas from the power plants is expected to reduce the temperatures for the implication of adsorbents to capture CO_2. Moreover, the flue gas often needs to be pretreated for removing any gases that can poison the adsorbent or reduce the selectivity by undesirable binding to the active sites.

Temperature swing and electric swing adsorptions are other regeneration processes where the adsorption capacities are altered with the change in temperature. Steam or hot air is used as the regeneration medium. In electric swing adsorption, the temperature of adsorbent is increased with the help of electricity (Joule effect) to liberate the adsorbates from adsorbents [56]. In situ heating of adsorbents in electric swing adsorption results in lower energy demand, fast kinetics, and dynamics without constraints on flow rate or heat transfer unlike temperature swing adsorption and vacuum swing adsorption [67]. The demand for longer regeneration time (i.e., in hours) in the case of temperature swing adsorption process compared to seconds in pressure swing adsorption makes the latter an economical option. A combination of pressure and temperature swing adsorption has also been attempted for CO_2 capture [47]. Adsorption also finds application in precombustion processes especially in steam reforming of hydrocarbons and water-gas shift reaction [68]. With the capture of CO_2, the equilibrium shifts toward the right and thereby enhances the productivity of H_2.

Membrane separation

Selective membranes have found versatile applications in the separation of gas mixtures such as in CO_2 capture from natural gas, separation of H_2 from CO_2 in synthesis gas and fuel cells. The two key parameters that decide the performance of the membranes are permeability and selectivity. The volume of gas passing through the membrane per unit area in unit time is termed as permeability. On the other hand, selectivity is defined as the ratio of permeability of the chief component to the other component in mixture. In the postcombustion processes, the flue gas stream comprises of NO_x and SO_x that have adverse impacts on the membranes. Prior to the application of membranes for separation of CO_2 from N_2, it is mandatory for the flue gas to be free of impurities. The flue gas needs to be cooled and compressed to create sufficient pressure differential for CO_2 transport.

The performance of the membranes depends on the driving force, that is, pressure differential (ΔP) for the transfer of the solute from feed to permeate side. The flux of a component (J_i) across the membrane of thickness (δ) can be obtained by Fick's law as shown in equation 1 [63].

$$J_i = \left(\frac{P_i}{\delta} \right) \Delta P \qquad (1)$$

There is always a constraint on the pressure differential in terms of economics and selectivity since the increase in pressure decreases the purity of the CO_2 stream on the permeate side. The mechanism for transfer of the solute from feed to permeate depends on the type of membranes. For example, in porous membranes the solute passes through the membrane by pore diffusion. In contrast, in nonporous and facilitated transport membranes, the solute follows the solution-diffusion mechanism and reacts with carrier molecules [69].

A variety of organic (i.e., polymeric), inorganic, mixed membrane, and hybrid systems can be applied to capture CO_2 in postcombustion processes [63, 69]. Different polymeric organic membranes are synthesized for separation of CO_2 from N_2. The organic membranes such as polyimides, polysulfones, polyarylates, polyacetylenes, polycarbonates, polyaniline, etc. exhibit satisfactory permeability (85–450 Barrer) and selectivity (5–55) [63]. The polymeric membranes with reactive carrier molecules (either amines or carbonates) are flexible to fabricate providing high selectivity due to charging. In a recent investigation, researchers have developed a polyvinylamine membrane with CO_2 permeance of 1000 GPU (Gas Permeance Unit) and selectivity of 200 [70]. Moreover, with materials having a selectivity of 50 and permeance of 1000 GPU, the CCS cost has been estimated to be as low as $23/ton [71].

Inorganic porous membranes such as zeolites, microporous silica, and carbon follow the pore diffusion mechanism for CO_2 transport [69]. Inorganic membranes have low CO_2 permeance with high selectivity. Microporous silica membranes exhibit comparable permeance and selectivity but are restricted for application due to the pore blocking with water and poisoning by SO_x [63, 69]. Membranes with inorganic particles dispersed in the continuous matrix of polymers, known as mixed matrix membranes, have also gained importance. The inorganic materials enhance the permeance and selectivity with good mechanical and thermal stability. In addition, the polymeric materials provide defect-free films at low processing cost. Research efforts are being invested to synthesize thin mixed matrix

membranes with high inorganic loading to improve both permeance and selectivity. A hybrid membrane system is the combination of membrane and absorption process to enhance the selective capturing of CO_2. Hydrophobic membranes are usually employed to avoid the interactions between gas and absorption liquid. CO_2 from the gas stream diffuses through the membrane and gets absorbed into the liquid.

Similar to postcombustion processes, membranes are also applied to precombustion processes to selectively separate CO_2 from H_2. High purity CO_2 can be attained with CO_2-selective membranes. Polymeric rubber materials with ether groups, polyethylene glycol polymeric blends and polyethylene oxide block copolymers are considered for high selectivity of CO_2 in the precombustion process [69]. Cross-linked hydrophilic membrane matrix composed of stationary and mobile carriers facilitates high CO_2/H_2 selectivity showing excellent mechanical stability and processing flexibility.

The performance of the facilitated membranes depends on the carrier molecules as well as the highly reactive and stable carriers. Mixed matrix membranes for precombustion processes are targeted to improve CO_2 selectivity and enhance the stability characteristics. The incorporation of nanofillers in polymeric matrix offers high sorption of CO_2 and increases free volume leading to high selectivity and permeability. Carbon nanotubes are found to be potential fillers with high separation and excellent mechanical stabilities [72]. Although mechanical stability can be easily achieved with mixed matrix membranes, yet their low CO_2 separation efficiency is still a challenge to overcome.

The membrane separation processes usually have high energy (i.e., electricity) requirement. For instance, in the case of a target energy requirement of 2 GJ/ton for heat regenerated process, nearly 0.5–0.7 GJ/ton of heat is allowed for membranes [73]. Ho et al. [74] attempted to pressurize the flue gas stream from a coal-fired power plant up to 0.15 MPa while maintaining the permeate stream pressure at 0.008 MPa. About 35% reduction in the cost was achieved in this trail study. Compared to U.S. $82/ton CO_2, the capture cost was U.S. $54/ton CO_2 using membrane separation at the pressurized feed conditions.

Cryogenic distillation

Cryogenic separation of flue gases is based on the freezing points of the components to condense the desired product by reducing the temperature. From the flue gas mixture, CO_2 is separated by cooling the gas mixture below $-73.3°C$ at atmospheric pressure to obtain it in liquid form [75]. The required cryogenic temperatures can be obtained by employing compressor, multi-stage heat exchangers, Joule–Thomson valve and cold traps. The composition of CO_2 in flue gas plays a key role on the operating temperature since the greater purity of CO_2 is attained at lower de-sublimation temperatures. For example, 2% CO_2 is obtained at de-sublimation temperature of −116°C, whereas 15% CO_2 is obtained at −99.9°C [63].

Currently, cryogenic separation has been performed in two ways for postcombustion processes. In the separation method proposed by Clodic and Younes [76], CO_2 is de-sublimated to solid CO_2 on the fins of heat exchangers, which is further heated and pressurized to obtain liquid CO_2 in the recovery stage. In another method by Tuinier et al. [77], packed beds are used for de-sublimation of CO_2. CO_2 is recovered from the packing material by feeding fresh gas stream to increase the temperature and enhance the concentration of CO_2 recovered from the packed bed. Though the cryogenic process is energy intensive, it does not involve any additional chemicals in the separation process.

Cryogenic distillation finds its application for the separation of O_2 from the air. The air separation unit employs cryogenic distillation for separation of O_2 from N_2 and other inert gases (mainly Ar). The energy requirement is proportional to the required purity of O_2. The condensation temperatures less than −182°C are applied to air for obtaining a pure stream of O_2 for oxy-fuel combustion [34]. Due to the oxy-fuel combustion, CO_2 concentration in flue gas stream reaches up to 89 vol% that is very high compared to other combustion processes. The efficiency of oxy-fuel CO_2 capture technology majorly depends on the method to obtain O_2. Typically, O_2 purity of 99.5 vol% is in demand, and an excess of 15% is supplied to the power plants for complete combustion [78]. The unreacted excess O_2 is recycled back, and the product stream is condensed to remove the product water and obtain pure CO_2 for compression and storage.

Some new routes to obtain the cold duty required for cryogenic separation of gases have been deployed. The cold duty from liquefied natural gas at its sites has been found to be an option to decrease the economics of cryogenic processes [56]. Cryogenic distillation is also applied in the separation of acid gases from natural gas. Controlled freeze zone process separates CH_4 and acid gases in a single unit [79]. CO_2 and other sulfur containing compounds along with heavier hydrocarbons result in the liquid product providing an economical route for transport and sequestration.

Hart and Gnanendran [80] have reported a cryogenic separation technology called CryoCell® that uses the distinctive solidification property of CO_2 as the basis of its separation from natural gas. The CryoCell® technology reduces water and chemical consumption as well as any corrosion-related issues. The cost savings in a CryoCell® plant are associated with gas treatment, CO_2 disposal, reduced electricity consumption, elimination of solvent pumping, and abridged heat requirement for amine reboiler duty.

Biological Routes for CO_2 Capture and Sequestration

Algal and bacterial systems

Microalgae, macroalgae, and cyanobacteria can be used to utilize the flue gas CO_2 as their carbon source along with sunlight and other nutrients to produce biofuels, value-added chemicals and byproducts. For the flue gas with 4% CO_2 and flow rate of 0.3 L/min, the carbon fixation rate of 15 g carbon/m^2 per day by microalgae has been attained [81]. The biofuels produced from algae are considered to be carbon-neutral as the CO_2 emitted from their combustion is consumed by fresh algae and plants during photosynthesis. Algae are simple photosynthetic aquatic organisms with estimated 300,000 species [82], that is, a diversity much greater than that of terrestrial plants. Microalgae photosynthesis can also result in the precipitation of $CaCO_3$ that is a potential long-lasting carbon sink [83].

The algal cells are made up of polysaccharides, especially hemicellulose (xylose, rhamnose, arabinose, mannose, and glucose); glucosamine; vitamin; proteins; fatty acids; lipids; and amino acids [84, 85]. Some typical green algae include *Botryococcus braunii*, *Chlorella* spp., *Chlamydomonas reinhardtii*, and *Dunaliella salina*. Algae also include diatoms such as *Phaeodactylum tricornutum* and *Thalassiosira pseudonana* as well as heterokonts such as *Nannochloropsis* and *Isochrysis* spp. [82]. Algae are used in BTL (biomass-to-liquid) and BTG (biomass-to-gas) conversion systems. They are attractive feedstock for the production of biofuels such as biodiesel, bioethanol, biobutanol, biogasoline, methane, hydrogen, and jet fuels. The lipids or the oily portion of algae are extracted and converted to biodiesel. *Botryococcus braunii*, although slow growing, is a rich source of hydrocarbons and ether lipids as it contains 60 wt% lipids within its cell wall [86]. The annual yield of oil from algae per unit area is estimated to be between 4.6 and 18.4 L/m^2 (5000 and 20,000 gallons per acre), which is about 30 times greater than the best oil-producing crop [87, 88].

The algal biomass is harvested and processed to release triacylglycerides for transesterification to produce biodiesel. However, extensive research is being done in recovering the lipids from the intracellular location of algae by energy-efficient and economical ways [89]. In addition, the primary consideration should be to convert most of the carbon from the algal biomass to biofuel with potential high-value byproducts. Biodiesel is one of the primary fuels of interest from algae because of: (1) its higher productivity than other plants; (2) higher accumulation of large amounts of triacylglycerides; and (3) lower cultivation cost with no arable land requirement. Today the top five biodiesel-producing countries in the world are USA, Germany, Argentina, Brazil, and France (Fig. 6).

After lipid extraction from algal cells, the *green* waste residue obtained is carbohydrate content of algae that can be fermented into bioethanol or biobutanol. Potts et al. [90] have demonstrated the production of butanol from macroalgae (*Ulva lactuca*) through fermentation by *Clostridium beijerinckii* and *C. saccharoperbutylacetonicum*. From the reducing sugar content of 15.2 g/L in the algal hydrolyzates, about 4 g/L of butanol was recovered as a result of fermentation. Furthermore, Ellis et al. [91] have also demonstrated the production of butanol from wastewater algae via fermentation using *C. saccharoperbutylacetonicum*. Algae such as *Gelidium amansii*, *Laminaria japonica*, *Sargassum fulvellum*, and *Ulva lactuca* have been used as raw materials for ethanol production by Kim et al. [92]. *Escherichia coli* KO11 employed for the fermentation of algal hydrolyzates resulted in ethanol yield of about 0.4 g/g of carbohydrate. The top five bioethanol-producing countries in 2012 were USA, Brazil, China, Canada, and France (Fig. 7).

Although algae utilize CO_2 during the growth cycle, their anaerobic digestion can result in the production of CH_4 as a gaseous fuel. The solar energy and CO_2 stored in the algal biomass in the form of carbohydrates could be released as CH_4 through anaerobic digestion. Yen and Brune [93] performed anaerobic co-digestion of *Scenedesmus* spp. and *Chlorella* spp. along with waste paper to generate CH_4. It is also suggested that the conversion of lipid extracted-algal biomass into CH_4 can recover more energy than the energy from the lipids alone [94]. Thermochemical liquefaction of microalga *Spirulina* using subcritical and supercritical ethanol has shown to produce high-quality bio-oil due to esterification of organic acids, ethanol, fatty acid methyl/dimethyl, and ethyl esters [95].

Although the benefits of algal cultivation are binary, particularly for CO_2 capturing and biofuel/biochemical production, yet there are some challenges to be addressed. Some disadvantages of microalgal cultivation in carbon sequestration are the high operating cost that makes the algal biofuels more expensive than fossil fuels. Sunlight is another limiting factor for algal growth. The artificial light source in the photobioreactors can be an alternative to sunlight, but their installation consequently adds to the overall production cost. Nevertheless, as the microalgal cultures also result in commercially viable products such as natural pigments, dyes, food additives, health

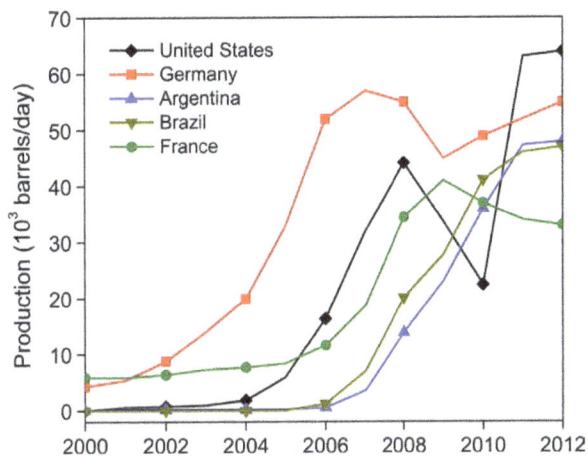

Figure 6. Top five biodiesel-producing countries in 2012 (Data source: [8]).

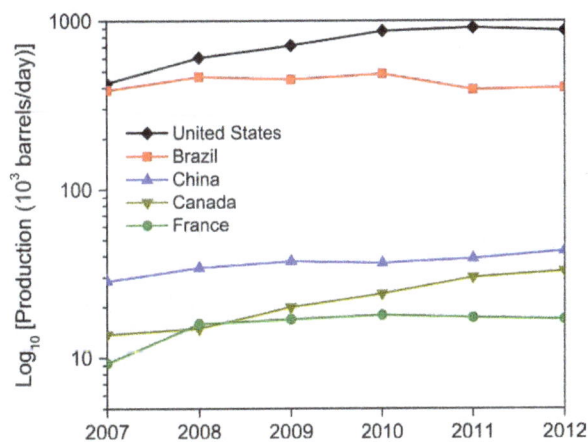

Figure 7. Top five bioethanol-producing countries in 2012 (Data source: [8]).

supplements, antioxidants, and bioactive compounds [96], their high market value may offset the high capital investment. Conversely, macroalgae have gained more attention than microalgae due to their higher biomass yield and utilization of flue gas as a direct source of CO_2, which significantly reduces the production cost [97]. The macroalgae such as *Gracilaria chilensis*, *Hizikia fusiforme*, and *Porphyra yezoensis* demonstrated nearly 2–3 times increased growth at elevated levels of CO_2 compared with the atmospheric CO_2 levels [98]. The use of flue gases containing 12–15% CO_2 is also found to maintain the desired pH range optimal for the macroalgal growth [99].

Cyanobacteria, also known as blue-green algae, are photoautotrophic bacteria that use CO_2 as the carbon source along with light and water as energy and electron sources, respectively. Cyanobacteria are photosynthetic in nature and require anaerobic growth conditions and sunlight. In addition to CO_2 fixation, cyanobacteria can also fix N_2 with the aid of nitrogenase enzymes. A few cyanobacteria that have been used for carbon fixation include *Anabaena* [100], *Aphanothece microscopica Nägeli* [101], *Chlorogleopsis* [102], *Fischerella* [103], and *Synechococcus elongates* [104]. *Synechococcus* is a promising cyanobacterium for CO_2 mitigation because of its high CO_2 uptake rate (i.e., 0.025 g/L/h or 0.6 g/L/day) at a cell concentration of 0.286 g/L [97]. Equation 2 gives the overall reaction of photosynthesis by cyanobacteria.

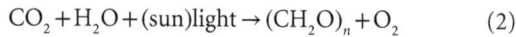

$$CO_2 + H_2O + (sun)light \rightarrow (CH_2O)_n + O_2 \qquad (2)$$

Cyanobacteria are also referred to as microbial fuel cells due to the ability to generate H_2 as a result of their metabolic pathway. A few cyanobacteria that can produce H_2 belong to the genus *Anabaena, Aphanocapsa, Calothrix, Chlamydomonas, Chroococcidiopsis, Cyanothece, Gloebacter, Gloeocapsa, Microcoleus, Microcystis, Microcystis, Nostoc, Oscillatoria, Rhodovulum, Synechococcus,* and *Synechocystis* [97, 105]. Both nitrogenase and hydrogenase enzymes in cyanobacteria can aid in biological H_2 production using CO_2 as the carbon source.

Dedicated energy crops

A dedicated energy crop is a non-food plant variety grown for the primary purpose of harvesting energy or biofuel production. The energy crops are mostly lignocellulosic in nature, and their cultivation is relatively less expensive with low-maintenance farming. The energy crops are used to generate biofuels such as bioethanol, biobutanol, bio-oil, biodiesel, and jet fuels, or combusted to generate electricity and heat. The energy crops can be woody (e.g., willow and poplar) or herbaceous in origin (e.g., *Miscanthus*, elephant grass, switchgrass, and timothy grass). These crops can be genetically modified for faster growth, greater biomass yields as well as less water and nutrient requirements. In addition to biofuel production, energy crops have tremendous potentials in carbon sequestration. In a study, *Miscanthus* and switchgrass have shown high yields with low production cost and less environmental impacts compared to conventional agricultural crops [106].

The pyrolysis of an energy crop can result in the production of combustible bio-oil and gases as well as biochar. Nanda et al. [107] and Mohanty et al. [108] have performed the carbon and energy (calorific value) balance of biochar and bio-oil from the pyrolysis of an energy crop – timothy grass. Timothy grass contains 43.4 wt% carbon (C), 6.1 wt% hydrogen (H), 1.3 wt% nitrogen (N), 0.1 wt% sulfur (S), and 45.4 wt% oxygen (O) [109]. The biochar from fast pyrolysis of timothy grass contained 63.7 wt% C, 3.6 wt% H, 1.9 wt% N, 0.04 wt% S and 30.8 wt% O [108], whereas bio-oil from the same process contained 49.2 wt% C, 9.3 wt% H, 2.2 wt% N, 0.9 wt%

S, and 38.4 wt% O [107]. However, slow pyrolysis biochar from timothy grass contained 67.5 wt% C, 2.3 wt% H, 1.9 wt% N, 0.1 wt% S and 28.2 wt% O [108], whereas slow pyrolysis bio-oil contained 44.9 wt% C, 8.4 wt% H, 1.8 wt% N, 0.5 wt% S and 44.4 wt% O [107]. As the biochar had more carbon than the biomass, it can make the biorefinery process carbon-negative upon integration with soil amendment for carbon sequestration [110]. The thermochemical conversion of energy crops can also result in energy dense bio-oil and synthesis gas. The calorific value of timothy grass biomass was determined to be 15.9 MJ/kg [109]. While fast pyrolysis bio-oil from timothy grass had high yields, it also demonstrated higher calorific value (23.8 MJ/kg) compared to low yielding slow pyrolysis bio-oil (20.2 MJ/kg) [107].

The soil contains over 2000 Gt of carbon, and the vegetation cover helps in sequestering up to 3 Gt of carbon per annum making it a significant natural carbon pool [97]. Since, the carbon captured in living vegetation is estimated to be around 550 Gt, the trees work as a good carbon sink. The energy crops are lignocellulosic in composition as they contain cellulose, hemicellulose and lignin. The biofuels produced from energy crops are considered carbon-neutral (or near-zero CO_2 accumulation) as the CO_2 emitted from their combustion is recycled by new plants for photosynthesis. In addition, energy crops do not compete with cash crops in terms of arable lands as they can grow in low-value soil without the need for intensive agricultural practices [11].

The carbon-neutral biofuels generated from energy crops can also offset some amount of fossil fuel-derived CO_2. In addition, the carbon sequestered into the soil represents supplementary benefit for CO_2 mitigation that no alternative energy source could provide [111]. Figure 8 gives the trend for the worldwide production of biofuels, especially biodiesel and bioethanol. The continual proliferation in the generation of biofuels is a good indication for the alleviation of CO_2 emissions from fossil fuels [112–114].

The physiological stimulus of plants to CO_2 has received considerable attention because CO_2 is a substrate for photosynthesis. The chief indications of elevated CO_2 on plants include: (1) stimulated photosynthesis; (2) accumulation of non-structural carbohydrates; (3) reduced tissue N_2 concentration; and (4) increased root-to-shoot ratio [115]. In plants, a substantial portion of photosynthate (i.e., 12–54% of carbon fixed by photosynthesis) is allocated to stimulate the root growth [116]. Much of this additional photosynthate is made available to soil microorganisms as root exudates. Hence, elevated atmospheric CO_2 levels could have a significant impact on the energy flow through microbial food webs in the soil. Nevertheless, studies have shown that the rhizosphere of

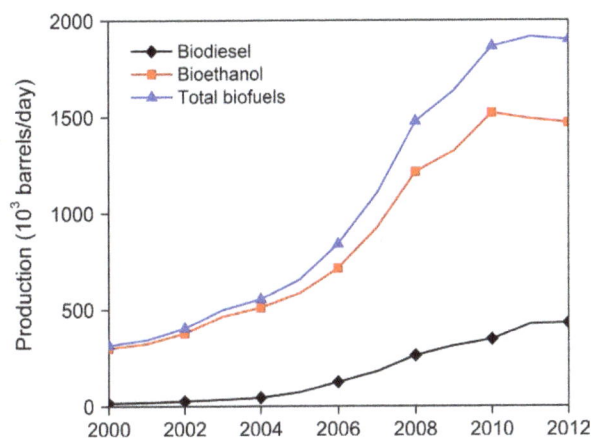

Figure 8. Worldwide production of biofuels from 2000 to 2012 (Data source: [8]).

the plants grown under elevated CO_2 contained 60% of soluble carbon [117]. In most terrestrial energy crop systems, much of the total carbon is below the ground because the soil contains two-third of the total terrestrial carbon in the biosphere [118].

As a response to elevated CO_2, the carbon assimilation in plants can increase through enhanced carbon fixation. This can also lead to higher carbon storage in the soil which is believed to be the largest and most stable terrestrial carbon pool. The high CO_2 levels lead to greater CO_2 fixation in energy crops through photosynthesis and higher carbon storage in the soil through soil-microbial-plant interactions. In addition, soil stores 2–3 times more carbon compared to the atmosphere. The temperature sensitivity of soil organic matter also determines the amount of soil carbon storage and emission over time. The long-term rise in atmospheric CO_2 may stimulate biomass production from energy crops through enhanced root growth resulting in greater carbon input in the soil through rhizodeposition and root exudates [119].

It should also be noted that increased soil microbial activity can lead to a loss of carbon from the soil by respiration and decomposition. However, increased CO_2 concentration can suppress soil organic matter degradation as microorganisms would primarily decompose plant root exudates as readily available carbon source before gearing up to utilize other organic matter from the soil via the priming effect [120]. The soil microorganisms are the chief regulators of soil organic matter and any alteration in its availability impedes their priming effect. In a study with an increase in CO_2 concentration, there was about 85% increase in rhizosphere bacterial population and 170% rise in respiring rhizosphere bacteria [121].

The natural decomposition and mineralization processes transforming soil organic matter into inorganic forms are affected by soil temperature and moisture content [122]. Soil temperature and moisture affect the soil microbial activity at microscale levels, which eventually impacts the cycling of soil organic carbon at local, regional and even global scenarios. The soil organic carbon decomposition and mineralization should be minimized to make an energy crop management system efficient for carbon sequestration. Soil disturbances such as change in the soil temperature, moisture, nutrients, and mineral matter maintain an optimal level of soil organic carbon and prevent its premature decomposition. The farming practices such as tillage is another soil disturbance factor that can disrupt soil aggregates that otherwise occludes soil organic matter [123]. Tillage can increase aeration and land surface area for triggered bacterial activity leading to the decomposition of soil organic matter.

Coalbed methanogenesis

The world coal reserve has dominated as one of the primary sources for fossil fuels. As mentioned earlier, there are significant consequences related to the direct combustion of coal concerning the GHG emissions. By reducing or eliminating the need for coal combustion, carbon emissions can be reduced dramatically, considering that a 500-megawatt coal-fired power plant produces 3 tonnes of CO_2 per year. This could potentially result in a net CO_2 emission reduction of 25% or more in the developed countries. A small but growing body of work related to coalbed methanogenesis exists today [124] including studies on methane-producing microbial communities from different coal seams in USA [125–127], Canada [128], China [129], and India [130].

Biogenic CH_4 is generated through anaerobic microbial process of methanogenesis and constitutes a significant fraction of the natural gas reserves. It is generally accepted that biogenic coalbed CH_4 is a product of coal biodegradation by methanogenic archaea and syntrophic bacteria inhabiting coal beds [129]. These microorganisms artificially stimulate the regeneration of biogenic coalbed CH_4. There have also been reports of field trials conducted in USA where stimulants have been added to coal seams to enhance biogenic CH_4 production [131]. Although the understanding of microbial coal conversion remains rudimentary, it is thought that both acetoclastic and hydrogenotrophic methanogens produce CH_4 by consuming small molecules generated from in situ biodegradation of coal components by other microorganisms [132]. It is reported that acetotrophic, hydrogenotrophic, and methylotrophic methanogenesis are the pathways for production of biogenic CH_4 in coalbed reservoirs [133, 134].

A study on the bioconversion of coal using core-flooding experiments revealed that coal can be economically converted into CH_4 and other value-added products [135]. Along with CH_4 generation, the methanogens produce CO_2, which can be pumped into a depleted oil reservoir for enhanced oil recovery [136], thereby enabling an efficient mechanism for geological carbon sequestration. However, anaerobic subsurface coal bioconversion science is in its infancy. While methanogenesis seems to be a feasible approach, other possibilities for biofuel generation are currently largely untapped. The understanding of microbial life in such coal seams, often refer to as the "microbial dark matter", is still being pursued vigorously due to the advancements of DNA sequencing and genomic analysis. New pathways for in situ bioconversion of coal to CH_4 would shed more light toward the overall reduction in GHGs.

Geological Routes for Carbon Sequestration

Biochar amendment

Increasing the soil carbon content is one of the best strategies to sequester carbon for longer timescales because more than 80% of organic carbon is preserved in the soil [137]. Biochar is one of the promising bioresources rich in recalcitrant carbon, thus making it a long-standing carbon pool. Biochar, a byproduct of biomass pyrolysis and gasification, is obtained along with crude bio-oil and gas components (e.g., H_2, CO, CO_2 and CH_4) [108, 138]. It comprises mostly of the recalcitrant aromatic forms of carbon that cannot be readily oxidized and released into the atmosphere as CO_2 [139]. The occurrence of biochar in the Amazonian *terra preta* soil is a historical indication that biochar can preserve carbon in the soil for hundreds to thousands of years. This makes biochar application a carbon-negative approach for CO_2 sequestration.

Biochar can also capture CO_2 through carbon sequestration and exclude NO_x and SO_x from flue gas through air purification. Biochar applied to soil can also help decrease emissions of N_2O and CH_4 whose respective potencies are 300 and 23 times higher than CO_2 [140]. The carbon sequestration by biochar is different to that of afforestation, conversion into trees, and no-tillage agriculture. While agricultural areas switched to no-tillage farming may lose carbon capturing efficiency within two decades, forestlands start maturing over the years and release CO_2 [141]. These divergences make biochar a long-term carbon sink for reducing CO_2 emissions. The non-charred carbonaceous materials such as dead plant or animal wastes decompose in the soil to release carbon slowly over time that might take a few decades. However,

the complete loss of carbon from biochar is very slow and difficult to estimate suggesting it to preserve carbon from centuries to millennia.

Integrating pyrolysis with carbon sequestration can help a biorefinery earn carbon credits. In other words, pyrolysis of biomass converts about 50% of the biomass carbon to bio-oil and gases, while the remaining 50% of carbon is stabilized in biochar [142]. Although burning of bio-oil and combustible gases (H_2 and CH_4) for energy can release CO_2, it is also consumed by plants for photosynthesis making the process carbon-neutral. Consequently, amending biochar to the soil can capture the stable biomass carbon for longer durations making the entire refinery process carbon-negative [11]. This helps in CO_2 abatement and earning carbon credits. In Brazil, carbonization of sugarcane bagasse has shown to optimize the pyrolysis process by producing biochar for use in household and industrial applications [143].

Carbon credits are based in the ratio of 1:1 in relation to the tons of CO_2 stored or removed through carbon sequestration. Nearly 8 Gt of CO_2 is being accumulated per year through burning of fossil fuels [144]. This represents around 8–10 billion carbon credits being created every year. At present, CO_2 trading price by Chicago Climate Exchange is set at U.S. $4/ton [141]. In such a scenario, the carbon credit economy would be comparable in size to the current fossil fuel economy. Hence, the Earth's capacity for storing biochar would be limitless. For evaluating the carbon budget and carbon credit of an integrated bioenergy system, a few factors should be taken into consideration such as: (1) GHG emissions associated with the production of biomass; (2) CO_2 emissions and energy input during biomass harvest, transportation, pretreatment, and conversion; and (3) energy input in product processing and upgrading, flue gas purification, and biochar handling. These factors could address the amount of GHG emissions and carbon balance in each step of the overall biomass conversion chain with carbon sequestration potential by biochar.

Biochar also has several advantages to the soil upon its amendment. Biochar enhances the available water holding capacity, plants' root development, and soil faunal population [145]. The micropores in biochar allow the diffusion of air and water, thereby reducing the overall bulk density of the soil. Biochar, when mixed with the soil, has the increased internal surface area for better adsorption of soil nutrients and humus. Biochar retains substantial quantities of mineral matter in the soil to make them available to plants through their roots. Lehmann [139] suggested two main routes through which biochar can reduce soil or groundwater pollution, that are: (1) by retaining nutrients and minerals in the soil, thereby lowering their chances of leaching into groundwater; and

(2) by improving nutrient availability in the soil thus reducing the use of chemical fertilizers.

Oceanic storage and mineralization

The oceanic CO_2 storage has an expected retention rate of several hundreds of years [146]. The formations of underground carbonate minerals prolong the residence time reducing the chances of CO_2 escape into the atmosphere [147]. The ocean contains approximately 40,000 Gt of carbon compared to 750 Gt of carbon in the atmosphere and 2200 Gt of carbon in the terrestrial ecosystem [56]. The carbon sequestration in oceans is expected at the rate of 5 Gt of carbon per year by 2100 [148]. In the scenario of projected climate change where high CO_2 saturation in oceans may enhance the sequestration rate due to greater carbon fixation by marine photosynthetic organisms, higher temperature might also decrease the solubility of CO_2 in ocean water [97].

Fung et al. [149] performed a series of experiments with the National Center for Atmospheric Research (NCAR) – Climate System Model (CSM1.4) to understand the influence of land and ocean as repositories and sinks of CO_2. The CSM1.4 model included the modified terrestrial biogeochemistry model (i.e., Carnegie–Ames–Stanford Approach or CASA) and the modified Ocean Carbon Intercomparison Project 2 (i.e., OCMIP-2) oceanic biogeochemistry model. The modeling results suggested that carbon sink strengths vary with CO_2 emission rates. In simple words, the terrestrial and oceanic carbon storage capacity can decrease with rapid emissions of CO_2 and other GHGs. According to the model, the magnitude of oceanic carbon storage will depend on the oceanic circulation and sensitivity of marine ecosystem processes at changing climatic conditions (i.e., high temperature and elevated CO_2 levels).

The CO_2 after being separated from the flue gas mixture is compressed to liquid or supercritical fluid state for storage deep under the oceans via pipeline [56]. The CO_2 can be transported and injected into the ocean as dry ice or through vertical injection, inclined pipe transfer, or pipe towed by ship [150]. The longevity of CO_2 captured and injecting in deep reservoirs under the ocean is around thousands of years. Under supercritical conditions, the CO_2 is less dense than water and tends to migrate up the storage formation. Water above its critical temperature ($T_c > 374°C$) and critical pressure ($P_c > 22.1$ MPa) is called supercritical water. Likewise, CO_2 above its critical temperature ($T_c > 31°C$) and critical pressure ($P_c > 7.4$ MPa) is called as supercritical CO_2. In the event of increasing pressure or weakening of the geological reservoir, the highly pressured CO_2 inclines to escape causing uncertainties on its deep water storability

[151]. Recently, saline aquifers were found to be attractive for geological CO_2 storage under water [152, 153]. However, the geological storage of CO_2 under the high-pressure system is under the developmental stage.

The inorganic carbon content in the oceans worldwide is estimated to be around 38,000 Gt with up to 2 Gt of carbon being sequestered annually [97]. The carbon sequestration in oceans can also occur naturally through several ways such as: (1) photosynthesis by deep water aquatic plants as well as photosynthetic bacteria and algae; (2) decomposition of dead aquatic plants and animals; and (3) carbonates formation. The carbon fixed by marine plants and microorganisms is also sequestered upon their decomposition and vertical flux for settlement in the ocean bed.

The photosynthesis by marine plants and microorganisms in the sunlit region and the turbulent diffusion of ocean water ensure the constant flux of carbon from the oceanic surfaces [154]. The biological carbon fixation and subsequent sequestration are enhanced through phytoplankton in the ocean by supplementing additional nutrients such as nitrogen and iron. Such a scheme can also help to earn carbon credits in response to the increased CO_2 emissions. The carbon fixed by the top surface photosynthetic organisms (e.g., aquatic plants, phytoplankton, algae, and photosynthetic bacteria) is eventually sequestered upon their death and descending to the ocean floor. However, any ecological disturbances and mutations in the plankton and photosynthetic microorganisms as a result of the additional nutrients cannot be ignored. The tumbling of surplus quantities of biomass formation from photosynthetic organisms and plankton can also lead to anaerobic digestion and CH_4 formation counteracting carbon sequestration [146].

The solubility pump is a technology used to dissolve CO_2 in ocean water for carbonate formation. This involves the conversion of CO_2 to inorganic carbonates through mineralization. The mineralization process offers an opportunity for the durable storage of CO_2, although it is subject to natural weathering process over longer durations. After the carbonates are generated, their complexes are carried away via water current towards the benthic ocean regions where they linger for longer periods [155]. In the oceans, the carbon is stored in the form of carbonates, especially $CaCO_3$ and $MgCO_3$, and their formation is summarized in equations 3 and 4.

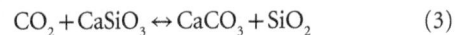

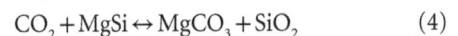

$$CO_2 + CaSiO_3 \leftrightarrow CaCO_3 + SiO_2 \qquad (3)$$

$$CO_2 + MgSi \leftrightarrow MgCO_3 + SiO_2 \qquad (4)$$

The carbonate formation acts as a long-term source of carbon that is driven by the metabolic activities by photosynthetic organisms including cyanobacteria and algae [156]. The formation of carbonates especially $CaCO_3$ in the form of shells is also found in coral reefs by many aquatic organisms such as benthic molluscs, echinoderms, corals, and plankton belonging to the group coccolithophore. However, the particulate organic matter and dissolved organic matter in oceans have a tendency to undergo microbial mineralization releasing most of the organic matter to dissolved inorganic carbon [156]. However, a small proportion of particulate organic matter escapes mineralization as it reaches the sediment where organic carbon remains buried for hundreds of years [157].

The geological CO_2 storage can also have some limitations in the case of unforeseen CO_2 leakage. Due to asphyxiation, CO_2 poses life-threatening risks for aquatic organisms as it can accumulate locally due to its density. Higher CO_2 levels may also change the water pH causing acidification of groundwater, thereby dissolving many toxic heavy metals [158]. The ocean acidification by CO_2 leakage has a direct impact on the growth of corals. It also alters the chemical speciation and biogeochemical cycling of many elements and compounds. Ocean acidification also lowers the saturation states of $CaCO_3$, which has adverse impacts on the shell-forming marine organisms such as plankton, corals, mussels, snails, clams, etc. [159]. The broader implications and adaptation of marine organisms against increasing CO_2 is not well understood for ocean ecosystems; hence future research is required to advance the innocuous geological CO_2 storage approach.

Conclusions

Carbon sequestration is considered as a feasible approach to mitigate the increased greenhouse gas emissions that lead to global warming and climate change. The currently available technologies for carbon capture and sequestration can be classified under physicochemical, biological and geological routes. The physicochemical route includes some fast-track techniques for the separation of CO_2 from flue gas mixtures with the use of absorbents, adsorbents, gas separation membranes and cryogenic distillation. In contrast, the biological route involves a long-term approach such as carbon fixation through the cultivation of microalgae, macroalgae, cyanobacteria, and energy crops. The biofuels derived from energy crops (e.g., temperate grasses and short-rotation woody biomass) are accounted as carbon-neutral as the plants recycle the CO_2 released from biofuel combustion. Coal, a geological fossil fuel resource, is also subject to biogenic degradation by methanogenic bacteria resulting in CH_4 generation. This biogenic CH_4 could be considered as a clean fuel gas compared to the CO_2 that would otherwise be released upon coal combustion. Finally, the geological storage of carbon includes amendment of biochar into the soil, pumping of CO_2 deep into the oceans, and mineralization of CO_2 in the form of carbonates in seawater. Although these routes seem to be promising, yet their commercial applications are required for implementation toward carbon credits and carbon trading. The current challenge is to develop an international regulatory framework that would help decide the viability of these carbon sequestering technologies based on a long-term lifecycle assessment.

Acknowledgments

The authors thank Natural Sciences and Engineering Research Council of Canada (NSERC) for the financial support toward this renewable energy research.

Conflict of Interest

None declared.

References

1. Metz, B., O. R. Davidson, P. R. Bosch, R. Dave, and L. A. Meyer. 2007. Climate change 2007: mitigation of climate change. Cambridge Univ. Press, New York, USA.
2. Rastogi, M., S. Singh, and H. Pathak. 2002. Emission of carbon dioxide from soil. Curr. Sci. 82:510–517.
3. Watson, R. T., M. C. Zinyowera, and R. H. Moss. 1996. Climate change 1995: impacts, adaptations and mitigation of climate change. Cambridge Univ. Press, New York, USA.
4. Battle, M., M. Bender, T. Sowers, P. P. Tans, J. H. Butler, J. W. Elkins, et al. 1996. Atmospheric gas concentrations over the past century measured in air from fin at the South Pole. Nature 383:231–235.
5. NOAA (National Centers for Environmental Information). 2015. Climate Monitoring – Global Summary Information April 2015. Asheville, North Carolina. Available at https://www.ncdc.noaa.gov/sotc/summary-info/global/201504 (accessed 30 June 2015).
6. U.S. Census Bureau. 2016. U.S. and World Population Clock. Available at http://www.census.gov/popclock (accessed 21 March 2016).
7. United Nations. 2007. World Population Prospects: The 2006 Revision, Population database. Economics and Social Affairs. United Nations, New York.
8. USEIA (U.S. Energy Information Administration). 2015. Independent Statistics & Analysis. Available at http://www.eia.gov (accessed 17 October 2015).
9. IEO (International Energy Outlook). 2011. U.S. Energy Information Administration, Washington, DC. Available at http://www.eia.gov/ieo/pdf/0484(2011).pdf. (accessed 01 March 2011).
10. Chu, S., and A. Majumdar. 2012. Opportunities and challenges for a sustainable energy future. Nature 488:294–303.

11. Nanda, S., R. Azargohar, A. K. Dalai, and J. A. Kozinski. 2015. An assessment on the sustainability of lignocellulosic biomass for biorefining. Renew. Sustain. Energ. Rev. 50:925–941.

12. The Guardian. 2011. Canada pulls out of Kyoto protocol. Available at http://www.theguardian.com/environment/2011/dec/13/canada-pulls-out-kyoto-protocol (accessed 30 June 2015).

13. Sutter, J. D., J. Berlinger, and R. Ellis. 2015. Obama: climate agreement 'best chance we have' to save the planet. Cable News Network (CNN). Available at http://www.cnn.com/2015/12/12/world/global-climate-change-conference-vote (accessed 17 January 2016).

14. UNCCC. 2016. United Nations Climate Change Conference COP21/CMP11. Available at http://www.cop21.gouv.fr/en/learn/what-is-cop21/the-phenomenon-of-climate-disruption (accessed 17 January 2016).

15. Solomon, S., D. Qin, M. Manning, M. Marquis, K. Avery, M. M. B. Tignor, et al. 2007. Climate change 2007: the physical science basis. Cambridge Univ. Press, New York, USA.

16. EEA (European Environment Agency). 2015. Atmospheric concentration of carbon dioxide, methane and nitrous oxide. Copenhagen, Denmark. Available at http://www.eea.europa.eu

17. Earth CO_2 Home Page. 2016. CO_2 Earth: Are we stabilizing yet?. Available at www.co2.earth (accessed 21 March 2016).

18. Pachauri, R. K., and A. Reisinger. 2007. Climate change 2007: synthesis report. Intergovernmental Panel on Climate Change (IPCC). IPCC, Geneva, Switzerland.

19. Smith, W. N., P. Rochette, C. Monreal, R. L. Desjardins, E. Pattey, and A. Jaques. 1997. The rate of C change in agricultural soils in Canada at the landscape level. Can. J. Soil Sci. 77:219–229.

20. Raich, J. W., and C. S. Potter. 1995. Global patterns of carbon dioxide emissions from soils. Glob. Biogeochem. Cyc. 9:23–36.

21. UNFCCC (United Nations Framework Convention on Climate Change). 2008. Kyoto protocol reference manual: on accounting of emissions and assigned amount. UNFCCC, Bonn, Germany.

22. Yang, S. S., C. M. Lui, C. M. Lai, and Y. L. Lui. 2003. Estimation of methane and nitrous oxide emission from paddy fields and uplands during 1990–2000 in Taiwan. Chemosphere 52:1295–1305.

23. Isermann, K. 1994. Agriculture's share in the emission of trace gases affecting the climate and some cause-oriented proposals for sufficiently reducing this share. Environ. Pollut. 83:95–111.

24. Yu, K. W., Z. P. Wang, A. Vermoesen, W. H. Jr Patrick, and O. Van Cleemput. 2001. Nitrous oxide and methane emissions from different soil suspensions: effect of soil redox status. Biol. Fertil. Soils 34:25–30.

25. Burton, M. R., and G. M. Sawyer. 2013. Deep carbon emissions from volcanoes. Rev. Mineral. Geochem. 75:323–354.

26. Carey, S., P. Nomikou, K. C. Bell, M. Lilley, J. Lupton, C. Roman, et al. 2013. CO_2 degassing from hydrothermal vents at Kolumbo submarine volcano, Greece, and the accumulation of acidic crater water. Geology 41:1035–1038.

27. Gerlach, T.. 2011. Volcanic versus anthropogenic carbon dioxide. Eos 92:201–208.

28. McHale, M. R., E. G. McPherson, and I. C. Burke. 2007. The potential of urban tree plantings to be cost effective in carbon credit markets. Urban For. Urban Green. 6:49–60.

29. Cairns, R. D., and P. Lasserre. 2006. Implementing carbon credits for forests based on green accounting. Ecol. Econ. 56:610–621.

30. Wang, B., Y. Li, N. Wu, and C. Q. Lan. 2008. CO_2 bio-mitigation using microalgae. Appl. Microbiol. Biotechnol. 79:707–718.

31. Nanda, S., J. Mohammad, S. N. Reddy, J. A. Kozinski, and A. K. Dalai. 2014. Pathways of lignocellulosic biomass conversion to renewable fuels. Biomass Conv. Bioref. 4:157–191.

32. Gibbins, J., and H. Chalmers. 2008. Carbon capture and storage. Energ. Policy 36:4317–4322.

33. Thiruvenkatachari, R., S. Su, H. An, and X. X. Yu. 2009. Post combustion CO_2 capture by carbon fibre monolithic adsorbents. Prog. Energ. Combust. Sci. 35:438–455.

34. Markewitz, P., W. Kuckshinrichs, W. Leitner, J. Linssen, P. Zapp, R. Bongartz, et al. 2012. Worldwide innovations in the development of carbon capture technologies and the utilization of CO_2. Energ. Environ. Sci. 5:7281–7305.

35. Rubin, E. S., H. Mantripragada, A. Marks, P. Versteeg, and J. Kitchin. 2012. The outlook for improved carbon capture technology. Prog. Energ. Combust. Sci. 38:630–671.

36. IPCC (Intergovernmental Panel on Climate Change). 2005. Special report on carbon dioxide capture and storage. Cambridge Univ. Press, New York, USA.

37. Kather, A., and G. Scheffknecht. 2009. The oxycoal process with cryogenic oxygen supply. Naturwissenschaften 96:993–1010.

38. Burdyny, T., and H. Struchtrup. 2010. Hybrid membrane/cryogenic separation of oxygen from air for use in the oxy-fuel process. Energy 35:1884–1897.

39. Richter, H. J., and K. F. Knoche. 1983. Reversibility of combustion processes. Pp. 71–85 in R. A. Gaggioli, ed. Efficiency and costing ACS symposium series 235. Oxford Univ. Press, Washington, D.C., USA.

40. Lyngfelt, A., B. Kronberger, J. Adanez, J. X. Morin, and P. Hurst. 2004. The GRACE project. Development

of oxygen carrier particles for chemical-looping combustion. Design and operation of a 10 kW chemical-looping combustor. Proceedings of the 7th International Conference on Greenhouse Gas Control Technologies, Vancouver, Canada, September 2004.

41. Kwon, J. H., and L. D. Wilson. 2010. Surface-modified activated carbon with β-cyclodextrin—Part II. Adsorption properties. J. Environ. Sci. Health A Tox. Hazard. Subst. Environ. Eng. 45:1793–1803.

42. Chabanon, E., B. Belaissaoui, and E. Favre. 2014. Gas–liquid separation processes based on physical solvents: opportunities for membranes. J. Membr. Sci. 459:52–61.

43. Padurean, A., C. C. Cormos, A. M. Cormos, and P. S. Agachi. 2011. Multicriterial analysis of post-combustion carbon dioxide capture using alkanolamines. Int. J. Greenh. Gas Control 5:676–685.

44. Padurean, A., C. C. Cormos, and P. S. Agachi. 2012. Pre-combustion carbon dioxide capture by gas–liquid absorption for Integrated Gasification Combined Cycle power plants. Int. J. Greenh. Gas Control 7:1–11.

45. Burr, B., and L. Lyddon. 2008. A comparison of physical solvents for acid gas removal. 87th Annu. Conv. Proc. 2–5.

46. Kaneco, S., H. Katsumata, T. Suzuki, and K. Ohta. 2006. Electrochemical reduction of carbon dioxide to ethylene at a copper electrode in methanol using potassium hydroxide and rubidium hydroxide supporting electrolytes. Electrochim. Acta 51:3316–3321.

47. Yu, C. H., C. H. Huang, and C. S. Tan. 2012. A review of CO_2 capture by absorption and adsorption. Aerosol Air Qual. Res. 12:745–769.

48. Aspelund, A., and K. Jordal. 2007. Gas conditioning— The interface between CO_2 capture and transport. Int. J. Greenhouse Gas Control 1:343–354.

49. Birley, R., A. Reichl, T. Schliepdiek, and O. Reimuth. 2014. Optimisation and integration of two post combustion capture plants to reduce CAPEX and OPEX. Power-Gen Europe, Cologne, Germany, June 3–5, 2014. Seimens AG, Germany: 1–15.

50. Knudsen, J. N., J. N. Jensen, P. J. Vilhelmsen, and O. Biede. 2009. Experience with CO_2 capture from coal flue gas in pilot-scale: testing of different amine solvents. Energ. Proc. 1:783–790.

51. Knudsen, J. N., J. Andersen, J. N. Jensen, and O. Biede. 2011. Evaluation of process upgrades and novel solvents for the post combustion CO_2 capture process in pilot-scale. Energ. Proc. 4:1558–1565.

52. Rochelle, G. T. 2009. Amine scrubbing for CO_2 capture. Science 325:1652–1654.

53. Freeman, S. A., R. Dugas, D. H. Van Wagener, T. Nguyen, and G. T. Rochelle. 2010. Carbon dioxide capture with concentrated, aqueous piperazine. Int. J. Greenh. Gas Control 4:119–124.

54. Darde, V., K. Thomsen, W. J. Van Well, and E. H. Stenby. 2010. Chilled ammonia process for CO_2 capture. Int. J. Greenh. Gas Control 4:131–136.

55. Reichl, A., G. Schneider, T. Schliepdiek, and O. Reimuth. 2014. CCS for enhanced oil recovery: Integration and optimization of a post combustion CO_2 capture facility at a power plant in Abu Dhabi. Society of Petroleum Engineers. Proceedings of Abu Dhabi International Petroleum Exhibition and Conference, Abu Dhabi, UAE, November 10–13, 2014. SPE 171692. SPE Int.: 1–11.

56. Pires, J. C. M., F. G. Martins, M. C. M. Alvim-Ferraz, and M. Simões. 2011. Recent developments on carbon capture and storage: an overview. Chem. Eng. Res. Des. 89:1446–1460.

57. Zhang, X., X. Zhang, H. Dong, Z. Zhao, S. Zhang, and Y. Huang. 2012. Carbon capture with ionic liquids: overview and progress. Energ. Environ. Sci. 5:6668–6681.

58. Zhang, Y., S. Zhang, X. Lu, Q. Zhou, W. Fan, and X. Zhang. 2009. Dual amino-functionalised phosphonium ionic liquids for CO2 capture. Chem. Eur. J. 15:3003–3011.

59. Samanta, A., A. Zhao, G. K. Shimizu, P. Sarkar, and R. Gupta. 2011. Post-combustion CO_2 capture using solid sorbents: a review. Ind. Eng. Chem. Res. 51:1438–1463.

60. Chester, A. W., and E. G. Derouane. 2009. Zeolite characterization and catalysis: a tutorial. Springer, Dordrecht, The Netherlands.

61. Siriwardane, R. V., M. S. Shen, and E. P. Fisher. 2003. Adsorption of CO_2, N_2, and O_2 on natural zeolites. Energy Fuels 17:571–576.

62. Coriani, S., A. Halkier, A. Rizzo, and K. Ruud. 2000. On the molecular electric quadrupole moment and the electric-field-gradient-induced birefringence of CO_2 and CS_2. Chem. Phys. Lett. 326:269–276.

63. Spigarelli, B. P., and S. K. Kawatra. 2013. Opportunities and challenges in carbon dioxide capture. J. CO2 Utilization 1:69–87.

64. Iwan, A., H. Stephenson, W. C. Ketchie, and A. A. Lapkin. 2009. High temperature sequestration of CO_2 using lithium zirconates. Chem. Eng. J. 146:249–258.

65. Venegas, M. J., E. Fregoso-Israel, R. Escamilla, and H. Pfeiffer. 2007. Kinetic and reaction mechanism of CO_2 sorption on Li_4SiO_4: study of the particle size effect. Ind. Eng. Chem. Res. 46:2407–2412.

66. Dawson, R., A. I. Cooper, and D. J. Adams. 2013. Chemical functionalization strategies for carbon dioxide capture in microporous organic polymers. Polym. Int. 62:345–352.

67. An, H., B. Feng, and S. Su. 2011. CO_2 capture by electrothermal swing adsorption with activated carbon fibre materials. Int. J. Greenh. Gas Control 5:16–25.

68. Mendes, D., A. Mendes, L. M. Madeira, A. Iulianelli, J. M. Sousa, and A. Basile. 2010. The water-gas shift

reaction: from conventional catalytic systems to Pd-based membrane reactors—a review. Asia-Pac. J. Chem. Eng. 5:111–137.

69. Ramasubramanian, K., Y. Zhao, and W. S. W. Ho. 2013. CO2 capture and H2 purification: prospects for CO2-selective membrane processes. AIChE J. 59:1033–1045.

70. Kim, T. J., H. Vrålstad, M. Sandru, and M. B. Hägg. 2013. Separation performance of PVAm composite membrane for CO_2 capture at various pH levels. J. Membr. Sci. 428:218–224.

71. Merkel, T. C., H. Lin, X. Wei, and R. Baker. 2010. Power plant post-combustion carbon dioxide capture: an opportunity for membranes. J. Membr. Sci. 359:126–139.

72. Murali, R. S., S. Sridhar, T. Sankarshana, and Y. V. L. Ravikumar. 2010. Gas permeation behavior of Pebax-1657 nanocomposite membrane incorporated with multiwalled carbon nanotubes. Ind. Eng. Chem. Res. 49:6530–6538.

73. Favre, E. 2011. Membrane processes and postcombustion carbon dioxide capture: challenges and prospects. Chem. Eng. J. 171:782–793.

74. Ho, M. T., G. W. Allinson, and D. E. Wiley. 2008. Reducing the cost of CO_2 capture from flue gases using membrane technology. Ind. Eng. Chem. Res. 47:1562–1568.

75. Olajire, A. A. 2010. CO_2 capture and separation technologies for end-of-pipe applications–a review. Energy 35:2610–2628.

76. Clodic, D., and M. Younes. 2002. A new method for CO_2 capture: frosting CO_2 at atmospheric pressure. Sixth International Conference on Greenhouse Gas Control Technologies, GHGT6. Kyoto, October 2002, 155–160.

77. Tuinier, M. J., M. van Sint Annaland, G. J. Kramer, and J. A. M. Kuipers. 2010. Cryogenic CO_2 capture using dynamically operated packed beds. Chem. Eng. Sci. 65:114–119.

78. Castillo, R. 2011. Thermodynamic analysis of a hard coal oxyfuel power plant with high temperature three-end membrane for air separation. Appl. Energ. 88:1480–1493.

79. Northrop, P. S., and J. A. Valencia. 2009. The CFZ™ process: a cryogenic method for handling high-CO2 and H_2S gas reserves and facilitating geosequestration of CO_2 and acid gases. Energ. Proc. 1:171–177.

80. Hart, A., and N. Gnanendran. 2009. Cryogenic CO_2 capture in natural gas. Energ. Proc. 1:697–706.

81. Watanabe, Y., and D. O. Hall. 1996. Photosynthetic CO_2 conversion technologies using a photobioreactor incorporating microalgae—energy and material balances. Energ. Convers. Manage. 37:1321–1326.

82. Scott, S. A., M. P. Davey, J. S. Dennis, I. Horst, C. J. Howe, D. J. Lea-Smith, et al. 2010. Biodiesel from algae: challenges and prospects. Curr. Opin. Biotechnol. 21:277–286.

83. Aresta, M., A. Dibenedetto, and G. Barberio. 2005. Utilization of macro-algae for enhanced CO_2 fixation and biofuels production: development of a computing software for an LCA study. Fuel Process. Technol. 86:1679–1693.

84. Toroghi, M. K., G. Goffaux, and M. Perrier. 2013. Observer-based backstepping controller for microalgae cultivation. Ind. Eng. Chem. Res. 52:7482–7491.

85. Abo-Shady, A. M., Y. A. Mohamed, and T. Lasheen. 1993. Chemical composition of the cell wall in some green algae species. Biol. Plantarum 35:629–632.

86. Metzger, P., and C. Largeau. 2005. Botryococcus braunii: a rich source for hydrocarbons and related ether lipids. Appl. Microbiol. Biotechnol. 66:486–496.

87. Atabani, A. E., A. S. Silitonga, I. A. Badruddin, T. M. I. Mahlia, H. H. Masjuki, and S. Mekhilef. 2012. A comprehensive review on biodiesel as an alternative energy resource and its characteristics. Renew. Sust. Energ. Rev. 16:2070–2093.

88. Yeang, K.. 2008. Biofuel from algae. Architect. Des. 78:118–119.

89. Hu, Q., M. Sommerfeld, E. Jarvis, M. Ghirardi, M. Posewitz, M. Seibert, et al. 2008. Microalgal triacylglycerols as feedstocks for biofuel production: perspectives and advances. Plant J. 54:621–639.

90. Potts, T., J. Du, M. Paul, P. May, R. Beitle, and J. Hestekin. 2012. The production of butanol from Jamaica bay macro algae. Environ. Prog. Sust. Energ. 31:29–36.

91. Ellis, J. T., N. N. Hengge, R. C. Sims, and C. D. Miller. 2012. Acetone, butanol, and ethanol production from wastewater algae. Bioresour. Technol. 111:491–495.

92. Kim, N.-J., H. Li, K. Jung, N. M. Chang, and P. C. Lee. 2011. Ethanol production from marine algal hydrolysates using Escherichia coli KO11. Bioresour. Technol. 102:7466–7469.

93. Yen, H. W., and D. E. Brune. 2007. Anaerobic co-digestion of algal sludge and waste paper to produce methane. Bioresour. Technol. 98:130–134.

94. Sialve, B., N. Bernet, and O. Bernard. 2009. Anaerobic digestion of microalgae as a necessary step to make microalgal biodiesel sustainable. Biotechnol. Adv. 27:409–416.

95. Huang, H., X. Yuan, G. Zeng, J. Wang, H. Li, C. Zhou, et al. 2011. Thermochemical liquefaction characteristics of microalgae in sub- and supercritical ethanol. Fuel Process. Technol. 92:147–153.

96. Singh, S., B. N. Kate, and U. C. Banerjee. 2005. Bioactive compounds from cyanobacteria and microalgae: an overview. Crit. Rev. Biotechnol. 25:73–95.

97. Farrelly, D. J., C. D. Everard, C. C. Fagan, and K. P. McDonnell. 2013. Carbon sequestration and the role of

biological carbon mitigation: a review. Renew. Sust. Energ. Rev. 21:712–727.

98. Wu, H. Y., D. H. Zou, and K. S. Gao. 2008. Impacts of increased atmospheric CO_2 concentration on photosynthesis and growth of micro- and macro-algae. Sci. China C: Life Sci. 51:1144–1150.

99. Israel, A., J. Gavrieli, A. Glazer, and M. Friedlander. 2005. Utilization of flue gas from a power plant for tank cultivation of the red seaweed *Gracilaria cornea*. Aquacult. 249:311–316.

100. Lopez, C. V. G., F. G. A. Fernandez, J. M. F. Sevilla, J. F. S. Fernandez, M. C. C. Garcia, and E. M. Grima. 2009. Utilization of the cyanobacteria *Anabaena* sp. ATCC 33047 in CO_2 removal processes. Bioresour. Technol. 100:5904–5910.

101. Jacob-Lopes, E., L. M. C. F. Lacerda, and T. T. Franco. 2008. Biomass production and carbon dioxide fixation by *Aphanothece microscopica Nägeli* in a bubble column photobioreactor. Biochem. Eng. J. 40:27–34.

102. Ono, E., and J. L. Cuello. 2007. Carbon dioxide mitigation using thermophilic cyanobacteria. Biosyst. Eng. 96:129–134.

103. Weissman, J. C., J. C. Radway, E. W. Wilde, and J. R. Benemann. 1998. Growth and production of thermophilic cyanobacteria in a simulated thermal mitigation process. Bioresour. Technol. 65:87–95.

104. Miyairi, S. 1995. CO_2 assimilation in a thermophilic cyanobacterium. Energ. Convers. Manage. 36:763–766.

105. Dutta, D., D. De, S. Chaudhuri, and S. K. Bhattacharya. 2005. Hydrogen production by Cyanobacteria. Microb. Cell Fact. 4:36.

106. Smeets, E. M. W., I. M. Lewandowski, and A. P. C. Faaij. 2009. The economical and environmental performance of miscanthus and switchgrass production and supply chains in a European setting. Renew. Sust. Energ. Rev. 13:1230–1245.

107. Nanda, S., P. Mohanty, J. A. Kozinski, and A. K. Dalai. 2014. Physico-chemical properties of bio-oils from pyrolysis of lignocellulosic biomass with high and slow heating rate. Energ. Environ. Res. 4:21–32.

108. Mohanty, P., S. Nanda, K. K. Pant, S. Naik, J. A. Kozinski, and A. K. Dalai. 2013. Evaluation of the physiochemical development of biochars obtained from pyrolysis of wheat straw, timothy grass and pinewood: effects of heating rate. J. Anal. Appl. Pyrol. 104:485–493.

109. Nanda, S., P. Mohanty, K. K. Pant, S. Naik, J. A. Kozinski, and A. K. Dalai. 2013. Characterization of North American lignocellulosic biomass and biochars in terms of their candidacy for alternate renewable fuels. Bioenerg. Res. 6:663–677.

110. Nanda, S., A. K. Dalai, F. Berruti, and J. A. Kozinski. 2016. Biochar as an exceptional bioresource for energy, agronomy, carbon sequestration, activated carbon and specialty materials. Waste Biomass Valor. 7:201–235.

111. Ragauskas, A. J., C. K. Williams, B. H. Davison, G. Britovsek, J. Cairney, C. A. Eckert, et al. 2006. The path forward for biofuels and biomaterials. Science 311:484–489.

112. Beerthuis, R., G. Rothenberg, and N. R. Shiju. 2015. Catalytic routes towards acrylic acid, adipic acid and ε-caprolactam starting from biorenewables. Green Chem. 17:1341–1361.

113. Moshkelani, M., M. Marinova, M. Perrier, and J. Paris. 2013. The forest biorefinery and its implementation in the pulp and paper industry: energy overview. Appl. Ther. Eng. 50:1427–1436.

114. Dhabhai, R., S. P. Chaurasia, K. Singh, and A. K. Dalai. 2013. Kinetics of bioethanol production employing mono- and co-cultures of *Saccharomyces cerevisiae* and *Pichia stipitis*. Chem. Eng. Technol. 36:1651–1657.

115. Chung, H., D. R. Zak, and E. A. Lilleskov. 2006. Fungal community composition and metabolism under elevated CO_2 and O_3. Oecologia 147:143–154.

116. Rogers, H. H., S. A. Prior, and S. V. Krupa. 1994. Plant responses to atmospheric CO_2 enrichment with emphasis to roots and rhizosphere. Environ. Pollut. 83:155–189.

117. Cheng, W., and D. W. Johnson. 1998. Elevated CO_2, rhizosphere processes, and soil organic matter decomposition. Plant Soil 202:167–174.

118. Lin, G., J. R. Ehleringer, P. T. Rygiewicz, M. G. Johnson, and D. T. Tingey. 1999. Elevated CO_2 and temperature impacts on different components of soil CO_2 efflux in Douglas-fir terracosms. Glob. Change Biol. 5:157–168.

119. Denef, K., H. Bubenheim, K. Lenhart, J. Vermeulen, O. V. Cleemput, P. Boeckx, et al. 2007. Community shifts and carbon translocation within metabolically-active rhizosphere microorganisms in grasslands under elevated CO_2. Biogeosciences 4:769–779.

120. Carney, K. M., B. A. Hungate, B. G. Drake, and J. P. Megonigal. 2007. Altered soil microbial community at elevated CO_2 leads to loss of soil carbon. Proc. Natl Acad. Sci. USA 104:4990–4995.

121. Montealegre, C. M., C. van Kessel, M. P. Russelle, and M. J. Sadowsky. 2002. Changes in microbial activity and composition in a pasture ecosystem exposed to elevated atmospheric carbon dioxide. Plant Soil 243:197–207.

122. Sartori, F., R. Lal, M. H. Ebinger, and D. J. Parrish. 2006. Potential soil carbon sequestration and CO_2 offset by dedicated energy crops in the USA. Crit. Rev. Plant Sci. 25:441–472.

123. Molina, J. A. E., C. E. Clapp, M. J. Shaffer, F. W. Chichester, and W. E. Larson. 1983. NCSOIL, a model

of nitrogen and carbon transformations in soil: description, calibration, and behavior. Soil Sci. Soc. Am. J. 47:85–91.

124. Wei, X. R., G. X. Wang, P. Massarotto, S. D. Golding, and V. Rudolph. 2007. A Review on recent advances in the numerical simulation for coalbed-methane-recovery process. SPE Reserv. Eval. Eng. 10:657–666.

125. Ulrich, G., and S. Bower. 2008. Active methanogenesis and acetate utilization in Powder River Basin coals, United States. Int. J. Coal Geol. 76:25–33.

126. Strąpoć, D., F. W. Picardal, C. Turich, I. Schaperdoth, J. L. Macalady, J. S. Lipp, et al. 2008. Methane-producing microbial community in a coal bed of the Illinois basin. Appl. Environ. Microbiol. 74:2424–2432.

127. Flores, R. M., C. A. Rice, G. D. Stricker, A. Warden, and M. S. Ellis. 2008. Methanogenic pathways of coal-bed gas in the Powder River Basin, United States: the geologic factor. Int. J. Coal Geol. 76:52–75.

128. Penner, T. J., J. M. Foght, and K. Budwill. 2010. Microbial diversity of western Canadian subsurface coal beds and methanogenic coal enrichment cultures. Int. J. Coal Geol. 82:81–93.

129. Guo, H., Z. Yu, R. Liu, H. Zhang, Q. Zhong, and Z. Xiong. 2012. Methylotrophic methanogenesis governs the biogenic coal bed methane formation in Eastern Ordos Basin, China. Appl. Microbiol. Biotechnol. 96:1587–1597.

130. Singh, D. N., A. Kumar, M. P. Sarbhai, and A. K. Tripathi. 2012. Cultivation-independent analysis of archaeal and bacterial communities of the formation water in an Indian coal bed to enhance biotransformation of coal into methane. Appl. Microbiol. Biotechnol. 93:1337–1350.

131. Pfeiffer, R. S., G. Ulrich, G. Vanzin, V. Dannar, R. P. DeBruyn, and J. B. Dodson. 2008. Biogenic fuel gas generation in geological hydrocarbon deposits. United States Patent No. 7,426,960 B2.

132. Bennett, B., and S. R. Larter. 2000. Quantitative separation of aliphatic and aromatic hydrocarbons using silver ion-silica solid-phase extraction. Anal. Chem. 72:1039–1044.

133. Strąpoć, D., M. Mastalerz, K. Dawson, J. Macalady, A. K. Callaghan, B. Wawrik, et al. 2011. Biogeochemistry of microbial coal-bed methane. Annu. Rev. Earth Pl. Sc. 39:617–656.

134. Shimizu, S., M. Akiyama, T. Naganuma, M. Fujioka, M. Nako, and Y. Ishijima. 2007. Molecular characterization of microbial communities in deep coal seam groundwater of northern Japan. Geobiology 5:423–433.

135. Stephen, A., A. Adebusuyi, A. Baldygin, J. Shuster, C. G. Southam, C. K. Budwill, et al. 2014. Bioconversion of coal: new insights from a core flooding study. RSC Adv. 4:22779–22791.

136. Baldygin, A., D. S. Nobes, and S. K. Mitra. 2014. Water-alternate-emulsion (WAE): a new technique for enhanced oil recovery. J. Pet. Sci. Eng. 121:167–173.

137. IPCC (Intergovernmental Panel on Climatic Change). 2000. Land use, land-use change, and forestry. Cambridge Univ. Press, Cambridge, UK.

138. Nanda, S., R. Azargohar, J. A. Kozinski, and A. K. Dalai. 2014. Characteristic studies on the pyrolysis products from hydrolyzed Canadian lignocellulosic feedstocks. Bioenerg. Res. 7:174–191.

139. Lehmann, J. 2007. Bio-energy in the black. Front. Ecol. Environ. 5:381–387.

140. Renner, R. 2007. Rethinking biochar. Environ. Sci. Technol. 41:5932–5933.

141. Lehmann, J.. 2007. A handful of carbon. Nature 447:143–144.

142. Lehmann, J., J. Gaunt, and M. Rondon. 2006. Biochar sequestration in terrestrial ecosystems - a review. Mitig. Adapt. Strategies Glob. Chang. 11:403–427.

143. Zandersons, J., J. Gravitis, A. Kokorevics, A. Zhurinsh, O. Bikovens, A. Tardenaka, et al. 1999. Studies of the Brazilian sugarcane bagasse carbonisation process and products properties. Biomass Bioenerg. 17:209–219.

144. Mathews, J. A. 2008. Carbon-negative biofuels. Energ. Policy 36:940–945.

145. Major, J., J. Lehmann, M. Rondon, and C. Goodale. 2010. Fate of soil-applied black carbon: downward migration, leaching and soil respiration. Glob. Change Biol. 16:1366–1379.

146. Stewart, C., and M. A. Hessami. 2005. A study of methods of carbon dioxide capture and sequestration—the sustainability of a photosynthetic bioreactor approach. Energ. Convers. Manage. 46:403–420.

147. Herzog, H. 2001. What future for carbon capture and sequestration? Environ. Sci. Technol. 35:148–153.

148. Cox, P. M., R. A. Betts, C. D. Jones, S. A. Spall, and I. J. Totterdell. 2000. Acceleration of global warming due to carbon-cycle feedbacks in a coupled climate model. Nature 408:184–187.

149. Fung, I. Y., S. C. Doney, K. Lindsay, and J. John. 2005. Evolution of carbon sinks in a changing climate. Proc. Natl Acad. Sci. USA 102:11201–11206.

150. Khoo, H. H., and R. B. H. Tan. 2006. Life cycle investigation of CO_2 recovery and sequestration. Environ. Sci. Technol. 40:4016–4024.

151. Xie, X., and M. J. Economides. 2009. The impact of carbon geological sequestration. J. Nat. Gas Sci. Eng. 1:103–111.

152. Michael, K., A. Golab, V. Shulakova, J. Ennis-King, G. Allinson, S. Sharma, et al. 2010. Geological storage of CO_2 in saline aquifers—a review of the experience from existing storage operations. Int. J. Greenh. Gas Control 4:659–667.

153. Yang, F., B. J. Bai, D. Z. Tang, S. Dunn-Norman, and D. Wronkiewicz. 2010. Characteristics of CO_2 sequestration in saline aquifers. Petrol. Sci. 7:83–92.

154. Shoji, K., and I. S. F. Jones. 2001. The costing of carbon credits from ocean nourishment plants. Sci. Total Environ. 277:27–31.

155. Anderson, L. G., T. Tanhua, G. Bjork, S. Hjalmarsson, E. P. Jones, S. Jutterstrom, et al. 2010. Arctic ocean shelf-basin interaction: an active continental shelf CO_2 pump and its impact on the degree of calcium carbonate solubility. Deep Sea Res. 57(Pt 1):869–879.

156. Raven, J. A., and P. G. Falkowski. 1999. Oceanic sinks for atmospheric CO_2. Plant, Cell Environ. 22:741–755.

157. Jiao, N., G. J. Herndl, D. A. Hansell, R. Benner, G. Kattner, S. W. Wilhelm, et al. 2010. Microbial production of recalcitrant dissolved organic matter: long-term carbon storage in the global ocean. Nature 8:593–599.

158. van der Zwaan, B., and K. Smekens. 2009. CO_2 capture and storage with leakage in an energy-climate model. Environ. Model. Assess. 14:135–148.

159. Doney, S. C., V. J. Fabry, R. A. Feely, and J. A. Kleypas. 2009. Ocean acidification: the other CO_2 problem. Ann. Rev. Mar. Sci. 1:169–192.

4

Low-grade heat recycling for system synergies between waste heat and food production, a case study at the European Spallation Source

Thomas Parker[1] & Anders Kiessling[2]

[1]WA3RM AB, Lund, Sweden
[2]Swedish University of Agricultural Sciences, Uppsala, Sweden

Keywords

Agriculture, aquaculture, cooling, horticulture, temperature, waste heat

Correspondence

Thomas Parker, WA3RM AB, Sandgatan 14F, 223 50 Lund, Sweden.
E-mail: thomas@wa3rm.se

Funding Information

Part of this work was substantially assisted by EU Grant Agreement No.: 312453 "EuCARD-2" for Enhanced European Coordination for Accelerator Research & Development

Abstract

At present food production depends almost exclusively on direct use of stored energy sources, may perhaps they be nuclear-, petroleum-, or biobased. Arable land, artificial fertilizers, and fresh water resources are the base for our present food systems, but are limited. At the same time, energy resources in the form of waste heat are available in ample quantities. The European Spallation Source (ESS) will require approximately 270 GWh of power per year to operate, power that ultimately is converted to heat. This multidisciplinary case study details an alternative food production cooling chain, using low-grade surplus heat, and involving fermentation, aquaculture, nutrient recapture, and greenhouse horticulture including both use of low-grade surplus heat and recycling of society's organic waste that is converted to animal feed and fertilizer. The study indicates that by combining the use of surplus energy with harvest of society's organic side flows, for example, food waste and aquatic-based cash crops, sustainable food systems are possible at a level of significance also for global food security. The effects of the proposed heat reuse model are discussed in a system perspective and in the context of the UNSCD indicator framework. The potential sustainability benefits of such an effort are shown to be substantial and multifaceted.

Introduction

The opportunity to recycle low-grade heat

In recent years, there has been a substantially increased research effort into the use of low-grade waste heat. The driver behind this interest seems to be the combination of concern of the climate impact from energy use and the substantial supply of low-grade heat. Low-grade heat is plentiful, because it is a by-product of thermal power production as well as various industrial processes in sectors such as metals and pulp and paper. In the United Kingdom, 11.4 TWh of recoverable heat was found to be wasted each year [1]. However, this figure only represents the wastage where there is a technically viable use available. The total amount of wasted heat is 48 TWh

per year [2]. Similarly, "In the USA, over two-thirds of the primary energy supply is ultimately rejected as low-grade waste heat" [3]. Sweden, considered a world leader in heat recycling with its well-developed district heating networks, reuses 4 TWh of 9.5 available industrial waste heat [4]. This figure does not include the considerably larger waste heat streams from nuclear power.

An emerging source of low-grade waste heat may be data centers. In these, "temperatures as high as 60°C are sufficient to cool microprocessors" and "switching to liquid cooling [is] inevitable" [5]. This would lead to two improvements, greater efficiency in the data center and the possibility to utilize the waste heat.

Certainly, the need to address the climate impact of world energy supply is well established. The concept of "energy poverty" also pinpoints a need to address energy efficiency from a viewpoint of economy and equity. The recovery of low-grade heat has been shown to be able to play a role in this and the opportunities for doing so seem to be increasing [6]. A low-temperature district heating network greatly facilitates heat recovery from industrial waste heat and leads to more efficient industry, cheaper heat, and lower emissions [7].

Purpose of the research

There is thus significant indication of an opportunity to create sustainability benefits by recycling low-grade heat. The purpose of research is to contribute to knowledge about how low-grade waste heat can be used in ways that are both practicable and sustainable. Target audiences for this article are sustainability managers, energy managers, government bodies involved in planning and energy systems, and researchers in the field.

To achieve the research goal, based on the identified case, the following research questions were posed:

1. What uses for industrial waste heat have been identified with development potential?
2. What are the identified sustainability benefits of the identified heat-recycling initiatives?
3. What potential sustainability benefits and costs might be associated with the heat recycling and how may they be evaluated?

Methods

This research is based on an in-depth case study. The organization in the case, the European Spallation Source (ESS) is large-scale, multinational research facility, a type of institution often called "research infrastructure." The study of research infrastructure offers some advantages. As organizations dedicated to facilitating scientific endeavor, research infrastructure tends to be default support research and to be open to the study. They tend to have an academic culture and publish design reports and other documents, and work with peer review as a management process [8]. Energy issues in research infrastructure have in recent years also attracted considerable interest [9].

The case is of special interest because ESS has committed to recycling its waste heat, and doing so in an as efficient way as possible. The commitment is strong, having been made between partner governments.

To achieve its goals, the ESS has formed a close collaboration with the Swedish University of Agricultural Sciences (SLU) and other interested parties, and has developed a proposal, based on the biological systems to reuse low-grade heat, which is detailed in the case study.

The case is analyzed in the perspectives of the interaction of the involved systems and sustainability impacts of the case proposal in these systems. In the systems perspective, the point of departure is the effect in the case on the energy system, and the analysis continues with connections to food and nutrient systems and water.

Many tools and indices to assess sustainability have been proposed, and these proposals have in turn been analyzed. In one such sustainability indices, the authors conclude that "We show that these indices fail to fulfill fundamental scientific requirements making them rather useless if not misleading with respect to policy advice" [10].

Indices do not express objective truths, but instead are powerful expressions of a chosen set of values [11]. "Indicators arise from values (we measure what we care about), and they create values (we care about what we measure)" [12] and further "from a scientific point of view, there cannot be such a thing as one comprehensive measure or index of sustainability" [13, 14].

Therefore, the sustainability assessment will focus on identifying the relevant categories of sustainability impacts and discussing the case from a systems perspective in these categories. Rather than attempting to assess the proposal into an index figure, the case is discussed in the context of each relevant theme in the framework put forward by the United Nations Commission for Sustainable Development [15].

Structure of the article

This first section of this article serves to introduce the issues and thereafter to present the purpose of the research and the methods employed to attain them. The rest of the article is structured as follows: "Use of Low-Grade Heat in the Literature" section contextualizes the study by giving a brief overview of uses of low-grade heat from the literature. Section "Case Study: The European Spallation Source" thereafter presents the case study, starting with the organization studied, its goals, and operating conditions as well as design solutions implemented at the point of study. Relevant local market and climate conditions are also briefly presented. Section "Proposal: waste heat for food production with a nutrient loop" contains a presentation of a proposal for improved heat recycling at lower temperatures that included integration to horticulture and aquaculture. Section "Analysis" is the analysis of the findings in the case, starting with the general applicability of the case study to industry, thereafter applying a systems perspective and lastly a sustainability perspective. And the final section presents the main conclusions of the study.

Use of Low-Grade Heat in the Literature

Potential uses of low-grade heat

The simplest use for low-grade waste heat is for space heating. Today, temperatures as low as 40°C can easily be used for heating purposes, either via ventilation or floor heating. The environmental benefits of this can be substantial, if the waste heat replaces burning of fossil fuels. Looking at waste heat in the United Kingdom, "one-third of all fossil fuels consumed in the UK to produce low-grade heat for buildings" and further that "district heating schemes can provide cost-effective and low-carbon energy to local populations" [16]. The word "local" is significant. Heating, as opposed to electrical power, can be relatively easily stored, but is difficult to transport. Therefore, "direct heat use will depend on whether [a] potential user can be found" [6].

Aside from residential space heating, low-grade heat may also be supplied to greenhouses. However, as a cooling source for industry, greenhouses are viable only in winter, or in very northern climate areas [17].

Heating demand for greenhouses in Sweden is around 0.5 TWh [18, 19]. Nonetheless, reducing the sustainability impact of energy use would be a significant improvement of the sustainability performance of greenhouses. Also, "energy is typically the largest over-head cost in the production of greenhouse crops." Counting indirect energy use, fertilizer is one of the most energy-consuming parts of greenhouse operation, accounting for 21% of energy use [18].

Greenhouses contribute to sustainable development by vastly increasing the yield for a given area. The increase can be by a factor of 10–20 times compared to outdoor horticulture [18].

An intriguing possibility is to generate electrical power with waste heat. The most common proposal for this to make use of the organic Rankine cycle (ORC), but many other proposals exist, such as Stirling engines or condensing boilers [1, 20]. Suppliers of ORC systems claim to be able to produce power at temperatures such as 80°C and even lower, but typically require a heat sink with a temperature difference to the supply of at least 40°C. Moreover, at these extreme levels the production is not very economical. Other systems tend to demand higher temperatures and/or higher temperature differences. According to Fang et al. [7], industrial uses of waste heat include desalination and power generation, but are difficult at temperatures under around 200°C.

It would seem more advantageous to find uses of heat not requiring conversion to mechanical or electrical power. There is a significant difference between *recovery* and use as heat or *upgrade* to work, electricity or cooling [3].

Aside from space heat, refrigeration and desalination allow direct use without conversion to mechanical work [5]. Desalination can be achieved at as low temperatures as 45–50°C using "near vacuum level pressures" [6]. Even freezing is possible [21]. The much-awaited hydrogen economy would open up new possibilities, such as bio-hydrogen via dark fermentation at 70°C [22], or biohydrogen and biomethane at 37°C [23]. Additional uses include bacteria growth, typically at 37–38°C, biogasification, drying biomass, and production of a variety of substances, including ammonia, hydrogen, and pure water [6].

The drying process can have a crucial impact on the total energy efficiency of the use of biomass as an energy carrier, because the drying requires significant amounts of energy. Similarly, in digestion processes, digestibility has been shown to increase by 5–10% when the temperature is increased from 35°C to 50°C, but the energy use for heating was higher than the gain [24]. Production of protein meals for constructed animal feed from biomass sources as fish, microbes, macroalgae, plants, and insects require energy for drying. For example, meal production from as varying sources as yeast or fish requires removal of more than 70–80% of the weight in water [25]. Traditionally, high-temperature systems with temperatures over 100°C have been used, but more recently it has become clear that lower drying temperatures improve the quality of the product [26]. Today more than 27 million tons of fish is processed into fishmeal annually [27, 28] requiring removal of nearly 20 million tons of water. Considering the prognosis of more than 90 million ton of farmed fish to 2030 [28] and a replacement of soy as the major protein source in their diet with single cell protein (as yeast, microalgae etc.), macroalgae, feed mussel meal, and/or insects, more than 300 million tons of water needs to be evaporated by low-temperature drying techniques in order to form a transportable commodity to be used in the aquatic feed production. Even if fish has an uniquely high protein need, the amounts will be similar or even larger in the terrestrial farmed animals considering their larger volumes and that also in these animals the use of human grade protein sources (mainly soy and corn) must be replaced by alternative sources like the ones mentioned earlier and thereby require more energy for evaporating water than presently is the case. Considering that 1 kg of water requires 2.3 MJ for evaporation, it becomes clear that alternative techniques to the present ones based on fossil energy are urgently needed.

Production of fish and microalgae in more closed systems will require heating, especially in temperate climates. The need of energy will vary with farming temperature, exchange of water, and ventilation. Salmon requires 14°C as the optimal temperature, while tilapia prefer close to 30°C. Most fish will have a Q10 of more than a doubling and, for example, will salmon smolt double their growth already when temperature is increased from 9°C to 14°C [29, 30]. On the other hand, fish, in parity with all biological organisms, are more sensitive to a temperature above their optimum than to a lower one [31]. In a complete flow through system roughly 70 m³ of water is needed per kg of salmon production, while about 2/3 is needed in less oxygen-demanding and CO_2-resistant species [32]. Given input of oxygen and CO_2 stripper, this requirement could be reduced to 50% and with high-technology filtration, protein skimming, and biological filters a reduction to 2.5% is possible [33]. However, in most food production system an exchange below 15% of farming volume per day is unusual due to quality reason. A quick calculation then gives with hand that a standing biomass of, for example, 1000 ton, with a density of 50 kg/m³, will require a total farming volume of 20,000 m³. An ambient water temperature of 10°C and a farming temperature of, for example, 25°C would require at an exchange of 15% heating of 3000 m³ at 15°C per day.

The importance of temperature

Temperature is fundamental to heat recycling: "The temperature of the low-grade heat stream is the most important parameter, as the effective use of the residual heat or the efficiency of energy recovery from the low-grade heat sources will mainly depend on the temperature difference between the source a suitable sink, for example another process or space heating/cooling" [6].

Within the literature discussing recycling and use of low-grade heat, there is variance in the definition "low grade," including the "widely accepted threshold temperature": 250°C [6, 20], 260°C [1], 60°C, and 120°C [3], and a typical heat from a solar collector (70°C) [21].

Case Study: The European Spallation Source

The European Spallation Source

The European Spallation Source (ESS) is a large-scale research facility in construction in Lund, Sweden. The facility will supply neutrons and a suite of neutron instruments for use in research in materials, life science, energy, and other disciplines. The facility will generate a far stronger neutron flux than existing facilities. To generate the flow of neutrons, a linear proton accelerator, the most powerful in the world, will propel bunches of 10^{12} protons into a target in the form of a large wheel of tungsten. In this target, the spallation process takes place, generating 30 neutrons for every proton.

Neutrons, being neutral particles, can penetrate into materials and can be used to create images of the insides of materials and substances on a nanometer scale and with nanosecond resolution. Neutrons are particularly useful for investigating light atoms, such as hydrogen, carbon, and oxygen found in organic molecules. This makes neutrons an important tool within life sciences, sustainability, and energy research. Within energy specifically, neutrons can help study the movement and structure of ions in batteries and fuel cells. The storage of hydrogen in metal substrates is another active research area, as is carbon capture and storage. With somewhat different research methods, neutrons can also be used to investigate photovoltaics and photosynthesis. The ESS and similar facilities can be used for in situ studies, typically looking at ongoing combustion processes to explore mechanical and chemical process improvements. A highly specialized area of research is superconductivity. This research makes use of another property of neutrons, their magnetic spin, to explore how magnetic properties of materials change with temperature.

Development of a sustainability strategy

The decision to build ESS in Lund was preceded by a competition between countries to host the facility. It was in this competition that the host governments of Sweden and Denmark committed to building a sustainable research facility, by implementing and energy strategy called Responsible, Renewable, and Recyclable. This meant that the facility was to be energy efficient, use energy from renewable sources, and recycle the waste heat resulting from activities. Importantly, this trio of goals was given as a hierarchy, so that energy efficiency was of higher priority than heat recycling [34].

The three parts of the energy strategy were given specific requirements in an Energy Policy [35]. The target for energy efficiency was set for a maximum of 270 GWh total annual energy use at full operation. The target for Renewable was that all energy used would be from renewable sources, and for Recyclable, it was that all "recuperated" waste heat would be reused. "Recuperated" meant heat captured in cooling systems.

ESS Scandinavia, as the Swedish and Danish bid to host was called, also proposed a shift to an almost completely superconducting linear accelerator, which was a significant gain in overall efficiency at ESS. A superconducting accelerator does not suffer from losses due to resistance in the accelerator, but requires cryogenics to chill the accelerator to approximately 2 K, in itself an energy-intensive process. Despite this, superconducting was still a net gain. This technological leap also had an important effect on heat recycling, as the cryogenics is required to run constantly, in order to preserve the cryogenic helium and to avoid thermal expansion in the accelerator, which would require time-consuming retuning. Therefore, a superconducting accelerator requires a constant minimum cooling, and thus supplies a constant minimum flow of heat.

The original energy concept envisioned heat recycling to the local district heating system. This system supplies Lund, as well as neighboring townships with a total of around a TWh of heat per year. Additionally, the operator has embarked on a project to connect this system with more distant heating systems in neighboring cities. However, district heating systems typically operate at temperatures of 80–120°C, whereas cooling systems for accelerator-based research facilities typically have cooling loops at two levels, one cooling-tower level of 30–40°C and one chilled water level of 5–20°C. Technically, this gap could easily be bridged with the use of heat pumps, but to do so for the full heat load at ESS would have directly conflicted with the "Responsible" goal of energy efficiency.

Since before the ESS Scandinavian proposal, ESS Scandinavia, and subsequently ESS has been collaborating in various agreements with the district heating operator Lunds Energi (now Kraftringen) to pursue a sustainable research facility. Throughout this long-term collaboration Kraftringen has been pursuing an independent effort to reduce temperatures in it local district heating system to reduce losses. As of December 2013, the two parties have reached a formal agreement to connect ESS to the district heating system. This agreement requires ESS to provide a temperature of 80°C, which is deemed sufficient for heating needs. The ESS receives back a temperature of under 50°C.

The evolution of the ESS energy strategy has been analyzed in sustainability strategy research [36].

Energy inventory

ESS has implemented a program to raise cooling temperature levels, in order to make use of the district heating system directly as cooling, an effort that has led to attention in the field [9]. High-temperature cooling is being implemented primarily for the klystrons providing the accelerating power and for the helium compressors providing the cryogenic cooling. Nevertheless, a significant amount of cooling is still necessary at lower temperature levels. The energy inventory, a biannual exercise conducted at ESS shows the projected energy use as well as the cooling needs at the three temperature levels.

The latest available energy inventory [37] shows a total annual power use of 265 GWh, of which 60 GWh is for the heat pumps that provide the lower temperature cooling and eject heat to the district heating system. The estimated total recovered heat amounts to 253 GWh per year, divided into temperature levels according to Table 1.

Table 1. The cooling temperature levels at ESS.

Temperature levels	Supply temperature (°C)	Return temperature (°C)	Part of total (%)
Low	5–15	30–35	30
Medium	30–35	40–50	35
High	40–50	75–80	35

The possible savings from recycling heat at lower temperatures than the 80°C required for the district heating system are thus significant.

Sustainability issues aside from energy at ESS

The indirect effect of energy was considered the most important and variable sustainability issue at ESS, but of course there were others as well. The facility would generate substantial volumes of radioactive waste, albeit mostly at rather low levels. Radiation protection was an important issue. Radioactive waste handling and radiation safety issues were dealt with a regulatory framework.

An issue that came up in relation to local inhabitants and in conjunction with the regulatory process was the use of 60 ha of prime agricultural land for the construction of the facility. This had led to some local opposition in the licensing process from the local farmer's organization and a local nature conservation organization. Placing this large facility on such excellent soil was perceived as unsustainable and a threat to future food security.

Indicative prices in the case

A full business case for heat recycling alternatives was not yet available, but some indicative prices could be uncovered. The long-term gross power price on the Nordic market is estimated at 5 ¢/kWh (eurocents per kilowatt hour). ESS is exempt from energy tax, but this would otherwise amount to 3 ¢/kWh. The average price paid for heat recovered to a district heating system in Sweden was 2 ¢/kWh. The average price of district heating to large companies in Sweden was 7 ¢/kWh. At an estimated COP for a heat pump to cool at 40°C and eject heat to district heating at more than 80°C would be around 4, indicating a cost of electrical power for the process of ¼ times 5 ¢ or 1.25 ¢/kWh.

Climate conditions in the case

The conventional cooling solution for research infrastructure facilities like the ESS is to either make use of a local body of water, if available and allowed, or to use cooling towers. Heating a local ecosystem has an associated impact on that system. Cooling towers consume electrical power and substantial amounts of water and chemicals.

A closed loop system based on heat pumps requires more energy than cooling towers, but does not consume the cooling water or the chemicals, and much of the chemical use is avoided completely as the system is not open to contamination.

Specific conditions in Sweden, compared to global averages, are that the climate is cool, so that heating is required for most of the year, whereas cooling is not in great demand, and that the supply of fresh water is ample.

Collaboration and comparison with other facilities

A study conducted within the EU-sponsored EuCARD2 (Grant Agreement 312453) project for accelerator development examines 12 large-scale research facilities (Research Infrastructure), of which 10 were in operation and two in construction. The average annual energy consumption for the facilities (including estimates for the two in construction) was 180 GWh per year, with considerable variation in the group. Discounting the outlier in each end, the average fell to somewhat under 100 GWh. The cooling requirement varied from 40% to 60% of the electricity use. Cooling at operating facilities was at low temperature (up to approximately 40°C). The facilities in construction had included a high-temperature (up to approximately 80°C) cooling loop for part of the cooling demand [38].

The study examines a number of technologies for reuse of surplus heat, divided into high-temperature technologies, meaning those requiring 80°C or more, and low-temperature technologies. The high-temperature technologies examined were district heating, heat-driven cooling, and power production using the organic Rankine cycle (ORC). Both the heat-driven cooling technologies examined and the ORC required cooling to function and thereby produced a flow of lower temperature heat. These options were therefore only of interest for facilities with a low-temperature heat sink available. District heating, on the other hand, was only of interest for facilities located close to a significant heat demand, preferably with an existing infrastructure for distribution.

The low-temperature option studies included low-temperature district heating, heat storage, food/fodder production, biological/chemical purification/separation techniques, wastewater treatment, and ground heating (e.g., for ice and snow removal).

The EuCARD2 project involves 40 partners from 15 European countries. The energy efficiency effort within this collaboration stems from "the need to increase the efficiency of energy use during operation for cost and sustainability reasons is common in all accelerator facilities in research and industry" [39]. Another European

collaboration within a similar area is the workshop series "Energy for Sustainable Science at Research Infrastructure" hosted by CERN, ESS, and ERF (European Association of National Research Facilities) (https://indico.cern.ch/event/245432/, http://europeanspallationsource.se/energy-workshop). After the first workshop, the hosts published an executive summary highlighting the value for society of the efforts at research infrastructure in energy and sustainability management. The value was created first by the direct effects in energy efficiency at the facilities, but potentially more important effects could be created by using research infrastructure as innovation hubs, testing grounds, and training grounds, roles that research infrastructure is created to fulfill [40].

Proposal: waste heat for food production with a nutrient loop

Based on the estimated heat flows at various temperatures, the value of heat sold for district heating purposes, and the estimated cost of electricity to drive the heat pumps, the estimated value of heat at the different temperature levels can be derived. This varies over the year with the value of the heat and electricity prices and may be a positive or negative value. These calculations form an economic basis for the development of alternative uses of waste heat.

The first such use to be developed in the case was for onsite space-heating needs. Space heating can easily be achieved in buildings so designed with water temperatures of 40°C. ESS has developed an internal heat distribution network with a supply temperature of 55°C and return 25°C. The annual heat use was estimated at 5 GWh. The remaining heat would therefore amount to 265 GWh.

Since existing demand required heat pumps to augment temperature, and internal demand was limited, ESS sought opportunities to create new demand, by offering the heat to users that would establish nearby specifically to utilize the offered heat. This was done in an open call for proposals published on 10 of September 2013.

Biological systems offer an opportunity to make use of heat. Fish, plants, algae all have in common that, within limits, growth is stimulated by an increase in temperature. As one example, an increase in temperature from 8.6°C to 13.7°C has been shown in a specialized research facility to double the growth rate in salmon smolt [29].

An existing use of surplus heat for biological systems exists only 125 km northeast of ESS, at the Elleholm greenhouse facility, which at 8000 m^2 greenhouse area is Sweden's second largest producer of tomatoes. Elleholm uses waste heat from nearby Södra Cell, a pulp and paper plant.

The details of the Elleholm case differ from ESS, in that the supplying facility at all times has a significant

Table 2. Identified uses for excess heat under 60°C in a cooling chain from heat to food.

Temperature (°C)	Food and fodder production process
40–60	Low-temperature drying
32–40	Fermentation of microbes (yeast, bacteria, microalgae)
22–32	Warm water fish farming (tilapia, shrimp, perch, turbot)
18–22	Green house hydroponics (tomatoes, cucumbers)
10–18	Cold water fish farming (salmonids, white fish, sturgeon, crayfish)

excess of heat of prime district heating temperature. There is therefore no incentive to explore low-temperature heating systems. Despite this, the Elleholm management has estimated minimum temperature needed at 38°C.

In collaboration with the Swedish University of Agricultural Sciences, SLU, and based on the results of the abovementioned open call, ESS has identified a number of potential biological uses of its excess heat, shown in Table 2.

A vital aspect of the SLU-ESS proposal is that the components of the cooling chain also could be linked in a nutrient chain. The heat recycling at the Elleholm facility is a significant gain in sustainability, but the facility still uses commercial fossil fertilizer, which represents a significant indirect energy use. If a greenhouse facility were colocated with fish farming, the fish excrements could be directly used as fertilizer in a hydroponic system. Fish fodder could be made from yeast, based on a substrate of food waste and agricultural waste (including plant waste from the greenhouses). This would require a drying process

to create fodder, which could also be achieved with low-grade waste heat.

A graphic of the SLU-ESS initial proposal taken from the ESS Energy Design Report [41] is shown in Figure 1.

Analysis

Generality of the case

The case of the European Spallation Source shows an example of the phenomenon that surplus heat has become an economic and environmental cost. The cost is first, a direct cost for operating cooling systems, and second, the cost of a lost opportunity, and third a presently undefined cost for the ecosystem in coping with an increased ambient temperature due to the release of large volumes of cooling water. The case thus shows some characteristics that may be common to energy-intensive activities, these being that (i) large quantities of excess heat are produced, (ii) well-managed, the heat can be conserved at a temperature that can be useful for heat and biological processes, and (iii) conventional management of this resource represents a cost burden to the organization. In EU it is estimated that this low-temperature heat loss equals 500 billion Euro in petrol equivalents [25].

Identification and interaction of relevant physical systems

In the case, a hierarchy of energy forms was established by the relative prices of heat and electricity, and the electricity demand for heat pumps to supply appropriate temperatures. The electrical power system represented the

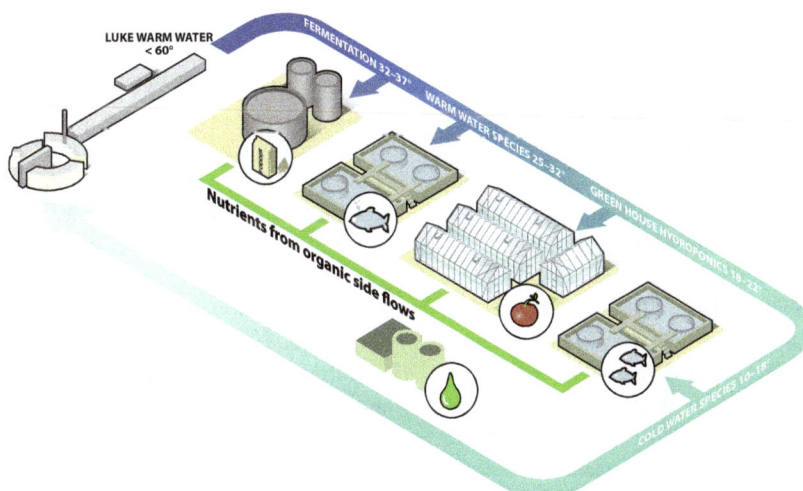

Figure 1. UNCSD theme indicator framework [15].

highest value systems, and cooling systems of various temperatures represented falling value with temperature level. This is a hierarchy of monetary value, but the same outcome would result from an analysis based on the second law of thermodynamics. The theoretical limits in Carnot cycle conversions function can be used to assign relative values to heat of various temperatures compared to electrical power [42].

A low-temperature heat-recycling scheme is therefore also a low-value scheme. This can enable uses that require only low-grade heat, but are not competitive if they must purchase heat at the high-value price.

Food is energy

Agriculture plants convert solar energy into energy for human consumption in the form of food. However, modern agriculture methods depend heavily on fossil fuels. Fertilizer represents a significant, indirect part of energy consumption for food production. For example, in Europe, farmers use about 10.5 million tons of nitrogen fertilizer 2010 and the trend is an increase. Phosphate fertilizer 2011/2012 was 23 million tons. Of this, EU imports roughly 25% (IFA, Fertilizer Europe.Com, 2014). The energy consumption to produce fertilizer varies considerably with the fertilizer type; nitrogen fertilizer requires much more than phosphorus or potash, but in order of magnitude this may represent an energy consumption of 100–150 TWh, or an eighth of a PWh. On the other hand, mining of organically available phosphorus, besides fossil fuels, do include a fraction of heavy metals, especially cadmium and aluminum, adding to its environmental load [43]. The total electrical consumption of Sweden in 2012 was 142 TWh; replacing mineral fertilizer is therefore a significant energy efficiency gain for agriculture.

Peak oil use, peak land, and peak fisheries

The term "peak oil" is used to describe a moment when the supply of oil would begin to decline, and when that might happen has been periodically much debated. More recently, "peak oil consumption" is being discussed, meaning the event that oil consumption would begin to decline.

It is clear that the event "peak arable land" has repercussions for future food supply. This moment might already have past. Loss of arable land is caused by degradation, and by constraints imposed by climatic, environmental, and human activity. The United Nations' Food and Agricultural Organization (FAO) track land degradation and publish a "Global Land Assessment of Degradation." If indeed "peak arable land" has been passed, it follows that all increased food production must come from greater output per land unit.

Furthermore, it is clear that "peak wild fish harvest" has passed. Nearly 85% of our wild fish stocks are near, at, or over its maximum harvest [27]. Increased fish consumption must therefore come from aquaculture.

Food and fodder

Within the world's food production system, basic food for human consumption competes with animal feed, and to a lesser extent also for energy production. In animal feed, 47% of soy and 60% of corn produced in the United States is used and at global scale more than 40% of these crops are used for animal feed [28, 43]. In 2007, EU produced less than 1 million tons of soy per year, and imported around 25 million ton of soybean meal (EU-27) (http://www.indexmundi.com/agriculture/?country=eu&commodity=soybean-meal&graph=imports). Today, only around 6% of soybeans are eaten directly as whole beans or in products like tofu and soy sauce [44]. Approximately 75% of all produced soybean is used for animal feed [44, 45]. To produce this amount of soy almost 15 million ha of agricultural land area is needed, nearly equal the total agricultural land of Germany [45]. In fact, soy fields now cover more than 1 million square kilometers of the world – the total combined area of France, Germany, Belgium, and the Netherlands [44]. Also, approximately 80% of that soy is genetically modified [44]. Based on the present increase in animal-derived food products, FAO [28] estimates that world soy production will double to 2050. Ever since soy production began increasing in South America in the 1960s, soy has been associated with clearance of some of the world's most crucial ecosystems, such as the Amazon and Cerrado, leading to loss of biodiversity. This loss of valuable forests and other native vegetation means that the carbon storage services they provide are lost forever, contributing to global climate change. Soy production is also linked with unsafe and excessive use of pesticides, violation of land rights, and unfair labor conditions [44]. The import of soy to EU has remained more or less stable over the last 15 years (EU-27) (http://www.indexmundi.com/agriculture/?country=eu&commodity=soybean-meal&graph=imports). In addition to plant-based nutrients, about a quarter of world catch of fish is used as animal feed [27] underlining the enormous amount of high-quality nutrients of human food quality presently used in animal feeds.

Horticulture is water intensive; the plants consume water and much is also lost in evaporation, one of the issues driving the idea of closed greenhouses. In the case, water usage was seen as an important sustainability advantage for location of greenhouses in Sweden, as the supply of fresh water is ample and the cost of often negligible. The cool climate and the resulting need for heat in greenhouses was a competitive disadvantage, to the extent that Sweden's horticulture industry was in a steady state of decline. Only 10% of tomatoes in Sweden are produced domestically [46]. An inexpensive, sustainable heat source would therefore potentially shift greenhouse production to a place with an abundance of water and thereby lessen the pressure on scarce resources elsewhere.

The indicative prices uncovered show that value of cutting out the middle man in the district heating system and establishing a direct relationship between the waste heat producer and the consumer was as much as 5 ¢/kWh. Additionally, a supply and demand at 40°C rather than 80°C would mean a savings of 1.25 ¢/kWh. With a total heat supply of 250 GWh per year, this would mean a total added value of 16 million Euros a year, quite likely enough to negate the competitive disadvantage of the heating need in Sweden.

Microbes and the protein chain

In cells, RNA relays the information of DNA to the protein synthesis. Microbes have high levels of RNA (10–15%) because of the high level of protein synthesis. Living cells metabolize the N (nucleotides) in RNA to uric acid. In mammals uric acid leads to kidney stones and gout [47]. Fish, however, have retained their ability to eat microbes and have no problem with uric acid [47–50]. Using microbes to produce fish fodder offers an opportunity to profitably use the fantastic growth rates of microbes. Microalgae tend to multiply at a rate of once per day, that is, a daily rate of 2^1. Yeast doubles every 2 h, that is, 2^{12}, whereas, for example, *Escherichia coli* bacteria the pace is as high as every 20 min during the exponential growth phase, leading to a daily rate of 2^{72}.

Taking the example of yeast, a study at Swedish yeast factory demonstrated a in favorable conditions a start culture of 10 mg of yeast developing into 150 tons in a week given free access to short carbon chains and ample supply of minerals.

Protein is important because around 40% of fish fodder is protein, in the average (tilapia/carp 30%, marine species 50%, salmon 40%). Table 3, below, details some of the main similarities and differences between protein chains for food production based on soy compared with yeast. For fodder production using yeast, the dominant energy use is for the drying process.

Creation of new dependency

From a systemic view, recycling is usually seen as an inferior option to prevention. This is also the case in the ESS program, where the "Responsible" goal of energy

Table 3. Protein chain comparison with energy use.

Protein chain	Soy	Yeast
Product	Soy meal	Protein meal
Protein content	50% dry matter	50% dry matter
Suitability	Terrestrial farmed animal[1] and humans	Fish and shrimp[2]
Main energy use	Farm/harvest, processing, transport	Drying
Energy use, kWh/kg	2.8	1.46
Energy supply	Fossil based	Surplus low-grade heat

[1]Not suitable for most fish, but if used to fish needs further processing into soy concentrate which require further energy in alcohol extraction and heat treatment to reduce antinutrient and endothelial inflammatory factors. In this process protein is concentrated from under 50% dry matter to over 65% demanding a parallel increase in amount of soy bean raw material per kg of concentrate, that is, in production energy.
[2]Suitable for monogastric terrestrial farmed animals at low inclusion level.

efficiency was the superior goal. A related issue is the question of whether heat recycling might cause a dependency on a fossil-based and/or wasteful process. For example, in this case, further efficiency gains were envisioned in the accelerator.

In comparison, heat recovery from incineration of waste streams or from electrical power production from fossil fuel sources may be seen as an unsustainable subsidy to these practices and as a risk if the heat source should cease.

In this case, ESS, backed by 17 democratic governments and a major potential source of new knowledge enabling new sustainability solutions, may be seen as a sustainable activity in its own right. The explicitly planned life span of 40 years also adds a measure of certainty of supply.

More generally, source dependency is lessened by lower temperature level. This is because the lower the temperatures, the more easily the heat source can be replaced by solar or geothermal heat. Therefore, heat recycling at low temperature can be seen as an enabler for restructuring of the energy system.

Sustainability assessment

The UNSCD sustainability framework [15] is shown in Figure 2. The framework is established for indicators on for a nation or region, not for a specific technology. It is chosen here as a reflection of a consensus view of the main international sustainability challenges.

In the CSD framework, the Health theme includes the subtheme Drinking Water. Climate Change is a subtheme of Atmosphere. Agriculture is a subtheme of Land and includes the indicators Arable and Permanent Crop Land Area and Use of Fertilizers. Oceans, Seas, and Coasts

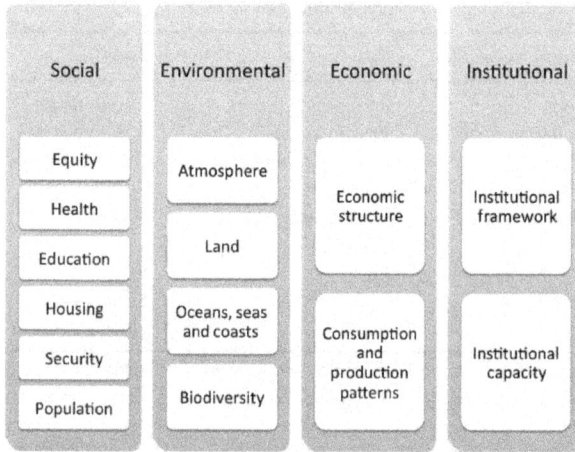

Figure 2. The proposed cooling chain for ESS with food and fuel production (*source:* [41]).

include issues of the impacts of nutrient flows into bodies of water as well as fishing yields. Among the Economic themes, Energy Use is a subtheme of Consumption and Production Patterns as is Waste Generation and Management, which includes the indicator Waste Recycling and Reuse.

The following themes, subthemes, and indicators are therefore relevant as categories of analysis to the case: groundwater and coastal waters; fishing, arable, and permanent crop land area; use of fertilizers; energy use; climate change; and waste recycling and reuse. The Institutional themes are specific to the national or regional level, and not applicable to a specific technology.

To ensure relevancy, a comparison was made with an assessment method developed specifically for energy [51]. As a result, four additional categories were added: resource depletion, cost/benefit, security and diversity of supply, and public acceptability. A summary of the assessments according to these categories is shown in Table 4.

The positive climate effect of replacing red meat with fish has been quantified in studies. At global level, meat production accounts for 18% of released climate gases [27, 28]. Gonzales et al. [52] in parity with Pelletier and Tyedemers [53] show a much lower energy use and release of CO_2 equivalents in producing 1 kg fish compared to red meat. Naturally this varies with species and production system. For example, Troell et al. [54], Tyedemers et al. [55], and Pelletier and Tyedemers [53] showed that cultured carp yielded an "industrial energy" return in edible food of 94%, while fishery harvested fish reached 8% and farmed salmon using fabricated diet based on fish and plant protein meals varied between 8% and 17% pending conventional or organic sources, while feed lot beef reached 2.5%. The differences mainly lie in the feed compartment. In these studies, the origin of the energy was not considered.

The increased predictability of supply from land-based fish farms is because unpredictable factors such as weather, diseases, etc., will be possible to control to a much higher extent in a closed system.

Main Conclusions

In this case, the value of energy quality is demonstrated. The preservation of temperature in cooling system was

Table 4. Summary of sustainability assessment by category.

Theme	+/−	Motivation
Groundwater and coastal waters	+	On-land closed-system fish farms can materially reduce the impact on inland and coastal waters from open-water fish farms.
Fishing	+	On-land aquaculture has the potential to alleviate pressure on wild stocks, allowing these to recover.
Arable and permanent crop land area	+	The production from greenhouses is 40–80 times higher (in monetary value) than from farmland per unit area.
Use of fertilizers		By combining the appropriate amount of fish farming with horticulture, nutrient flows from fish excrements can replace commercial fertilizer in greenhouses, provided the feed is based on nonplant materials.
Energy use	+	Use of waste heat for greenhouses and land-based fish farms can substantially reduce their energy use.
Climate change	+	A positive climate change effect comes from reduced energy use, reduced fertilizer use, and a replacement of red meat by fish due to avoided methane release and great energy efficiency.
Waste recycling and reuse	+	Waste nutrient streams are proposed as a basis for producing fish fodder. If implemented, this would be a significant valorization of a waste stream. However, technical hurdles remain.
Resource depletion	+	Fossil fuel, fresh/ground water, and fossil nutrient use is reduced.
Security and diversity of supply	+	The diversity of energy supply would increase. Heat storage capacity would be necessary and could contribute to stability. Diversity and predictability of food supply would increase.
Public acceptability	−	Industrial-scale fish and greenhouse farming may be considered unsightly. Light pollution may be an issue. Traditional farming and fishing is culturally ingrained. Animal feed based on recycled nutrients need strict food security control.

shown to add significant value. Conversely, using as low quality as possible can significantly lower costs. The indicative figures provided in the case, for example, 40°C for heating, 60°C for cooling data centers, and 70–80°C for cooling power electronics corroborate some earlier research. With additional verification, these benchmarks could inform future energy performance efforts.

Heat recycling was shown to be a considerable enabler for food production. The analysis indicates that waste heat resources available that are comparable to those in the case are abundant. If these can be combined with waste nutrient streams and converted to food and energy, there is potential for noticeable impact on global food supply. Additionally, if the efforts to integrate nutrient and cooling chains are successful, a large-scale rollout of this technology could supplant substantial amounts of fossil fertilizer and thus significantly lower the environmental impact of food supply.

Making use of low-grade heat in biological systems may also enable future development by lowering the threshold for renewable heat sources such as solar and geothermal.

The study is limited to a single case, and although the studied case organization is well established, the energy processes described are still being designed and will not be observable for some years. Further studies are therefore urgently needed, particularly studies of real energy flows and demonstration facilities for food production.

Conflict of Interest

None declared.

References

1. Law, R., A. Harvey, and D. Reay. 2013. Opportunities for low-grade heat recovery in the UK food processing industry. Appl. Therm. Eng. 53:188–196.
2. Element Energy, Ecofys, Imperial College, P. Stevenson, and R. Hyde. 2013. The potential for recovering and using surplus heat from industry. Department of Energy & Climate Change, London.
3. Little, A. B., and S. Garimella. 2011. Comparative assessment of alternative cycles for waste heat recovery and upgrade. Energy 36:4492–4504.
4. Swedish Government Official Reports. 2005. SOU 2005:033 Fjärrvärme och kraftvärme i framtiden. Fritzes offentliga publikationer.
5. Zimmermann, S., M. K. Tiwari, D. Poulikakos, I. Meijer, S. Paredes, and B. Michel. 2012. Hot water cooled electronics: exergy analysis and waste heat reuse feasibility. Int. J. Heat Mass Transf. 55 no. 23–24:6391–6399.
6. Ammar, Y., S. Joyce, R. Norman, Y. Wang, and A. P. Roskilly. Jan. 2012. Low grade thermal energy sources and uses from the process industry in the UK. Appl. Energy 89:3–20.
7. Fang, H., J. Xia, K. Zhu, Y. Su, and Y. Jiang. 2013. Industrial waste heat utilization for low temperature district heating. Energy Pol. 62:236–246.
8. Parker, T. 2013. The view from below – a management system case study from a meaning-based view of organization. J. Clean. Prod. 53:81–90.
9. Parker, T. 2011. Sustainable energy: cutting science's electricity bill. Nature 480:315.
10. Böhringer, C., and P. E. P. Jochem. 2007. SURVEY: measuring the immeasurable – a survey of sustainability indices. Ecol. Econ. 63:1–8.
11. Parker, T. 1998. Total cost indicators – operational performance indicators for managing environmental efficiency. International Institute of Industrial Environmental Economics, Lund.
12. Meadows, D. 1998. Indicators and information systems for sustainable development – a report to the Balation Group. The Sustainability Institute, Hartland, WI.
13. 06/02001 Will the information society be sustainable? Towards criteria and indicators for a sustainable knowledge society. Spangenberg, J. H. International Journal of Innovation and Sustainable Development, 2005, 1, (1–2), 85–102. Fuel Energy Abstr. 47: 300, 2006.
14. Singh, R. K., H. R. Murty, S. K. Gupta, and A. K. Dikshit. Jan. 2012. An overview of sustainability assessment methodologies. Ecol. Indic. 15:281–299.
15. UNCSD. 2001. Indicators of sustainable development: framework and methodologies. Division for Sustainable Development, UN Department of Economic and Social Affairs, Background Paper No. 3.
16. Swithenbank, J., K. N. Finney, Q. Chen, Y. B. Yang, A. Nolan, and V. N. Sharifi. 2013. Waste heat usage. Appl. Therm. Eng. 60:430–440.
17. Leffler, R. A., C. R. Bradshaw, E. A. Groll, and S. V. Garimella. 2012. Alternative heat rejection methods for power plants. Appl. Energy 92:17–25.
18. Vadiee, A., and V. Martin. 2014. Energy management strategies for commercial greenhouses. Appl. Energy 114:880–888.
19. Statens Jordbruksverk. 2012. The 2011 horticultural census. Statens Jordbruksverk, Stockholm.
20. Walsh, C., and P. Thornley. 2013. A comparison of two low grade heat recovery options. Appl. Therm. Eng. 53:210–216.
21. Le Pierrès, N., D. Stitou, and N. Mazet. 2007. New deep-freezing process using renewable low-grade heat: from the conceptual design to experimental results. Energy 32:600–608.
22. Markowski, M., K. Urbaniec, A. Budek, M. Trafczyński, W. Wukovits, A. Friedl, et al. 2010. Estimation of

energy demand of fermentation-based hydrogen production. J. Clean. Prod. 18(Suppl. 1):S81–S87.

23. Liu, D. 2008. Bio-hydrogen production by dark fermentation from organic wastes and residues. Ph.D. Thesis, Department of Environmental Engineering Technical University of Denmark.

24. Lakaniemi, A.-M., O. H. Tuovinen, and J. A. Puhakka. 2013. Anaerobic conversion of microalgal biomass to sustainable energy carriers – A review. Biorefineries 135:222–231.

25. Langeland, M., A. Kiessling, and O.-I. Lekang. 2014. Baltic Aquaculture Innovation Centre (BIC), Aquaculture Center East, 1.

26. Halver, J. E., and R. W. Hardy. 2002. Fish nutrition, 3rd ed. Elsevier Science and Academic Press, San Diego, CA.

27. FAO. 2014. The state of World Fisheries and Aquaculture. Opportunities and challenges. Food and Agriculture Organisation of the United Nations, Rome.

28. FAO. 2014. Sustainable fisheries and aquaculture for food security and nutrition. Food and Agriculture Organisation of the United Nations, Rome.

29. Terjesen, B. F. 2012. Forskning på miljökrav og produktionsmetoder I RAS for Atlantisk laks. *in Proceedings*, Sunndalsöra, Norway.

30. Terjesen, B. F., S. T. Summerfelt, S. Nerland, Y. Ulgenes, S. O. Fjæra, B. K. Megård Reiten, et al. 2013. Design, dimensioning, and performance of a research facility for studies on the requirements of fish in RAS environments. Aquaculture 54:49–63.

31. Randall, D., W. Burggren, and K. French. 2001. Eckert: animal physiology, 5th ed. W.H. Freeman, New York, NY.

32. Lekang, O. I. 2013. Aquaculture hatchery water supply and treatment systems. Pp. 3–22 *in* X. Allan and X. Burnell, eds. Advances in aquaculture hatchery technology. Woodhead publishing, Cambridge, U.K.

33. Bergheim, A., H. Thorarensen, A. Jøsang, O. Alvestad, and F. Mathisen. 2013. Water consumption, effluent treatment and waste load in flow-through and recirculating systems for salmonid production in Canada – Iceland – Norway *in Proceedings*.

34. Malm, M., K. McFaul, P. W. Carlsson, C. Vettier, and C. Carlile. 2008. Focus Lund. The ESS Scandinavia submission to the ESFRI Working Group on ESS siting. European Spallation Source Scandinavia.

35. Parker, T. 2013. Energy policy. European Spallation Source ESS AB. 13-June-2013.

36. Peck, P., and T. Parker. 2015. The "sustainable Energy Concept" – making sense of norms and co-evolution within a research facility's energy strategy. J. Clean. Prod. doi: 10.1016/j.jclepro.2015.09.121.

37. Lindström, E. 2014. Energy inventory. European Spallation Source ESS AB, 24-Feb-2014.

38. Torbentsson, J. 2014. Cooling related inventory. EuCARD2 Consortium, Lund, Sweden, Deliverable Report EuCARD2-Del-D3-1, 2014.

39. Stadlmann, J., R. Gehring, E. Jensen, T. Parker, and P. Seidel. 2014. Energy efficiency of particle accelerators – a networking effort within the EuCard2 program, presented at the 5th International Particle Accelerator Conference, Dresden, Germany, Pp. 4016–4018.

40. Bordry, F., T. Parker, and C. Rizzuto. 2011. Main findings of the first joint workshop on Energy management for large-scale research infrastructures, presented at the Energy for Sustainable Science workshop, Lund, Sweden.

41. Indebetou, F. 2013. Business plan energy recycling. Pp. 104–117 *in* T. Parker, ed. ESS energy design report. ESS, Lund, Sweden.

42. Lebrun, P. 2014. Energy consumption and savings potential of CLIC, presented at the 55th ICFA Advanced Beam Dynamics Workshop on High Luminosity Circular e+e- Colliders – Higgs Factory, Beijing.

43. Brown, L. R. 2012. Full planet empty plates. The new geopolitics of food scarcity. W.W. Norton and Company, New York, NY.

44. WWF Soy Report Card. 2014. Assessing the use of responsible soy for animal feed in Europe. World Wildlife Found, International, 56 pp.

45. Gelder, J. V., K. Kammeraat, and H. KroesSoy. 2008. Consumption for feed and fuel in the European Union. Friends of the Earth Netherlands, 22 pp.

46. Ekelund, L. L., L. Johnson, S. Lundqvist, B. Persson, H. Sandin, H. Schroeder, A. Sundin, I. Christensen, G. Larsson, and L.-L. Björkman. 2012. Branschbeskrivning Trädgård – område hortikultur, utemiljö och fritidsodling, Sveriges lantbruksuniversitet, Fakulteten för landskapsplanering, trädgårds- och jordbruksvetenskap, Omvärld Alnarp 2012, ISBN:978-91-576-9114-9.

47. Rumsey, G. L., R. A. Winfree, and S. G. Hughes. 1992. Nutritional value of dietary nucleic acids and purine bases to rainbow trout (*Oncorhynchus mykiss*). Aquaculture 108:97–110.

48. Kinsella, J. E., B. German, and J. Shetty. 1985. Uricase from fish liver: isolation and some properties. Comp. Biochem. Physiol. B 82:621–624.

49. Andersen, Ø., T. S. Aas, H. Takle, S. van Nes, B. Grisdale-Helland, and B. F. Terjesen. 2006. Purine-induced expression of urate oxidase and enzyme activity in Atlantic salmon (*Salmo salar*). FEBS J. 273:2839–2850.

50. Langeland, M., A. Vidakovic, J. Vielma, J. Lindberg, A. Kiessling, and T. Lund. 2015. Digestibility of microbial and mussel meal for Arctic charr (*Salvelinus alpinus*) and Eurasian perch (*Perca fluviatilis*). Aquac. Nutr. *In press*.

51. Santoyo-Castelazo, E., and A. Azapagic. 2014. Sustainability assessment of energy systems: integrating environmental, economic and social aspects. J. Clean. Prod. 80:119–138.

52. Gonzales, A., B. Frostell, and A. Carlsson-Kanyama. 2011. Protein efficiency per unit energy and per unit greenhouse gas emissions: potential contribution of diet choices to climate change mitigation. Food Policy 36:562–570.

53. Pelletiers, N., and P. Tyedemers. 2007. Feeding farmed salmon: is organic better? Aquaculture 272:399–416.

54. Troell, M., P. Brunsvik, N. Kautsky, and P. Ronnback. 2004. Aquaculture and energy use. Pp. 97–108 *in* C. Cleveland, ed. The encyclopedia f energy Vol. 1. Elsevier, St. Louis, MO.

55. Tyedmers, P. H., R. Watson, and D. Pauly. 2005. Fueling global fishing fleets. Ambio 34:635–638.

5

Amorphous single-junction cells for vertical BIPV application with high bifaciality

Nies Reininghaus, Clemens Feser, Benedikt Hanke, Martin Vehse & Carsten Agert

NEXT ENERGY EWE Research Centre for Energy Technology, University of Oldenburg, Carl-von-Ossietzky-Str. 15, 26129 Oldenburg, Germany

Keywords
a-Si, bifacial, multijunction solar cell, PECVD, thin film

Correspondence
Nies Reininghaus, NEXT ENERGY EWE Research Centre for Energy Technology, University of Oldenburg, Carl-von-Ossietzky-Str. 15, 26129 Oldenburg, Germany.
E-mail: nies.reininghaus@next-energy.de

Funding Information
No funding information provided.

Abstract

Solar cells used in building integration of photovoltaic cells (BIPV) are commonly made from crystalline wafer cells. This contribution investigates the challenges and benefits of using bifacial solar cells in vertical installations. We show that those cells get up to 13% more irradiance compared to optimum tilted south facing monofacial modules in Germany. The role of the n-layer in thin amorphous bifacial single-junction cells intended to be used as bifacial cells in BIPV applications is investigated. In contrast to the superstrate cell design, a transparent n-layer and back contact play a key role to achieve high bifaciality. We therefore increased the transparency of the n-layer by adding CO_2, increasing the PH_3 flow in the deposition gas and tested different thicknesses. With those measures, we reached a bifaciality of 98% for short-circuit current density and 99% for open-circuit voltage.

Introduction

The changing situation in the social awareness of climate change and the resulting need for renewable energy facilitates the intention to implement photovoltaic devices into already established products to generate additional benefit. Especially, the thin film silicon solar technology has some great advantages compared to conventional silicon wafer solar cells concerning the integration into existing products:

- Large area with homogenous optical appearance
- Nonvisible circuitry by laser interconnection
- Can be deposited on different substrates (glass, ceramic, plastic, metal, etc.)
- Small temperature coefficient
- Very short energy payback time
- Nontoxic in case of recycling

An area with huge potential which has seen almost no attention until now [1], is the integration of solar cells into the traffic infrastructure. This would have the benefit of no additional intrusion into nature because the traffic infrastructure already exists. Cost-cutting efforts in the area of infrastructure in the EU lead to a lack of investments that have to be redeemed in the future. Adding photovoltaic modules to building elements in traffic infrastructure has the benefit of return of those investment costs in the long term, making them viable investment projects.

Bifacial cells in vertical installation on highways and rail ways have the potential to generate 42 TWh/a [2] without changing the appearance of the landscape because the noise barrier walls already exist.

The combination of sustainable energy generation and noise barriers next to highways would increase the public acceptance of structural modifications as an additional benefit.

Integrating photovoltaic power generation into noise barrier walls has been tested before [3] by integrating silicon wafer modules. The bifacial concept leads to an increase in power generated per module [4]. It also pointed out some disadvantages of using silicon wafer technology: They cannot be integrated in every object and every position because of the restricted dimensions of those modules. Integrating semitransparent wafer modules has the optical disadvantage of casting a shadow like a chess board [4] that can be very distracting.

With wafer technology, it is therefore not possible to use an existing substrate.

As we demonstrated in a previous paper [5], there is more than one way to fabricate functional bifacial thin film silicon solar modules. As we have shown, multijunction cells are not feasible in bifacial applications. On the other hand, single-junction thin film solar cells do not suffer from this problem. Compared with μc- silicon thin film modules, amorphous silicon thin film modules have some advantages:

- Higher absorption coefficient makes very thin cells and short fabrication times possible
- Stable deposition regime

These are the reasons why we believe that amorphous silicon thin film modules are a promising technology for this application.

Irridiance and Bifaciality

Using a conventional wafer-based solar cell in vertical configuration has the disadvantage of using only one-half space and therefore only a reduced part of the incoming sunlight.

The efficiency characterization of solar modules is dependent on the solar spectrum. The solar spectrum can be divided into three parts: The direct, indirect (diffuse), and reflected solar irradiance. The direct sunlight reaches the module surface after crossing the atmosphere with no further changes. Indirect sunlight is scattered on clouds or particles and far away objects before reaching the solar modules. The reflected part of the irradiation spectra results from the reflection of the near environment and back scattered sunlight of the ground. For monofacial modules on slanted rooftops, the reflected sunlight of the near environment does not play a significant role. Figure 1 depicts the composition of those parts for three German cities. Even the southernmost city Munich has an indirect solar irradiance of over 50% in average over the year. This is true for the northern European area in general.

To get a deeper understanding, in what way the mounting orientation affects the module generation, we calculated the irradiance on a plane in relation to the angle of tilt and orientation. Data for this simulation was extracted from the meteonorm [6]. The results are depicted in Figure 2. To get the optimum yield in the yearly average for a conventional one-sided module in Oldenburg Germany, it should face to the south and slightly to the west. In this optimum tilt, the average irradiance would reach 1137 kWh/m²/a.

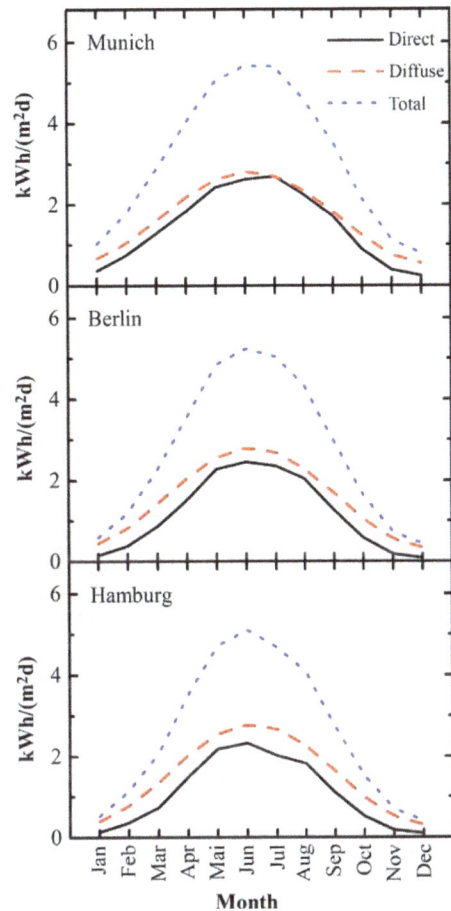

Figure 1. Composition of the solar spectrum in terms of direct and indirect irradiance for three German cities.

The same calculation can be done for a vertical plane facing east, south, west, and north. Because of the before mentioned diffuse irradiation, the average irradiance on those vertical planes per year is still very high. The results of the calculations can be seen in Figure 3. A vertical one-sided module facing south would generate the highest average yearly yield with an irradiance of 844 kWh/m²/a, which is 74% of the irradiance of the optimum tilted installation.

We propose to use bifacial modules for the vertical installation. Those modules are able to use both half spaces of the light hitting the back- and the front side of the module and are able to compensate the negative effect of partial shadowing to a certain extent. Furthermore, the vertical installation increases the naturally occurring cleaning effects and reduces the negative soiling problems caused by leaves, bird drops, and similar impurities. Thinking about noise barrier walls lining the motorways and railways going from north to south, we can use the data of Figure 3 to calculate the irradiance shining on a

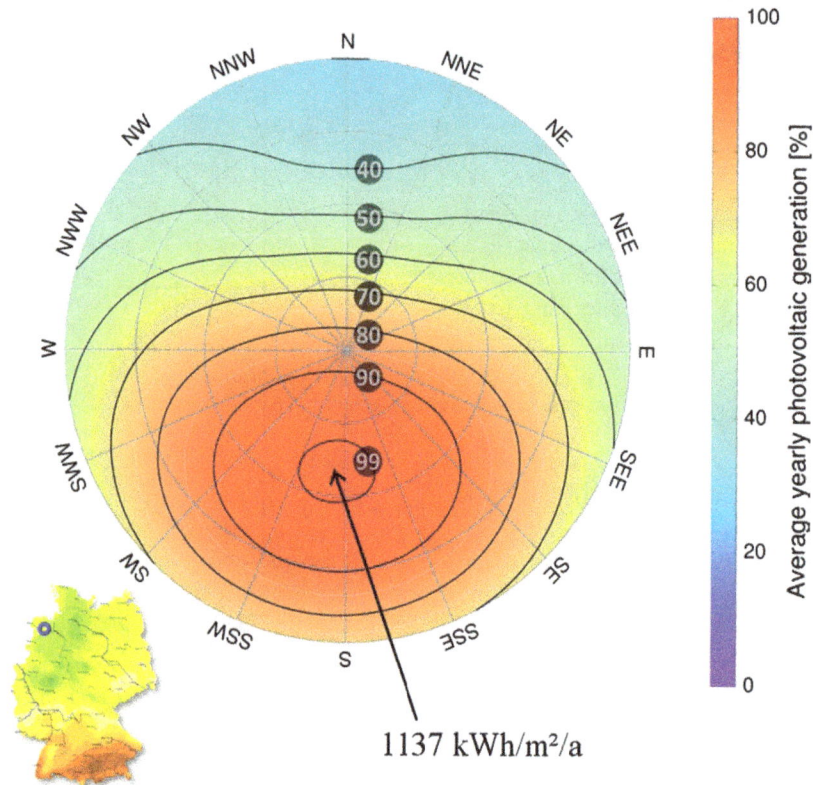

1137 kWh/m²/a

Figure 2. Tilt in relation to average yearly photovoltaic irradiance for Oldenburg, Germany. Black lines represent the isoline in percentage from the maximum irradiance.

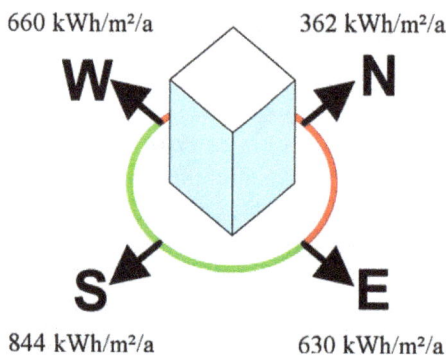

Figure 3. Photovoltaic yearly average irradiance for vertical planes facing north, east, south and west in Oldenburg with an albedo of 0.25.

bifacial module (100% bifaciality, vertical installation facing east and west) which it could convert to power. In yearly average it would face 1290 kWh/m²/a, which is 113% of a conventional module in the optimum tilt position would face. Even for transport infrastructure going from east to west the yearly average irradiance is still higher than for the modules in optimum tilt position (1206 kWh/m²/a vs. 1137 kWh/m²/a).

When using modules in vertical installation, the daily generation profile also changes. To give an example, we

used measured irradiance data to simulate the irradiance on different planes of orientation and tilt on an average day in July. The results are depicted in Figure 4.

Using bifacial solar cells in vertical installation could shift the power generation peaks to the morning and evening hours. This would reduce the demand for power storage and increase the self-consumption rate since the demand profile for northern European regions is not dominated by air-conditioning.

The shift in the irradiance profile exists also on a yearly timescale. Using the same irradiance data, we calculated the same graph for those tilts and orientations on a monthly timescale. The results are shown in Figure 5.

Bifacial cells facing in north/south direction have the benefit of being able to generate more power in the winter month because of the sun being lower in the sky. In contrast, bifacial modules facing east and west could generate more power in the sunny months compared to conventional tilted modules.

Experimental

The ratio between rear- and front-side efficiency, and therefore the current generation density, is called bifaciality

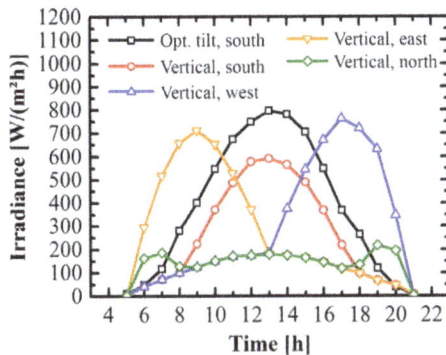

Figure 4. Simulated irradiance of an average day in July in Oldenburg for different tilts and orientations of the module plane.

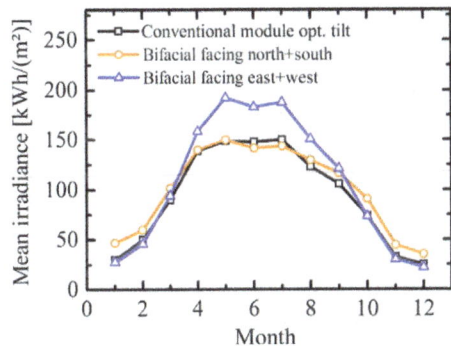

Figure 5. Monthly irradiance for three different planes in Oldenburg in different orientations.

[7]. In our previous work, we reached a bifaciality of 50% using a standard process [5] for the deposition of single a-Si thin-film solar cells in combination with a transparent back contact. The i-layer thickness was fixed at 250 nm and showed the expected light-induced degradation of 8% in average (1000 h, @STC). No further optimizations where done. To improve the bifaciality of our single-junction amorphous solar cells, we focused on the n-layer of the cell-stack. In superstrate pin configuration, amorphous silicon thin film solar cells see light only from the p-side. Therefore, there was no need for blue transparent n-layers. In the situation of a high bifaciality, the n-layer has to fulfill the same requirements as the p-layer. This means, a high transparency, high doping, less surface recombination, and good enough cross conductivity [8].

To improve the transmittance of the amorphous n-layer for low wavelengths, we varied:

• the CO_2 concentration
• the doping concentration
• the layer thickness

We decided to use an amorphous n-layer instead of the commonly used $\mu c\text{-}Si_{1\text{-}X}O_X$:H layer [9,10] because we expected a better long-term stability of highly doped amorphous n-layers compared to highly doped $\mu c\text{-}Si_{1\text{-}X}O_X$:H layers [work in prep.]. Furthermore, the process stability is better.

All presented samples were produced with a LEYBOLD Phoebus three chamber cluster PECVD-System (plasma enhanced chemical vapor deposition) at 13.56 MHz under a pressure of 4 mbar and substrate temperature between 190° to 220°. As substrates, glass with rough etched commercial sputtered ZnO:Al from the company Solayer were used. We also tested (results not shown) our cells on flat substrates and found no difference in terms of efficiency. In all samples the hydrogen flow (300 sccm), the silane flow (37.2 sccm), and the plasma power (70W) were the same. All samples were prepared with ITO-back contacts and the cell area of 1×1 cm² was defined by a laser isolation process.

IV-curves of solar cells under illumination (AM 1.5 g, 1000 W/cm², 25°C) were measured under a continuous light (DC) sun simulator of class A (WACOM WXS-155S-L2-AM1.5GMM). For determining the external quantum efficiency (EQE) of the solar cells, monochromatic probe beams using RERA System equipment with wavelength from 300 to 800 nm were used. Data for the irradiance graphs were extracted from the meteonorm [6].

Bifaciality Optimization

CO_2 variation

To increase the transparency of the n-layer, especially in the 300–500 nm wavelength regime, a common approach is to alloy the layer by adding CO_2 to the deposition gas mixture [9,10]. When adding too much CO_2 gas to the mixture, a decrease of the n- conductivity [8] can be observed. To prevent such an effect, we increased the CO_2 flow in small steps until we reached a gas flow of 10 sccm. As shown in Figure 6, the addition of CO_2 did not change the behavior of the cell when illuminated from the p-side. Therefore, we conclude that the used max gas flow of 10 sccm is small enough to neglect the influence on the n-layer conductivity. In contrast to the results of the p-side illumination, the short circuit current density increases slightly with the increasing gas flow. As expected, an increase in the transparency of the n-layer can be observed by increasing the CO_2 proportion at the n-layer. The maximum benefit of 0.6 mA/cm² is reached by adding a flow of 10 sccm CO_2 to the deposition gas mixture. To see where the additional current is generated, we measured the external quantum efficiency and extracted the results depicted in Figure 7. The reflectance (not shown) does not change measurably for all samples.

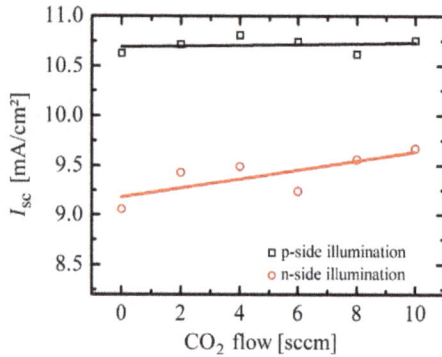

Figure 6. Short circuit current density of amorphous single-junction cells with oxygenated n-layers. Lines are fits of the measurement points.

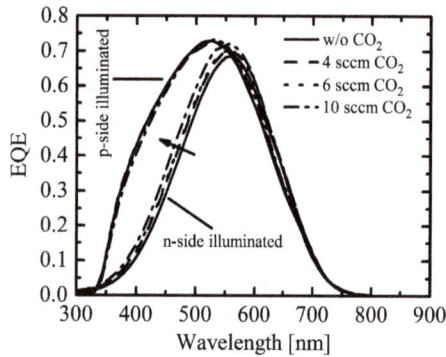

Figure 8. Open circuit voltage values of bifacial amorphous single-junction cells with decreased n-layer thickness. Lines are fits of the measurement points.

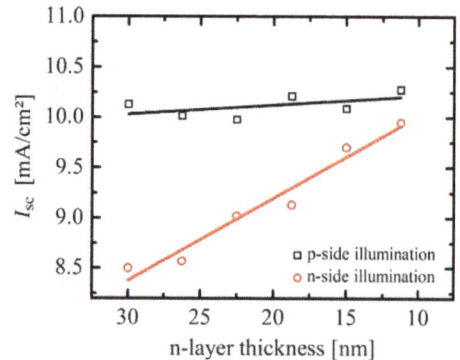

Figure 7. External quantum efficiency measurement of bifacial amorphous single-junction cells with CO_2 doping in the n-layer. The arrow indicates the tendency.

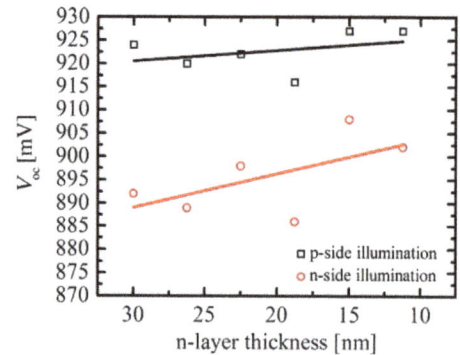

Figure 9. Short circuit current density of bifacial amorphous single-junction cells with decreased n-layer thickness. Lines are fits of the measurement points.

While the addition of CO_2 to the n-layer did not change the behavior, a distinct increase in conversion efficiency in the wavelength area between 400 and 600 nm can be observed (indicated by the arrow), when the cells are illuminated from the n-side. This can be attributed to the increased transparency due to the incorporation of oxygen and/or carbon to the layer. V_{oc} is not affected by the variation and remains nearly constant at an average of 920 mV for p-side illumination and 900 for n-side illumination.

Thickness variation

Besides mixing the n-layer deposition gas with CO_2 to increase its transparency, a thickness variation in the n-layer was performed to investigate the influence of the n-layer thickness on the absorption behavior, cell efficiency, and the bifaciality. The standard p-layer used in our cells is between 5 and 10 nm thick, whereas the default n-layer is 20–30 nm thick to ensure a strong built-in field for

the improved carrier extraction and the homogeneity of the field. Figure 8 depicts the V_{oc} values of the series, proving that the used n-layer of approximately 10 nm is thick enough to conserve the strong built-in field.

As shown in Figure 9, the decreasing thickness of the n-layer does not influence the current generation density when the cell is illuminated from the p-side.

In case of the illumination from the n-side, an increase in short-circuit current density generation can be observed, reaching a short-circuit current generation bifaciality value of 96% for the cell with an n-layer of approximately 10 nm thickness.

As demonstrated in Figure 10, the additional current generated originates from the short wavelength regime. Compared to the CO_2 variation and the doping gas variation, the thickness variation shows the highest impact on the transparency for small wavelength.

The observed improvement in conversion efficiency for light from the high-energy regime is caused by a reduced absorption in the n-layer. Light absorbed in the n-layer does not contribute to the photocurrent. By reducing the

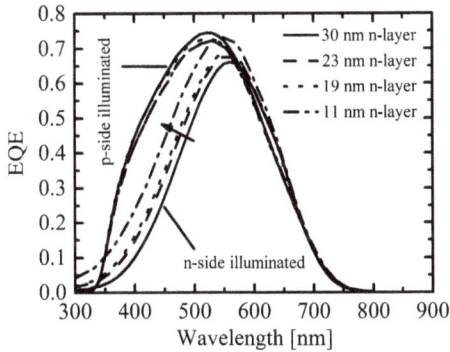

Figure 10. External quantum efficiency measurement of bifacial amorphous single-junction cells with reduced n-layer thickness. The arrow indicates the tendency.

thickness, more high energy light passes the n-layer to reach the i-layer where it is absorbed and converted into photocurrent.

Doping variation

Because a variation in the thickness of the n-layer might reduce the strength of the built-in field generated by the phosphorus doping, we deposited our bifacial a-Si:H cells with an increasing amount of PH_3 doping gas. For a better comparability to the other experiments, we used the same n-layer as a starting point as in the experiments before. The results of the doping variation are depicted in Figure 11.

With n-side illumination, an increase in the short-circuit current density of 0.4 mA/cm^2 can be seen when an additional gas flow of 5 sccm is added to the deposition gas mixture. The p-side formed by short-circuit current does not show such dependence.

The V_{oc} is not affected by the variation and remains constant at an average of 925 mV for p-side illumination and 900 for n-side illumination, respectively.

This asymmetric behavior is in good accordance to theory. The built-in field of amorphous silicon thin film cells largely originates from the boundary between i-layer and adjacent doped layer. Free charge carriers excited by absorption of light at the i-layer are forced toward the contacts by the built-in field (p for holes, n for electrons). Because of their lower mobility, free holes have a higher recombination probability in the i-layer compared to electrons [11]. This is why conventional a-Si:H modules are built in superstrat configuration, where the light enters the cell through the p-layer and the high density of exited holes on the first few nanometers can reach the contact before recombining.

The depicted n-doping increase in Figure 11 influences mainly the hole transport in the area of the n-i junction. This is why, only the short circuit current generation under n-side illumination increases. A faster excited charge transport has the additional benefit of removing the holes faster from the area where the highest concentration of free carriers is located in case of illumination from the n-side. This improved charge separation behavior resulted in a reduced recombination probability and consequently in an improved fill factor we also observed.

Combination of variations

The reference cell at the start of the investigations had a bifaciality in short-circuit current generation of only 77%. Due to a combination of the discussed improvements, we were able to fabricate a bifacial amorphous single-junction cell with improved bifaciality. This cell incorporated a 40% thinner n-layer an increased PH_3 and CO_2 deposition gas flow rate and was deposited on rough etched commercial ZnO:Al substrates. With that n-layer we managed to reach a bifaciality of 98% for short-circuit current density and 99% for open circuit voltage. The EQE in Figure 12 depicts the spectra of the optimized and the reference cell. The optimization leads to an increased

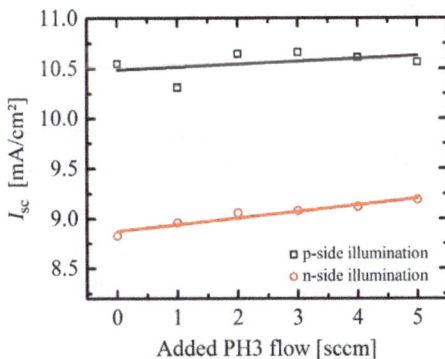

Figure 11. Short circuit current density of bifacial amorphous single-junction cells with increased doping gas flow. Lines are fits of the measurement points.

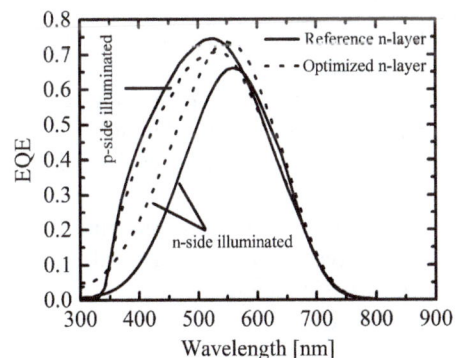

Figure 12. External quantum efficiency measurement of bifacial amorphous single-junction cells with reference and optimized n-layer.

conversion efficiency up to a wavelength of 600 nm when illuminated from the n-side while maintaining the performance when illuminated from the p-side.

The average power conversion efficiency of the monofacial reference cells was 7.0% while the optimized bifacial cells reached an efficiency of 7.4%.

Since the deposition time of the final n-layer is less than half of the deposition time of the reference layer and the only additional gas is cheap CO_2, the new n-layer would cost even less in industrial production than the reference n-layer.

Conclusion

Simulating the irradiance on different planes, we demonstrated that bifacial modules in freestanding vertical installation face more irradiance than monofacial modules in optimum tilt and orientation, regardless of the modules facing east-west or north-south.

We could show that a slight tuning of the n-layer of a standard amorphous single-junction cell in combination with the removal of the reflecting layer (Ag/white light reflector) can produce a cell with a bifaciality of 98%. Our improved bifacial amorphous single-junction solar cells show the same efficiency as our conventional cells under STC. Due to their bifacial nature this could lead to twice the output power generated under optimum conditions. We estimate that they could reach 40% to 70% improved power generation in a yearly average compared to monofacial solar cells under real-life conditions.

Outlook

To make measurements of bifacial modules comparable under STC, additional conditions like the reflectivity of the measurement block are being proposed [12]. To gage the performance of a plant with bifacial cells, additional measurement standards have to be defined [13]. To determine the plant yield, new factors like the albedo and the ratio between direct and indirect sunlight have to be taken into account.

Bifacial solar cell applications differ from classic rooftop or power plant installations demanding new solutions and different circuitry to compensate for shadowing by moving objects for example.

Current building material regulations hinder the incorporation of solar cells in building materials. To establish solar cells as part of building materials different hazards have to be evaluated:

- Danger of electrocution
- Fire hazards (ignition and toxicity)
- Blinding of vehicle drivers by reflections
- Breakage of glass

Since building material glass is a different type of glass compared to solar-grade glass, it has to be investigated how combination of thin film silicon solar cells and the building type glass used in noise barrier walls performs.

Conflict of Interest

None declared.

References

1. http://www.theguardian.com/environment/2014/nov/05/worlds-first-solar-cycle-lane-opening-in-the-netherlands (accessed 09.07.2015).
2. Quaschning, V. *Systemtechnik einer klimaverträglichen Elektrizitätsversorgung in Deutschland für das 21 Jahrhundert.* Fortschritts-Berichte, VDI Reihe 6, 2000; Nr. 437.
3. Pukrop, D. *Zur Modellierung großflächiger Photovoltaik-Generatoren* 1997; These; Carl von Ossietzky University Oldenburg.
4. Duran, C. *Bifacial Solar Cells: High Efficiency Design, Characterization, Modules and Applications* 2012; These; University Konstanz.
5. Reininghaus, N., C. Feser, and K. von Maydell. 2014. Investigation of bifacial thin-film silicon solar cells for building integration. Proc. EUPVSEC 29th:3837–3841. doi:10.4229/EUPVSEC20142014-6AV.5.39.
6. http://meteonorm.com (accessed 09.07.2015).
7. Kopecek, R., Y. Veschetti, E. Gerritsen, A. Schneider, C. Comparotto, V. D. Mihailetchi, et al. 2014. One small step for technology, one giant leap for kWh cost reduction. Photovolt. Int. 26:32–45.
8. Cuony, P. *Optical Layer for Thin-film Silcion* 2011, These, Ecole polytechnique federale de Lausanne.
9. Smirnov, V., A. Lambertz, and F. Finger. 2014. Bifacial microcrystalline silicon solar cells with improved performance due to µc-SiOx: H doped layers. Can. J. Phys. 92:913–916.
10. Veneri, P. D., L. V. Mercaldo, and I. Usatii. 2011. Improved micromorph solar cells by means of mixed-phase n-doped silicon oxide layers. Prog. Photovolt. Res. Appl. 21:148–155.
11. Shah, A. 2010. Thin-film silicon solar cells. EPFL Press, Switzerland.
12. Hohl-Ebinger, J., and W. Warta. 2010. Bifacial solar cells in STC measurement. Proc. EUPVSEC 25th:1358–1362. doi:10.4229/25thEUPVSEC2010-2CO.4.1.
13. Singh, J. P., A. G. Aberle, and T. M. Walsh. 2014. Electrical characterization method for bifacial photovoltaic modules. Sol. Energy Mater. Sol. Cells 127:136–142.

6

GIS- based multiregional potential evaluation and strategies selection framework for various renewable energy sources: a case study of eastern coastal regions of China

Yanwei Sun[1] [iD], Run Wang[2], Jialin Li[1] & Jian Liu[3]

[1]School of Architectural Civil Engineering and Environment, Ningbo University, Ningbo 315211, China
[2]School of Resources and Environment, Hubei University, Wuhan 430062, China
[3]Zhejiang Academy of Social Science, Hangzhou 310007, China

Keywords
Composite index, eastern coastal regions of China, multicriteria evaluation, potential assessment, renewable energy sources

Correspondence
Yanwei Sun, School of Architectural Civil Engineering and Environment, Ningbo University, Ningbo 315211, China.
E-mail: sunyanwei2002@126.com

Funding Information
Academic Discipline Project of Ningbo University (Grant/Award Number: '011-431501312'); K.C.Wong Magna Fund.

Abstract

Evaluation of available potential and suitable alternative options for various renewable energy sources (RES) requires the explicit consideration of inter-regional disparities and site-specific conditions (e.g., environmental, socio-economic, and resources features) for a larger scale region or country. This paper presents a novel multiregional analytical framework considering inter-regional suitability difference and inter-RES competitiveness at a given location in order to support efficient and sustainable RES spatial planning strategies. A GIS-based multicriteria evaluation technique was used to identify the geospatial potential/appropriate sites for various RES through combination of satellite-derived information, reanalysis data, ground measurements, and statistical data. The composite index (CI) considering location suitability and sustainability of renewable energy technology was then used to compare different RES options, and as a selection criteria to define suitable strategies for each specific region. A real case study concerning the China's eastern coastal 10 provinces of RES planning issue demonstrates the applicability of the proposed approach. The present results illustrate how to coordinate the competing relationships between inter-regions and inter-RES for a given location in the eastern coastal regions of China. This could provide useful insight into plan the long-term RES developing paths and facilitate the determination of optimal energy mix for the other same type area.

Introduction

Increasing the proportion of renewable energy power has been a consensus of the international community to mitigate global climate change and get rid of energy dependence on traditional fossil fuels. For the past three decades, China has experienced spectacular economic growth, and successively become the largest carbon emitter since 2007, the largest energy consumer since 2009 [1]. However, since China's energy mix is dominated by coal which accounted for 66.2% of its primary energy consumption in 2012 and cannot be substantially changed in the near future, the control of carbon emissions will be rather difficult [2]. These present key challenges to China to

balance between rising energy demands and potential environmental issues [3].

China's eastern coastal region is one of the main engine for country's economic development, especially after the reform and opening-up in 1978, which account for 13% of total area, but carried on 40% of population, and contributed about 60% of GDP [4]. With rapid economic growth and population agglomeration in this region, eastern coastal zone is also in the forefront of energy consumption, while having limited fossil fuels resources. Most of provinces along the coastal lines are currently heavily depending on energy import, such as west-east electricity transmission project, importing liquefied natural gas and raw coal from east and south-east Asia countries. In such a case, clean energy sources like renewable energy sources and nuclear energy are becoming potential and safety solution for eastern developed provinces to ensure the security of energy supply and reduce CO_2 emissions. Some of developed provinces have made reasonable development goals of renewable energy sources in their medium- and long-term energy development panning to promote the exploitation of clean energy. But there is still lack of adequate feasibility studies about clean energy development in such region.

Due to spatial and temporal variability in renewable energy sources, the development of renewable energy projects requires a thorough analysis of land use issues and lots of constraints [5]. GIS-based analyses have been used to evaluate different categories of renewable energy sources at various research scale, such as global [6, 7], national [8–13], regional and local scales [14–19]. The majority of such applications focus on potential assessment, environmental and ecological impacts, and preferable locations selection. Aydin et al. [17] suggested identification of preferable locations for renewable energy systems is a decision-making problem that requires evaluation of the potential of the resource together with economic and environmental limitations. Spatial multicriteria evaluation (MCE) integrated with GIS allows incorporation of the geographical data with the decision makers' preferences in order to provide overall assessment of multiple, conflicting, and incommensurate criteria [20]. Thus, GIS analysis might aid to determine appropriate zones according to specific criteria for future development [21].

The purpose of this study was to propose an integrated analytic framework for determine suitable utility technology and optimal site selection at the multiregional level. In this study, the structure of energy consumption and supply for eastern coastal regions of China were reviewed firstly; Furthermore, the large-scale exploitation potential of main categories of RES (i.e., offshore wind, onshore wind, larger scale PV, and biomass energy) and their contributions to regional energy system were estimated using GIS-based MCE method with respect to geographical and technical constrains; Subsequently, we introduced the appropriate development strategies for each coastal regions from a spatial planning perspectives. These present may help build a macroscopical vision for sustainable energy systems based on spatially explicit information.

Study Area and Datasets

Study area

The target region of this study consists of 10 provinces/municipalities along east coastal regions of China, including Liaoning, Jingjinji (taking Hebei, Beijing, and Tianjin as one region), Shandong, Jiangsu, Shanghai, Zhejiang, Fujian, Guangdong, Guangxi, and Hainan province (Fig. 1), which is the most developed part in China with the highest population density, large amount of GDP, and huge amount of energy consumption. This region is the forefront of reform and opening up in China, with total area of 129.4×10^4 km^2 (about 13.5%) and a population of 595.9 million (about 43.5%) in 2010, and contributes to over 54% of total GDP in China. Meanwhile, the amount of energy consumption for eastern coastal regions has been increasing rapidly, from 629.3 Mtoe in 1995 to 2175.2 Mtoe in 2012; the total amount of energy consumption has already cover 60% of that in China in 2012 (Fig. 2). On the other hand, China's energy import dependency was added up to 9% in 2013, and will reach to 26% in 2020. This trend is more serious for some eastern coastal regions, such as Zhejiang (>90%), Guangdong (>65%), Jiangsu (>90%), etc, which indicates eastern coastal regions will face great challenge of energy supply security for a long time.

Eastern coastal regions of China have limited fossil fuels sources, while have good endowment of renewable energy (e.g., onshore and offshore wind energy, biomass energy). Currently, coastal provinces have focused on development of solar, wind, and biomass energy by taking advantage of favorable geographical locations. Till the end of 2012, the total installed power capacity for wind and solar power generation in this region has, respectively, reached to 26,484.1 MW, and 2763 MW, which accounted for 34.9% and 34.7% of national wind and solar power installed capacity, respectively. As shown in Figure 3, the total amount of wind power installed capacity in Jingjinji, Liaoning, Shandong have dramatically increased, in particular after 2005. Similarly, solar power installed capacity also has experienced a steadily growth, whereas the share is still small due to higher generation cost comparison with onshore wind power. Jiangsu, Jingjinji, and Shandong, where endow with good solar energy resources, have larger solar power installed capacity. In addition, biomass power installed capacity is concentrated in eastern coastal provinces of China, and its grid-connection capacity has reached to 3514.84 MW accounting for 45.12% of national total installed capacity at the end of 2013, such as Jiangsu, Shandong, Guangdong, and Zhejiang (Fig. 4). The biomass power generation technology types focus on agriculture and forestry biomass direct combustion and municipal solid waste incineration power. As a whole, wind power has been at a higher development level than the other categories except Guangxi and Hainan provinces, where biomass power have high utility proportion.

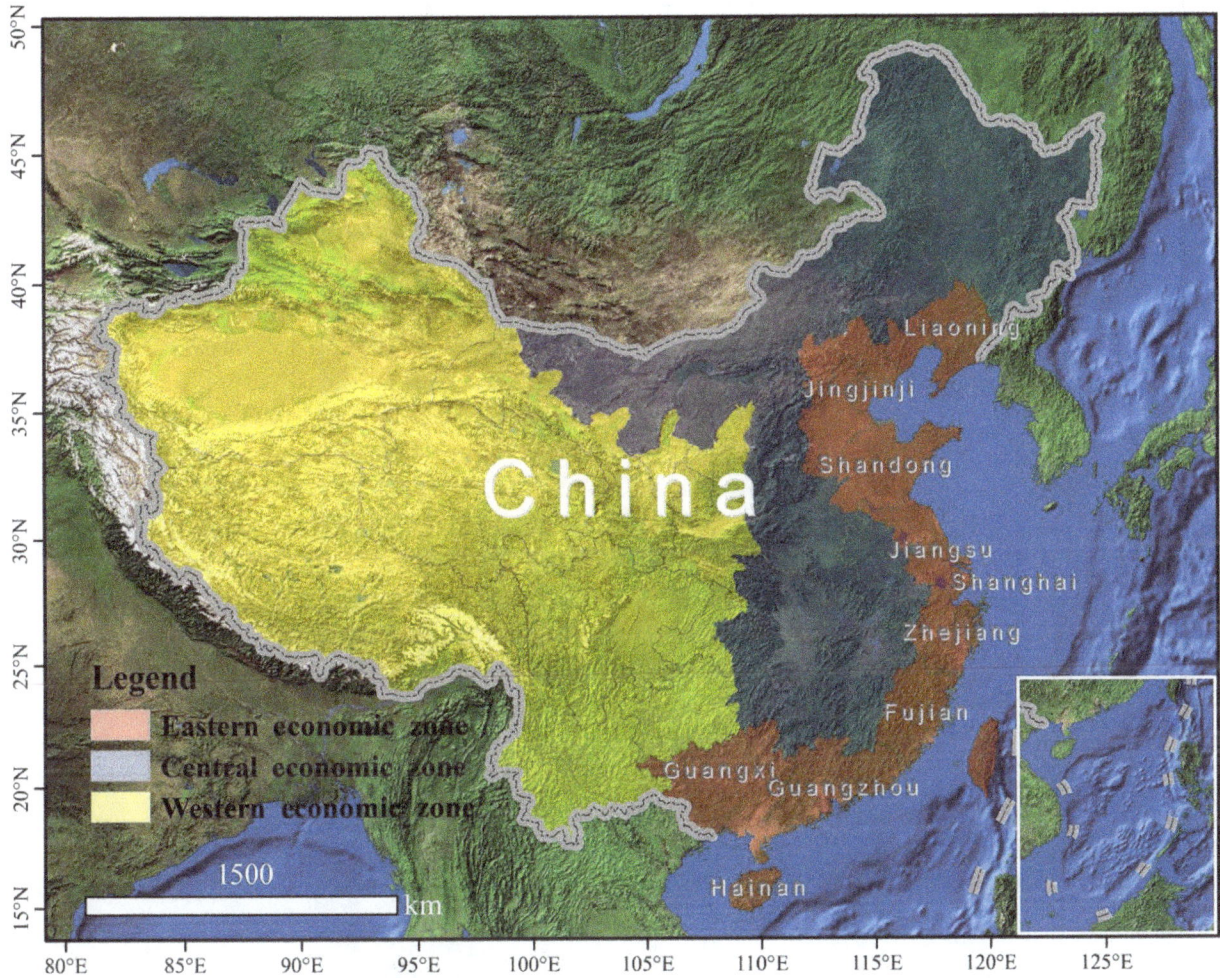

Figure 1. Study area of eastern coastal regions of China (Eastern economic zone).

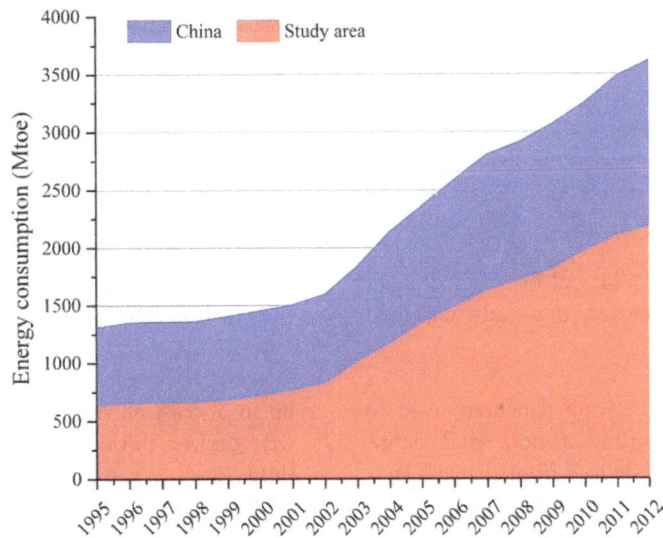

Figure 2. Growth of energy consumption in the study area (1995–2012).

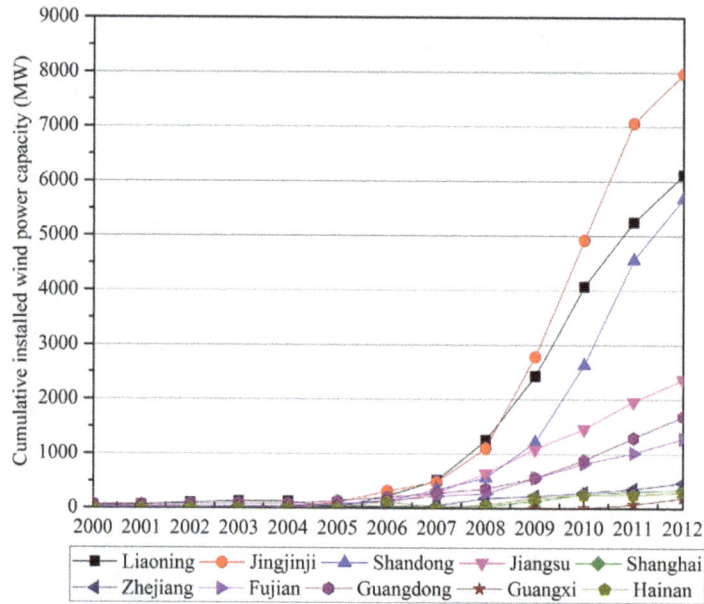

Figure 3. Cumulative installed wind power capacity in the study area from 2000 to 2012 [22].

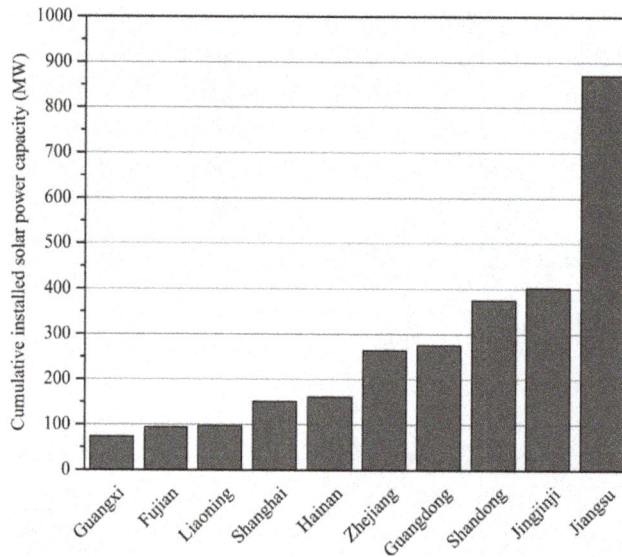

Figure 4. Cumulative installed solar power capacity for the study area in 2012.

Data sources

To exactly evaluate harnessed potential of regional wind, solar, and biomass energy sources, the study has used the datasets as below:

1. Wind field data: Two surface wind data were used to assess the potential of onshore and offshore wind energy. Ocean surface wind data are derived from spatial blending of high-resolution satellite data (SeaWinds instrument on the QuikSCAT satellite-QSCAT) and global weather center reanalyses (NCEP). The wind field data contain

the U and V wind components at a 25 km resolution. The global coverage datasets begin in July 1999 and ends in July 2009 [23]. The dataset was produced using improved Geophysical Model Function (GMF). The WindSat retrievals are believed to be accurate for winds up to at least 30 m/sec. Land surface wind speed data were derived through the School of Geography Oxford (http://www.geog.ox.ac.uk). New et al. [24] constructed a 10′ latitude/longitude dataset of mean monthly surface climate over global land areas, which includes eight climate elements: precipitation, wet-day frequency, temperature, diurnal temperature range, relative humidity, sunshine duration, ground frost frequency, and wind speed, and was interpolated from a dataset of station means for the period centered on 1961 to 1990.

2. Bathymetric data: The bathymetric data of eastern ocean of China are retrieved from the National Geophysical Data Center by the National Oceanic Atmospheric Administration (NOAA) (http://www.ngdc.noaa.gov/mgg/global/relief/ETOPO1/).

3. Solar radiation data: Solar radiation at latitude tilt in China was derived from GIS data of Solar and Wind Energy Resource Assessment (SWERA), which was developed by the National Renewable Energy Laboratory for the U.S. Department of Energy. This data provide monthly average and annual average daily total solar resource averaged over surface cells of approximately 40 km by 40 km in size [25].

4. Land use/land cover data: Multi-temporal Landsat images (Landsat TM/ETM+) were used to product land use/land cover map in the study area. The high-resolution Google Earth imageries were used as auxiliary information to improve the accuracy.

5. Net primary production (NPP) data: Annual MODIS17A3 NPP data for 2010 in sinusoidal projection were downloaded tile by tile from the Numerical Terradynamic Simulation Group (NTSG) at the University of Montana (http://hdfeos.net/). MODIS17A3 NPP data are formatted as a HDF EOS (Hierarchical Data Format – Earth Observing System) tile and have a resolution of 1 km.

6. Statistical data: In this study, we mainly consider the five food crop categories including rice, wheat, corn, legume crop and tuber crop, and three oil crops categories including peanut, rapeseed, sesame. Each type of annual crop yield at provincial level can be acquired from China Agricultural Statistical Yearbook 2012. In order to match the crop classes with the land-use dataset, the crops mentioned above are reclassified into two classes: paddy field crops (rice) and dry field crops. The Forest stock data are derived from eighth national forest resources inventory, which provides a source of forest inventory data from 2009 to 2013. The forest biomass includes fuel wood, forestry harvesting residues, wood processing waste, and forest pruning and brunches, etc. In this study, only the pruning and tending woody residues are considered. The related conversion parameters used in this study are mainly referenced from Shi et al. [26].

Methodology

A brief schematic of our study to determine the alternative RES solution for multiregion level is presented in Figure 5. First, we collect the required data of evaluating the RES potential, which include satellite-derived information,

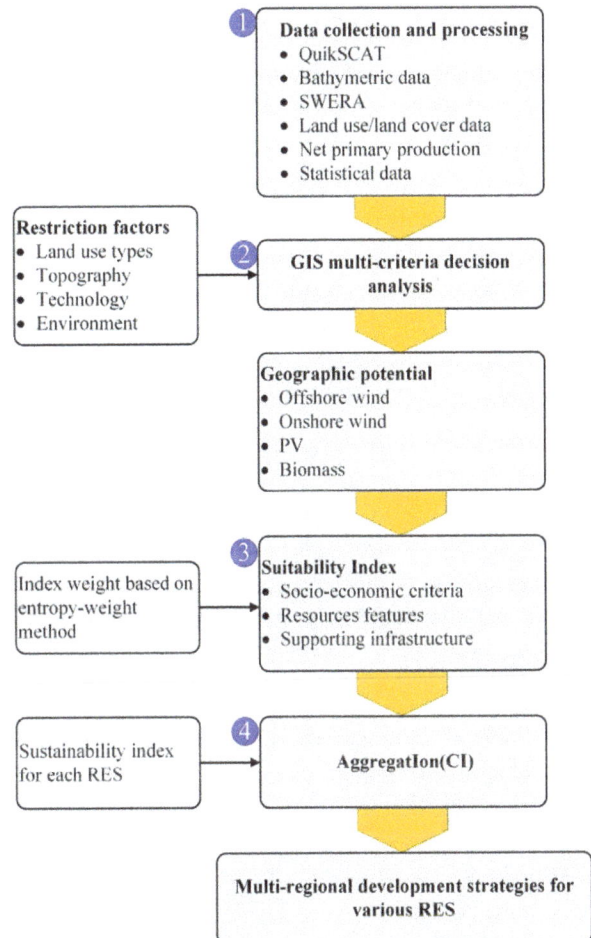

Figure 5. Procedures of decisions analysis for multiregional energy planning.

reanalysis data, ground measurements and statistical data. All available dataset are then processing in GIS to prepare the subsequent analysis. The second step is to identify the multiple constraint criteria for eliminating infeasible sites to various RES development. In this article, a set of technological, economic, social, and environmental criteria is selected. Then the mainstream technologies and evaluation methods of energy production for each RES options are selected, and the GIS procedures are conduced to produce various RES energy potential maps. Entropy-weight method and GIS multicriteria decision analysis are further used to calculate the suitability index for each prefecture level region. In the fourth step, a composite index (CI) considering location suitability and sustainability of renewable energy technologies is then used to compare different RES options, and as a selection criteria to define suitable strategies for each specific region. Finally, the composite decision making information including priority of RES options and optimal location for multiregion is determined through qualitative and quantitative evaluation for supporting energy planning.

Onshore wind energy

To obtain high spatial resolution wind speed surface, land surface wind speed data with a 10′ latitude/longitude was firstly interpolated to a grid size of 1 km × 1 km, and then masked by administrative boundary of study area using GIS spatial analysis tools. Since the wind speed changes with altitude due to frictional effects at the surface of the earth, monthly wind data at 10 m height should be adjusted to the hub height of chosen wind turbine model. The power law is used by many wind energy researchers to extrapolate the reference wind speed to the hub height [27], and its basic form as follows:

$$V = V_0 (z/z_0)^{\alpha}, \tag{1}$$

where V is the wind speed at height z, V_0 is the reference wind speed at the reference height z_0, and α is the wind speed power law coefficient, and take a value of 0.28, because the suitable areas of wind farm are always located in terrain uniformly and covered with obstacles 10–20 m, for example, residential suburbs, woodland [28].

The electric energy output of wind turbine depends on local wind regime and characteristics of selected wind turbine, such as rated power of the wind turbine generator, the swept area and the power curve of the turbine. Weibull and Rayleigh distributions are often used to derive the theoretical appearance frequency of wind speed [29]. However, the calculated process of such method is relatively complicated, and it need daily wind speed dataset. For the purpose of this study, we follow the estimation method which is proposed by Hoogwijk [6]. There is a correlation between full-load hours and average wind speed, so the full-load hours could be simplified as a linear function of the average annual wind speed. It is supposed that Repower 5 MW wind turbines are installed in the study area. We build a simple linear regression equation between full-load hours and average annual wind speed using a Weibull distribution and the power curve of Repower at 23 meteorological stations of China's eastern coastal provinces. The detailed results of this correlation analysis was performed by Ref. [30]. Based on this supposition, the electric energy potential of wind turbine in each grid cell (E_i) is estimated by the below formula:

$$E_i = P_R \cdot H_i \cdot \lambda \tag{2}$$

In the above equation, P_R is the power of selected wind turbine [MW], H_i is the full-load hours [h], and λ represents a conversion factor [0.86].

Not all the land surface are suitable to install wind turbines. In the determination of geographical potential of onshore wind power, some of land use categories, including forest, water bodies, farmland, and built-up area, are excluded. In addition, land with slopes >10° are also excluded, since such areas are difficult to access for heavy machinery to installing and maintaining wind turbines.

Recent studies have shown that for turbines that are spaced 10 rotor diameters, D, apart in the prevailing downwind direction and five rotor diameters apart in the crosswind direction, array losses are typically very low (<10%) [27]. The spacing parameters indicate a spacing factor of about 5 MW/km² if considering a rotor diameter D = 90 m. As a result, the regional onshore wind energy potential can be estimated with consideration of geographical constrains.

Offshore wind energy

The offshore wind energy potential was calculated using the similar method as onshore wind energy. Ocean surface wind data were interpolated to a grid size of 1 km × 1 km using kriging method, and then masked by administrative boundary of study area using GIS spatial analysis tools. The wind speed at the hub height was linearly interpolated from 10 m height through power law of wind profile, which the wind speed power law coefficient (α) take a value of 0.1. Water depth is the primary concern to determine the type of foundation. Bottom-mounted foundations have been widely used for the area where water depth is <20 m. For deeper water, floating foundations have to be used. Even with the floating foundation, the area where the water depth is deeper than 200 m is difficult to use due to economical reason [31, 32]. In this study, two scenarios of water depth were assumed: (1) 0~20 m, (2) 20~50 m.

Table 1. Selected indicators and weights for evaluating location suitability index of various RES options.

Category	Evaluation indicators	Weights for onshore wind	Weights for solar PV	Weights for biomass
Social-economic criteria	Per capital GDP (Yuan/person)	0.0859	0.0887	0.1200
	Total demand of electricity (kWh)	0.2054	0.2123	0.2871
Resources features	Available supply potential (GWh/year)	0.4070	0.4225	0.1133
	Average energy output density (MWh/km²)	0.0579	0.0246	0.1387
Supporting infrastructure	Road density (km/km²)	0.0370	0.0382	0.0517
	Marginal land area percentage (%)	0.2069	0.2137	0.2891

Solar energy

Generally, there are two kinds of application models for solar PV generation: (1) for the suitable area outside of built-up areas, it means for large-scale PV station; (2) for the built-up areas, the roof-top PV is the main system. In this study, we only estimate the potential of large-scale PV system in the study area. The PV production energy is determined by three main parameters, solar radiation of local

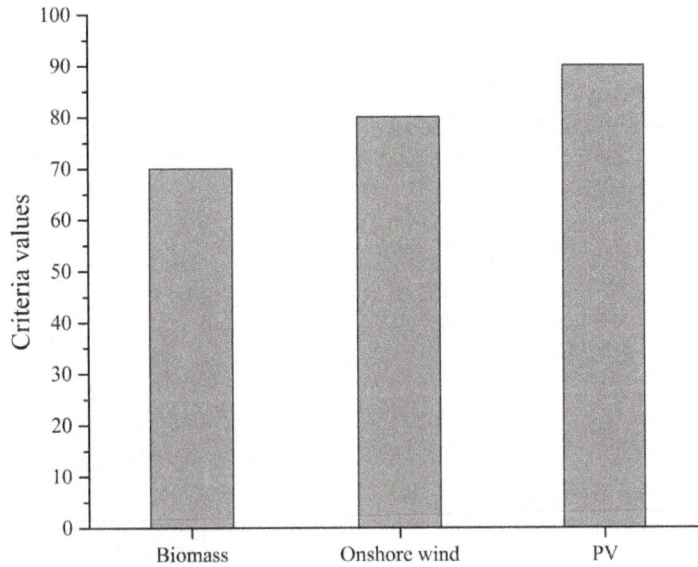

Figure 6. Ranking of the renewable energy technologies using fixed criteria values.

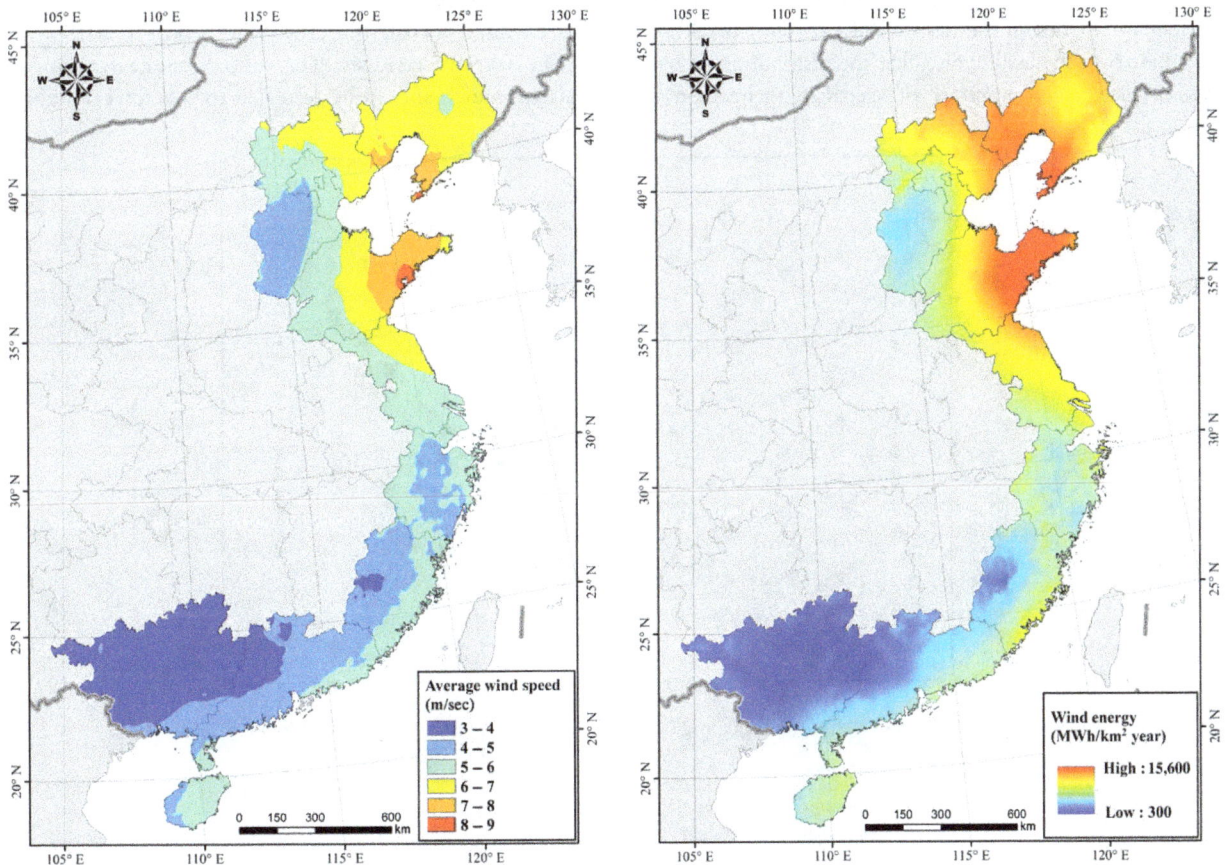

Figure 7. Annual wind speed (left) and annual wind energy (right) of the study area at 100 m height.

area, and size and performance ratio of PV systems. The annual total amount of PV generation electricity in the grid cell i, E_i, was calculated using the following equation [33]:

$$E_i = \frac{P_i G_i \eta_T}{1000 \text{ w/m}^2}, \tag{3}$$

where P_i is the peak power of PV system installed in grid cell i, G_i is the annual total amount of global radiation tilted at the latitude in grid cell i, η_T is the performance ratio of PV system. A typical value for PV system with modules from mono- or polycrystalline is around 0.75.

Based on the previous studies, geographical restriction areas are referred to forest, built-up area, water body, natural reserve, agriculture land, and land with slopes of more than 4°. Additionally, we defined a PV capacity density of 44 MW in each grid cell [34].

Biomass energy

Unlike the other renewable energy, biomass is distributed over large areas and biomass productivity varies greatly with site-specified conditions of climate and soil. Thus spatial variations of biomass production are crucial for the estimation of the available potential. In this study, we followed the method applied in the research of Gehrung and Scholz [35]. Based on the assumption that the distribution of biomass available for purposes of electricity generation is

influenced directly by biomass increment on hand as net primary productivity, the total amount of each type of biomass production was allocated to the whole regarded land cover weighting with NPP data. The weighted disaggregation methodology is expressed in the following equation:

$$P_{i,j} = P_R \times n^{-1} \times NPP_{i,j} \times \overline{NPP^{-1}}, \tag{4}$$

where $P_{i,j}$ is the amount of biomass production on pixel i,j; P_R is the total amount of biomass production at provincial level; n is the number of pixel of regarded land cover; $NPP_{i,j}$ is the net primary productivity of pixel i,j in region; $\overline{NPP}$ is the average net primary productivity of pixels of regarded land cover in the region.

Then the total amount of annual energy generation for biomass residuals in each 1 km^2 pixel can be calculated by using equation as follows:

$$P_{i,j}^T = P_{i,j} r_i e_i (1 - s - l) \eta_i, \tag{5}$$

where $P_{i,j}^T$ is the annual energy generation of biomass residues on pixel i,j; $P_{i,j}$ is the amount of biomass production on pixel i,j; The parameter r_i is the fraction of biomass residues which can be used to generate electricity; The parameter e_i is the energy content of biomass residues, the low heating value of 15 MJ/kg was adopted for all the crop types, and 16 MJ/kg for all the forestry residues; The ratio S is the fraction of residues which should be returned to soil for ecological/

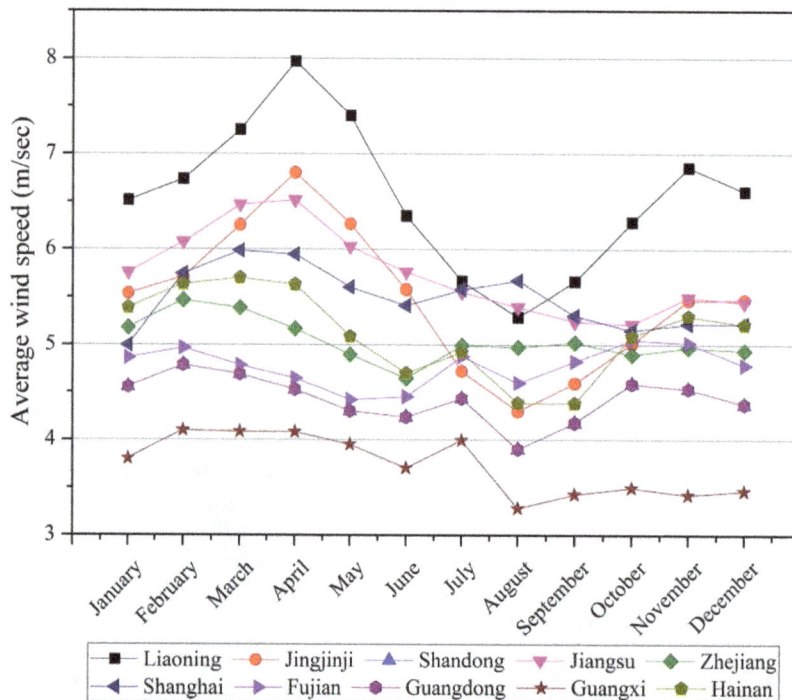

Figure 8. Seasonal variability in average onshore wind speed for the study area.

Table 2. The potential of onshore wind energy in the study area.

Name	Total area (km²)	Annual average wind energy (MWh/km²)	Min. value of wind energy (MWh/km²)	Max. value of wind energy (MWh/km²)	Suitable area (km²)	Energy Potential (TWh/year)
Liaoning	145,500	10,523	8163	14,771	5853	61.4
Jingjinji	215,200	7300	3499	11,700	24,940	181.4
Shandong	153,500	10,709	5943	15,586	4439	47.3
Jiangsu	101,000	7170	5789	11,455	295	2.0
Zhejiang	101,900	5485	4120	7403	1131	6.1
Shanghai	6254	7059	6195	7722	0	0.0
Fujian	121,700	4376	1255	7665	4740	20.6
Guangdong	176,700	3187	543	6595	1253	4.0
Guangxi	236,300	1556	359	3998	6026	9.4
Hainan	33,800	5495	4614	6296	316	1.7

environmental reasons, we defined the conservative value of 0.5; The fraction l is the loss coefficient of biomass residues during the collection and transportation process, and the value of 0.05 is taken for all the residues. A η_i is the efficiency of the convention biomass to power. For estimation of technical potential, we choose the high efficient power generation system with a convention efficiency of 40% as reference system.

Composite index by GIS-based multicriteria evaluation techniques

Location suitability and sustainability index of RES technologies were often used for assessing/comparing the comprehensive performance of various RES systems at the multiple scales. The evaluation of optimal RES alternative options for multiple regions requires the explicit consideration of inter-regional disparities and sustainability performance of

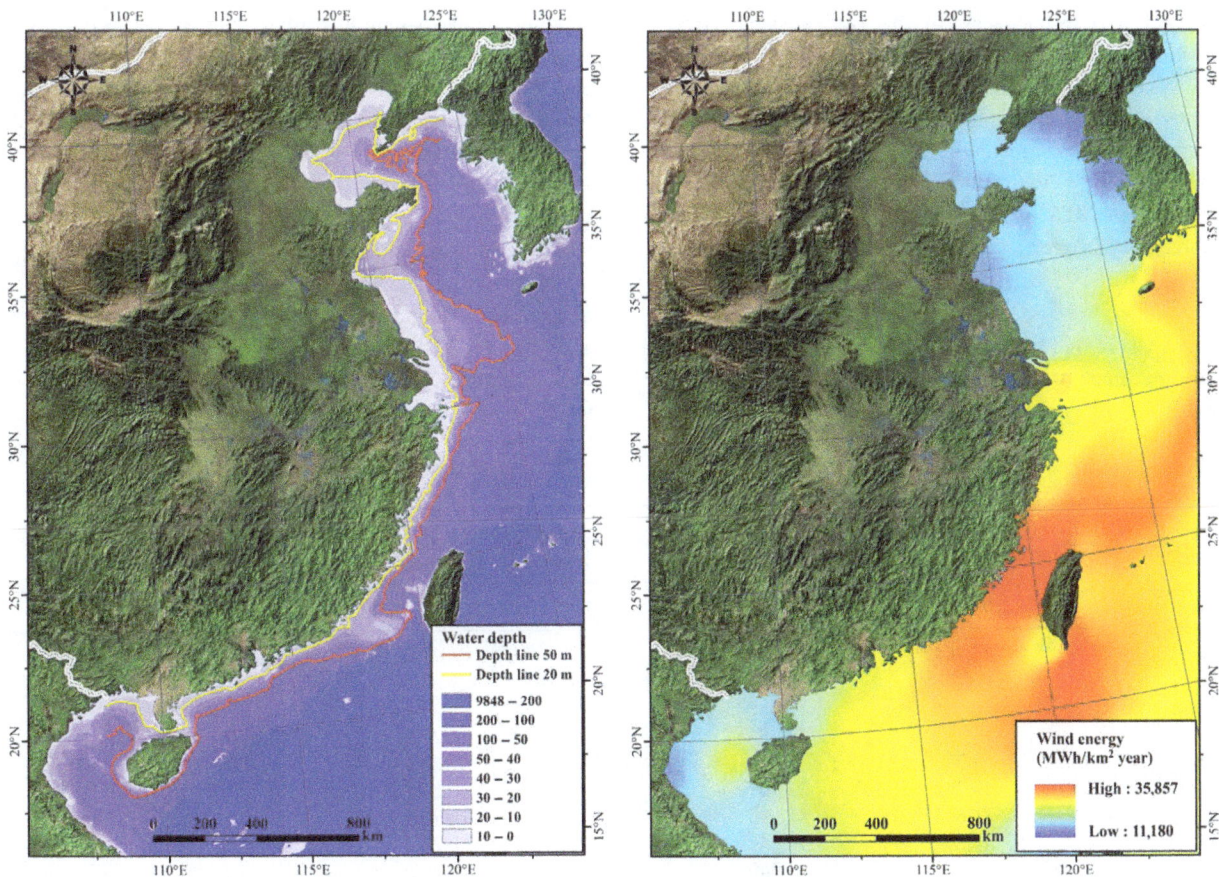

Figure 9. Water depth (left) and annual wind energy (right) of the study area at 100 m height.

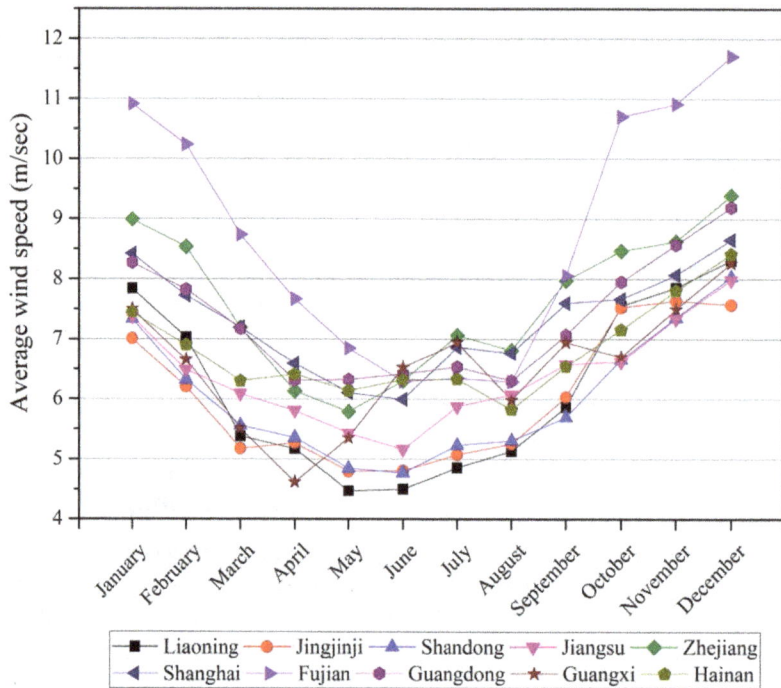

Figure 10. Seasonal variability in average offshore wind speed for the study area.

different RES technologies. In this context, we propose a novel composite index (CI) aggregated both location suitability and sustainability of RES technologies using GIS-based multicriteria evaluation techniques. CI can be calculated by the maximum selection method. We can take CI as a qualitative and quantitative evaluation criteria to determine the priority development RES option or optimal energy mix for each subregion. Therefore, the CI can be expressed as follows:

$$CI_j = Max\left[SI_1 \cdot UI_1, SI_2 \cdot UI_2, SI_3 \cdot UI_3 \cdots, SI_n \cdot UI_n\right], \quad (6)$$

where CI_j is the composite index for district j, SI_i is the location suitability index of RES type i (i = 1, ⋯, n) for district j, UI_i is the sustainability index of RES type i.

The location suitability index (SI) for various RES is introduced to define the priority sites in a larger scale area and produce the final suitability map. A weighted linear combination (WLC) is selected to combine multiple evaluation criteria, and is expressed as follows:

$$SI_j = \sum_{i=1}^{n} w_i x_{i,j}, \quad (7)$$

where w_i is the relative weight of the evaluation criteria i, and $x_{i,j}$ is the standardize score of district j for the evaluation criteria i. RES location selection action associated with local infrastructure conditions, social-economic and resources features aspects. In this section, six indicators are selected

Table 3. Offshore wind energy potential in different water depth.

Name	Total area (km²)		Annual average wind energy density (GWh/km²)		Offshore wind energy potential (TWh/year)	
	0–20 m	20–50 m	0–20 m	20–50 m	0–20 m	20–50 m
Liaoning	23,764	22,001	21.3	20.7	506.9	441.2
Jingjinji	10,230	13,093	20.2	20.1	206.2	263.7
Shandong	33,029	56,589	20.1	20.2	664.8	1149.9
Jiangsu	38,635	57,134	20.9	21.6	807.2	1265.5
Zhejiang	24,579	19,544	26.5	26.5	651.6	519.6
Shanghai	10,357	23,109	26.1	25.2	270.6	572.4
Fujian	11,608	42,596	31.9	30.9	369.8	1306.1
Guangdong	26,848	53,045	24.4	25.3	655.7	1368.3
Guangxi	6410	11,240	21.6	22.0	138.8	250.1
Hainan	–	19,388	–	23.1	–	448.8

aiming to evaluate the location suitability of each region (Table 1). The entropy-weight method is used to calculate the weights of each criteria for onshore wind, solar PV, and biomass energy. If there is a large difference between the objects for a criterion determined, this criterion can be regarded as an important factor for the analysis of alternatives [36]. Offshore wind is not considered in this evaluation process because of its location specificity. In equation (6), UI is another key factor for evaluating the sustainability of different RES technologies, and there is a range of important criteria that needs to be considered. Troldborg et al. [37] provided a ranking of the RES technology with MCA by selecting criteria and gathered information from extensive literature reviews. We assign directly a fixed criteria value as the UI value for each RES type in this study based on the raking order of Troldborg's paper (Fig. 6). As all above evaluation procedure, the priority RES alternative map was finally produced, and it can offer a comprehensive decision making information on optimal RES alternative solution for each district in the study area.

Results and Discussion

In this section, the results of onshore wind energy, offshore wind energy, solar PV, as well as biomass energy potential of each eastern coastal regions considering

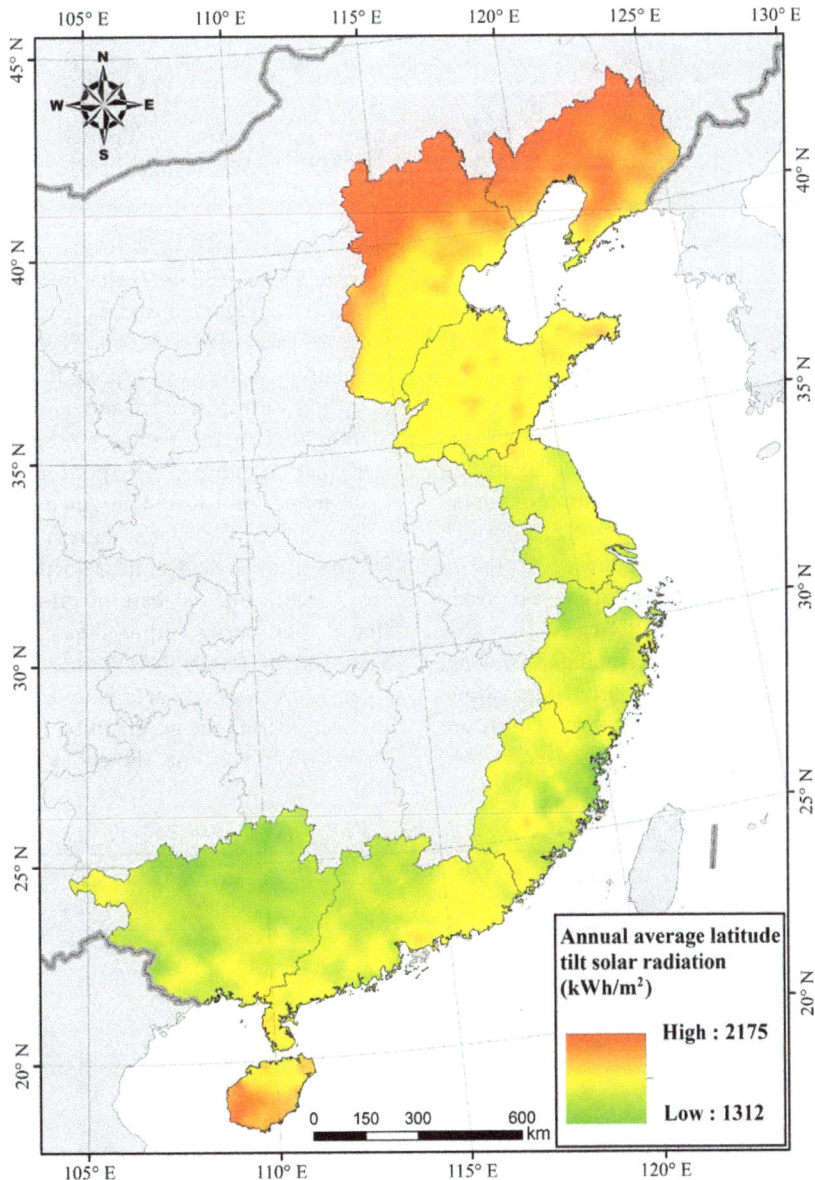

Figure 11. Spatial distribution of annual solar radiation for the study area (kWh/m²).

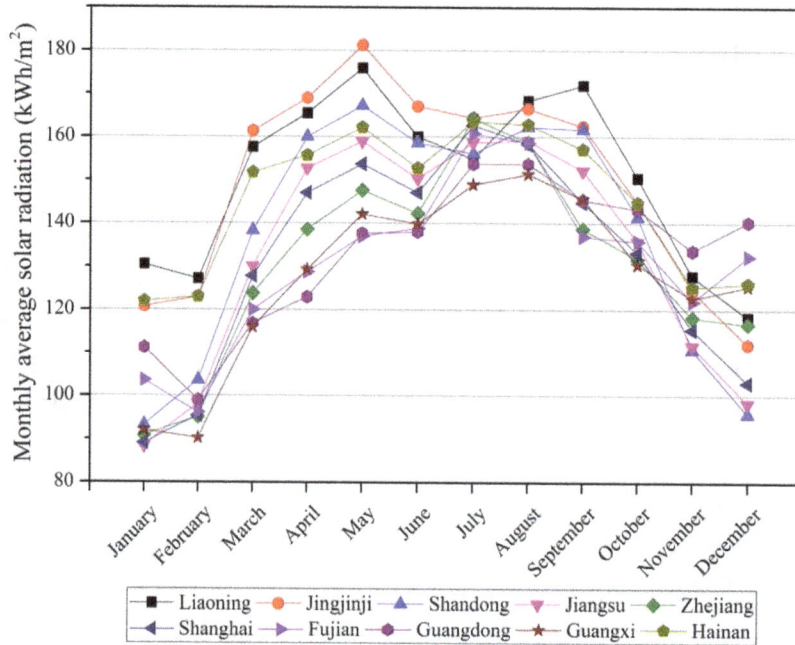

Figure 12. Seasonal variability in monthly average solar radiation for the study area.

geographical and technical restrictions are presented and discussed.

Onshore wind energy

Onshore wind power is one of the most promising GHG mitigation technologies with huge resource availability and a large deployment rate. The spatial distribution of annual wind speed and electricity output in each grid cell through eastern coastal regions is shown in Figure 7. At 100 m height, annual wind speed of the land surface varied greatly over the region from 3 to 9 m/sec. It can be seen that the provinces around Bohai rim like Liaoning, Shandong, Hebei, have the highest wind energy resources. Meanwhile, the coastal zones in the eastern and southern China are also endowed with excellent wind energy resources (i.e.,

Yangtze Delta region, Zhejiang, Fujian, Guangdong; see Fig. 7 left). Figure 8 shows the monthly average wind speed in each coastal region. There is a marked seasonal variation in average wind speed, for most of regions higher in spring-winter months and lower in summer-autumn months due to the typical monsoon climate. It can be seen that monthly wind speed is the highest in April (5.58 m/sec) and lowest in August (4.57 m/sec).

In addition, the spatial pattern of wind power output for the study area is similar to that of wind speed (see Fig. 7 right). On the base of primary estimation results, annual wind power output differs in value over the region from 300 to 15,600 MWh/km². The total wind energy potential in the study area is up to 7590 TWh/year, while after considering the geographical restrictions, it is reduced to 334 TWh/year. As shown in Table 2, the northern

Table 4. The potential of solar energy in the study area.

Name	Total suitable areas (km²)	Min. value of solar energy density (TWh/km²)	Max. value of solar energy density (TWh/km²)	Mean value of solar energy density (TWh/km²)	Solar energy potential (TWh/year)
Liaoning	2310	52.1	66.8	60.7	140.2
Jingjinji	8907	52.9	71.7	63.9	569.4
Shandong	1421	52.0	56.9	54.6	77.6
Jiangsu	172	51.2	55.7	52.8	9.1
Zhejiang	588	48.0	54.3	51.9	30.5
Shanghai	–	49.2	53.7	52.3	–
Fujian	1330	44.0	55.7	51.9	69.0
Guangdong	495	49.2	55.6	52.0	25.8
Guangxi	1963	46.2	54.9	50.2	98.5
Hainan	161	52.8	62.0	58.2	9.4

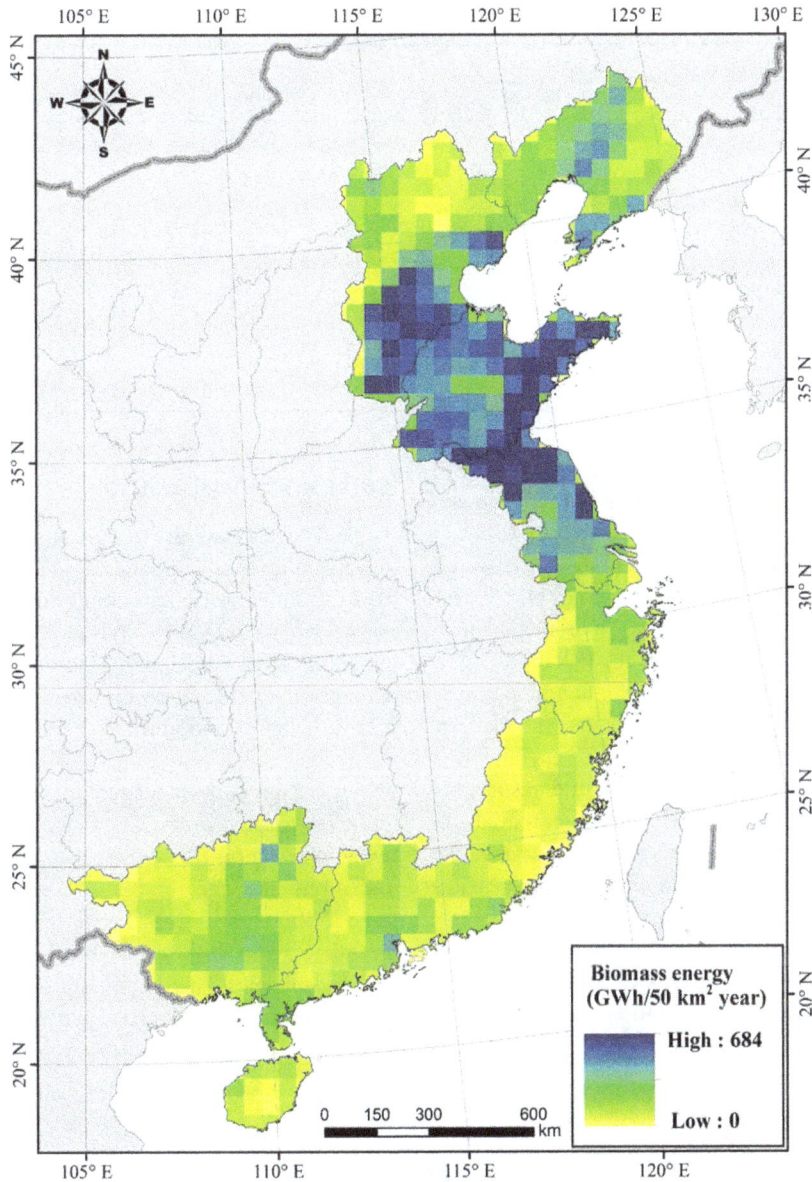

Figure 13. Spatial distribution of biomass energy for the study area.

part of coastal region (i.e., Jingjinji, Liaoning, Shandong) had higher wind energy potential due to the larger amount of suitable installed areas. Correspondingly, the eastern and southern parts of China had lower onshore wind energy potential. For example, the onshore wind energy potential for metropolis Shanghai is almost zero, since there is no suitable areas to install wind turbine with the process of urban sprawl and population agglomeration.

Offshore wind energy

The distribution of offshore wind speed in east coastal zone is shown in Figure 9. The highest annual wind speed is found in the coast of Fujian due to the funnel effect of Taiwan Strait. The northern Guangdong and southern Zhejiang are also endowed with good wind resources compared with the other coastal areas. It is essential to demonstrate the seasonal variation in wind speed, which has a significant impact on the electricity supply and adjustment. Figure 10 shows the monthly average wind speed profile in each coastal area. As shown in the data, the minimum monthly average wind speed appeared in summer months (5.8 m/sec), and the maximum in winter (8.3 m/sec). In general, higher wind speed variation located in the coast of Fujian, Zhejiang, and Guangdong.

Table 5. The potential of biomass energy in the study area.

Name	Available paddy crops (10^8 MJ/year)	Available dry crops (10^8 MJ/year)	Available forest residues (10^8 MJ/year)	Total amount of biomass residues (GWh/year)
Liaoning	85.42	329.59	0.48	11,541.2
Jingjinji	8.40	802.98	0.24	22,545.1
Shandong	17.39	977.74	0.17	27,647.1
Jiangsu	319.61	312.40	0.12	17,559.4
Zhejiang	102.32	41.08	0.41	3994.7
Shanghai	2.94	0.36	0.00	91.9
Fujian	84.74	31.87	1.16	3271.5
Guangdong	189.50	72.53	0.68	7297.5
Guangxi	191.93	81.14	0.97	7612.1
Hainan	26.17	8.31	0.17	962.7

To get a more comprehensive understanding of off-shore wind energy potential, the total annual power generation at different locations are presented in Table 3. Overall, the offshore wind energy potential in the water depth of 0~20 m and 20~50 m can reach to 4271.5 TWh/year and 7585.6 TWh/year, respectively. For the shallow water zones with the depth of 0~20 m, the higher potential is located in Jiangsu, Shandong, Guangdong, and Zhejiang, which have longer coastline. For the deeper water zones with the depth of 20~50 m, Guangdong, Fujian, and Jiangsu have good offshore wind resources. Spatial patterns of offshore wind energy potential in Hong's study [32] are consistent with our results. Currently, offshore wind energy in shallow waters with depths of 0–20 m are viable due to technically mature and economically feasibility. As a result, onshore and offshore wind energy are shown a complementary effect from the point view of spatio-temporal dimension.

Solar energy potential

Regional solar energy resources are related to latitude, terrain, and local climatic conditions. Figure 11 shows that the annual solar radiation differs in value over coastal region from 1312~2175 kWh/m². In general, annual solar radiation on latitude tilt in northeastern parts of China is larger than that in southeastern and southern parts of China. Liaoning and northern part of Jingjinji are endowed with good solar resources. The time series data within the year are also calculated for each coastal region in order to clarify the seasonal variation pattern of solar radiation. Figure 12 shows that monthly average solar radiation has the highest value in summer month (156 kWh/m²±5 kWh/m²), and the lowest value in winter month (114 kWh/m²±9 kWh/m²).

The available area and potential of large-scale PV station are shown in Table 4. Total amount of suitable area for

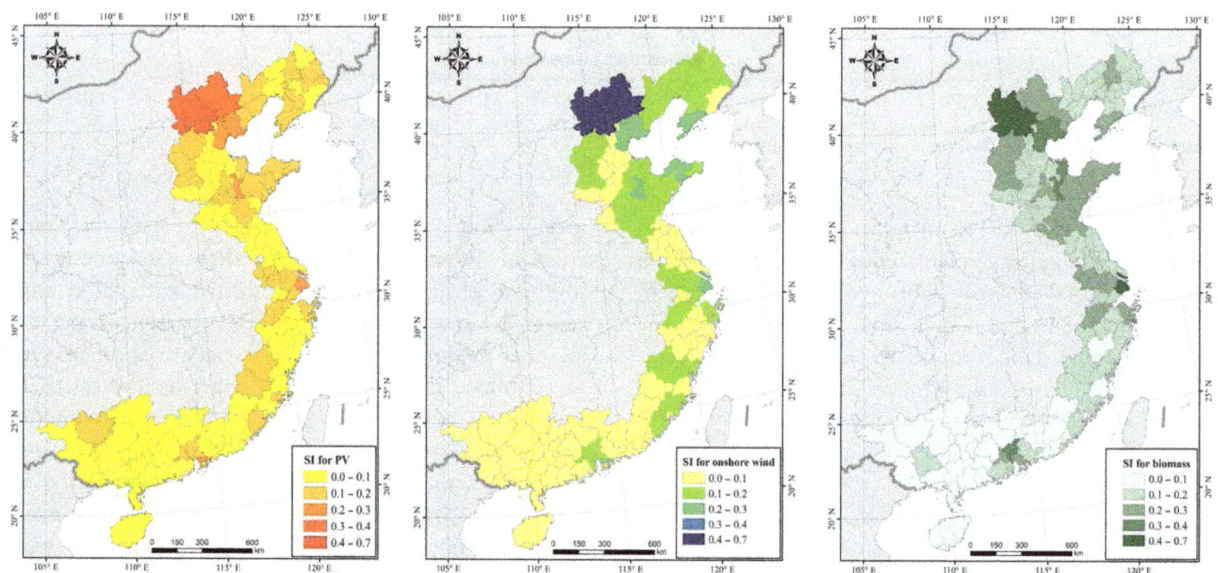

Figure 14. Spatial distribution of suitability index (SI) for PV, onshore wind and biomass in the study area.

large-scale PV is the largest in Jingjinji, which has a larger amount of grassland and flat ground. However, several developed regions like Jiangsu, Zhejiang, Guangdong, and Shanghai have limited available areas to install PV station. Thus, Jingjinji and Liaoning have larger solar energy potential (accounting for about 70% of total mount) compared with other coastal regions. Overall, the potential of large-scale PV in eastern region reaches to 1029.5 TWh/year.

Biomass energy potential

To get a spatial explicit result of biomass residues, the spatial allocation of available agricultural crop and forest pruning residues amount was conducted with spatial

analysis tools. In Figure 13, the total amount of biomass residues energy density in 50 km × 50 km supply regions are shown. The energy density of biomass residues varies greatly over region in the range between 0 and 684 GWh/year. The hotspot regions are located in major agricultural provinces such as Shandong, Jiangsu, and southern part of Hebei. Some regions of Liaoning also have larger biomass energy potential.

As shown in Table 5, the total amount of agriculture and forestry residues is 102.5 TWh/year for the study area, and the top four regions include Shandong, Jingjinji, Jaingsu, and Liaoning. Moreover, the available amount of dry crops residues is twice more than that of paddy residues. The potential production of four provinces

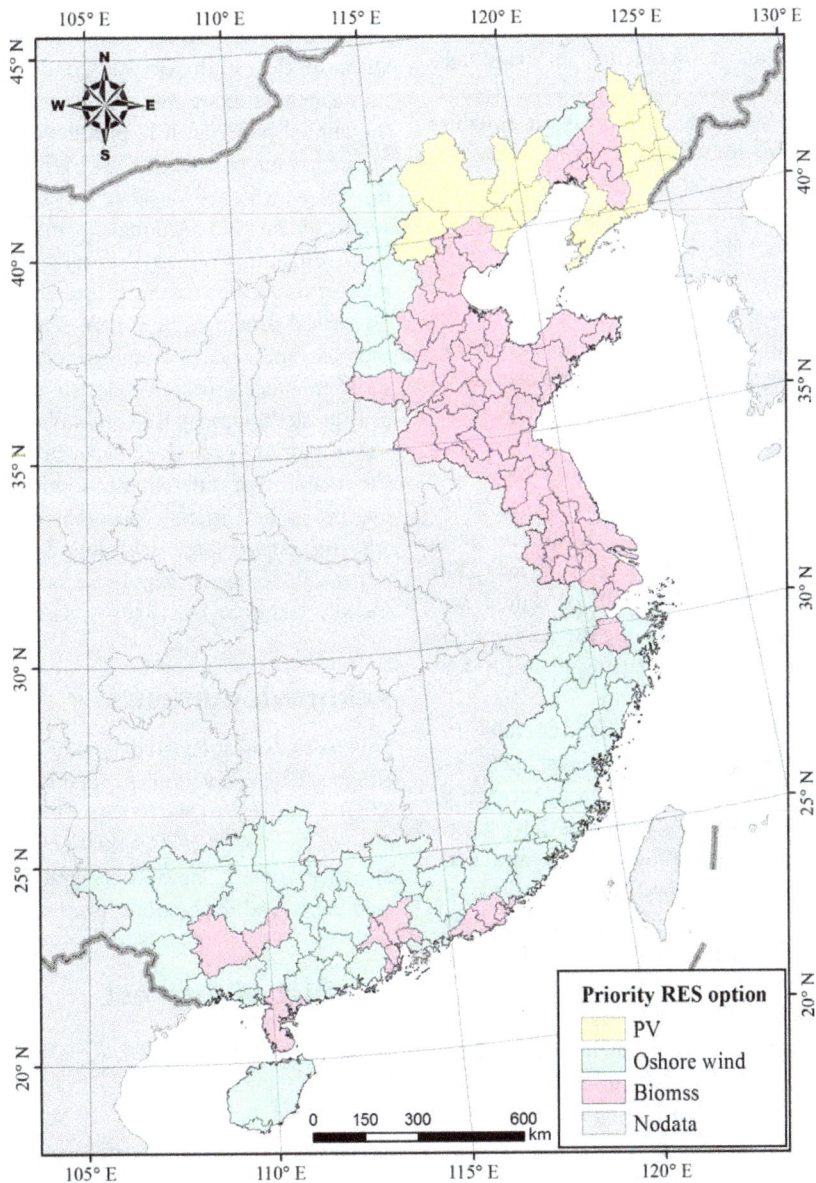

Figure 15. Spatial distribution of priority RES options for the study area.

occupies more than 70% of total across study area. Due to its superb climate and geographical conditions, it is as major agricultural production regions in China. In addition, plenty of marginal land are also available in such regions. Thus, biomass energy as a stable RES option should be priority exploration in the north part of China.

Renewable energy development strategies in eastern coastal region

Location suitability index maps for each RES type were determined by combining entropy-weight method and GIS multicriteria decision analysis in eastern coastal regions of China. In this procedure, we took the prefectural level regions as basic assessment units (a total of 115 subregions). As the results shown in Figure 14, the larger SI for three RES types are all emerged in the Jingjinji, Shandong, Liaoning, Yangtze River Delta and Pearl River Delta Region with about 15% of land coverage. Such areas are suitable to development of RES due to better economic foundation and larger marginal land. It is clearly seen that most of areas in southeastern parts of our country have the lower value of suitability index. The lack of marginal land is the largest constrain factor for these regions.

A ranking procedure was then conducted by a trade-off between SI and UI of different RES options to determine an overall RES alternative solution map. The priority RES technology was chosen according to the value of CI for each prefectural level region (Fig. 15). As the result, about 12% of the study area (including Beijing, part of Hebei and Liaoning) is suitable for solar PV farms, and 35% of the study area (including Shandong, Jiangsu, and part of Liaoning, Hebei) is suitable for biomass farms. The other 53% of the study area is found to be suited for onshore wind development. By optimizing the RES option for each district, the issue of spatial energy planning for multiple regions can be easy to deal with.

Eastern China's region has shown to be vulnerable to electricity supply and has demanded significant RES development in order to address its challenges. According to official statistics, eastern China's provinces have been increasing its RES installed generating capacity and had reach to 32.76 GW in 2013. Onshore and offshore wind play an important role in the past decades, and account for 80% of total installed capacity. Total amount of electric consumption for 10 provinces of study area had reach to 2885.13 TWh in 2013. Based on the evaluation of our study, current overall RES generation potential in eastern China is 5737.5 TWh, which is close to twice time of electric consumption in 2013. Considering technical and economic feasibility of RES technologies, onshore and offshore wind will continue to play an important role in future electric supply system of eastern China. Among the main sources of RES options, biomass and solar PV energy offer greater potential as supplementary sources of electric generation. Besides some technical, economic, and policy issues, an integrated spatial coordination and a scientific environmental cognition are two important factors for it. The spatial coordinated development should be implemented according to the spatial disparity and consistency among the major renewable energies, coal resources, energy consumption and its major influencing factors in China [38].

Conclusions

This study attempted to propose a new multiregional analytic framework to determine the location suitability and priority RES options for a larger scale region. A case study of this approach was conducted to 10 provinces/municipalities along east coastal regions of China. Overall, the annual potential RES production from wind, PV, and biomass sources was 5737.5 TWh in eastern China's 10 provinces, which is close to twice times of electric consumption in 2013. Among the main sources of RES options, offshore wind in the water depth of 0~20 m will play an important role in future electric supply, especially for coastal provinces of Fujian, Zhejiang, and Guangdong. Biomass and solar PV energy offer greater potential as supplementary sources of electric generation. Selection of priority development RES technology for specific location was a trade-off procedure between location suitability and sustainability of different RES options. A favorable incentive policy and spatial coordinated development strategy are also essential for achieving higher future targets in terms of electricity supply security and environmental friendly technologies.

Acknowledgments

This work was sponsored by K.C.Wong Magna Fund in Ningbo University and Academic Discipline Project of Ningbo University (No. 011-431501312). Run Wang thanks the support by the "Chutian Scholarship Program" by Province Hubei. The authors gratefully acknowledge the comments and suggestions from anonymous reviewers.

Conflict of Interest

The authors declare no conflict of interest.

References

1. Chen, S. Y. 2013. Energy, environment and economic transformation in china. Taylor & Francis Group, Routledge, London and New York.

2. Zhang, S. F., P. Andrews-Speed, X. L. Zhao, and Y. X. He. 2013. Interactions between renewable energy policy and renewable energy industrial policy: a critical analysis of China's policy approach to renewable energies. Energ. Policy 62:342–353.

3. Zhang, X., L. Q. Wu, R. Zhang, S. H. Deng, Y. Z. Zhang, J. Wu et al. 2013. Evaluating the relationships among economic growth, energy consumption, air emissions and air environmental protection investment in China. Renew. Sustain. Energy Rev. 18:259–270.

4. Zhao, X. T., W. Yang, D. Zhou, and S. L. Chen. 2008. The five major issues related to the sustainable development of China's estuary. Ocean Dev. Manag. 25:91–93.

5. Sun, Y. W., R. Wang, J. Liu, L. S. Xiao, Y. J. Lin, and W. Kao. 2013. Spatial planning framework for biomass resources for power production at regional level: a case study for Fujian Province, China. Appl. Energy 106:391–406.

6. Hoogwijk, M. M. 2004. On the global and regional potential of renewable energy sources. Universiteit Utrecht, Utrecht, Netherlands.

7. Delucchi, M. A., and M. Z. Jacobson. 2011. Providing all global energy with wind, water, and solar power, Part II: reliability, system and transmission costs, and policies. Energ. Policy 39:1170–1190.

8. Bekele, G., and B. Palm. 2009. Wind energy potential assessment at four typical locations in Ethiopia. Appl. Energ. 86:388–396.

9. Mondal, M. A. H., and M. Denich. 2010. Assessment of renewable energy resources potential for electricity generation in Bangladesh. Renew. Sustain. Energy Rev. 14:2401–2413.

10. Gómez, A., M. Rodrigues, C. Montañés, C. Dopazo, and N. Fueyo. 2010. The potential for electricity generation from crop and forestry residues in Spain. Biomass Bioenerg. 34:703–719.

11. Tucho, G. T., P. D. M. Weesie, and S. Nonhebel. 2014. Assessment of renewable energy resources potential for large scale and standalone applications in Ethiopia. Renew. Sustain. Energy Rev. 40:422–431.

12. Siyal, S. H., U. Mörtberg, D. Mentis, M. Welsch, I. Babelon, and M. Howells. 2015. Wind energy assessment considering geographic and environmental restrictions in Sweden: a GIS-based approach. Energy 83:447–461.

13. Wu, S. P., C. Y. Liu, and X. P. Chen. 2015. Offshore wave energy resource assessment in the East China Sea. Renew. Energy 76:628–636.

14. Aydin, N. Y., E. Y. Kentel, and S. Duzgun. 2010. GIS-based environmental assessment of wind energy systems for spatial planning: a case study from Western Turkey. Renew. Sustain. Energy Rev. 14:364–373.

15. Grassi, S., N. Chokani, and R. Abhari. 2012. Large scale technical and economical assessment of wind energy potential with a GIS tool: case study Iowa. Energ. Policy 45:73–85.

16. Sun, Y. W., A. Hof, R. Wang, J. Liu, Y. J. Lin, and D. W. Yang. 2013. GIS-based approach for potential analysis of solar PV generation at the regional scale: a case study of Fujian Province. Energ. Policy 58:248–259.

17. Aydin, N. Y., E. Kentel, and H. S. Duzgun. 2013. GIS-based site selection methodology for hybrid renewable energy systems: a case study from western Turkey. Energy Convers. Manag. 70:90–106.

18. Sarralde, J. J., D. J. Quinn, D. Wiesmann, and K. Steemers. 2015. Solar energy and urban morphology: scenarios for increasing the renewable energy potential of neighborhoods in London. Renew. Energy 73:10–17.

19. Mirzaei, A., F. Tangang, and L. Juneng. 2015. Wave energy potential assessment in the central and southern regions of the South China Sea. Renew. Energy 80:454–470.

20. Malczewski, J. 1999. GIS and multi-criteria decision analysis. John Wiley & Sons, New York.

21. Malczewski, J. 2006. Ordered weighted averaging with fuzzy quantifiers: GIS-based multi-criteria evaluation for land-use suitability analysis. Int. J. Appl. Earth Obs. Geoinf.. 8:270–277.

22. Chinese Wind Energy Association (CWEA). 2012. China's wind power installed capacity statistics (2000–2012). CWEA, Beijing, China.

23. Colorado Research Associates/Northwest Research Associates Inc. 2001. QSCAT/NCEP Blended ocean winds from Colorado Research Associates (version 5.0). National Center for Atmospheric Research, Computational and Information System Laboratory, Boulder, CO.

24. New, M., D. Lister, M. Hulme, and I. Makin. 2002. A high-resolution data set of surface climate over global land areas. Clim. Res. 21:1–25.

25. Solar and Wind Energy Resource Assessment (SWERA). 2015. Solar: monthly and annual average latitude tilt GIS data at 40 km resolution for China from NREL. NREL, Fort Collins, CO.

26. Shi, X., A. Elmore, X. Li, N. J. Gorence, H. M. Jin, X. H. Zhang et al. 2008. Using spatial information technologies to select sites for biomass power plants: a case study in Guangdong Province, China. Biomass Bioenerg. 32:35–43.

27. Manwell, J. F., L. Rogers Anthony, and J. G. McGowan. 2009. Wind energy explained: theory, design and application, 2nd ed. Wiley, Chichester.

28. Gao, X. X., H. X. Yang, and L. Lu. 2014. Study on offshore wind power potential and wind farm optimization in Hong Kong. Appl. Energy 130: 519–531.

29. Mentis, D., S. Hermann, M. Howells, M. Welsch, and S. H. Siyal. 2015. Assessing the technical wind energy potential in Africa a GIS-based approach. Renew. Energy 83:110–125.

30. Sun, Y. W., R. Wang, J. Liu, L. S. Xiao, and D. W. Yang. 2012. Assessment of onshore wind energy potential in Fujian province based on GIS. Resour. Sci. 34:1167–1174.

31. Yamaguchi, A., and T. Ishihara. 2014. Assessment of offshore wind energy potential using mesoscale model and geographic information system. Renew. Energy 69:506–515.

32. Hong, L. X., and B. Möller. 2011. Offshore wind energy potential in China: under technical, spatial and economic constraints. Energy 36:4482–4491.

33. Šŭri, M., T. A. Huld, E. D. Dunlop, and H. A. Ossenbrink. 2007. Potential of solar electricity generation in the European Union member states and candidate countries. Sol. Energy 81:1295–1305.

34. NREL. 2009. Potential for development of solar and wind resource in Bhutan. National Renewable Energy Laboratory, Fort Collins, CO.

35. Gehrung, J., and Y. Scholz. The application of simulated NPP data in improving the assessment of the spatial distribution of biomass in Europe. Biomass Bioenerg 2009; 33:712–720.

36. Delgado, A., and I. Romero. 2016. Environmental conflict analysis using an integrated grey clustering and entropy-weight method: a case study of a mining project in Peru. Environ. Model. Softw. 77: 108–121.

37. Troldborg, M., S. Heslop, and R. L. Hough. 2014. Assessing the sustainability of renewable energy technologies using multi-criteria analysis: suitability of approach for national-scale assessments and associated uncertainties. Renew. Sustain. Energy Rev. 39:1173–1184.

38. Bao, C., and C. L. Fang. 2013. Geographical and environmental perspectives for the sustainable development of renewable energy in urbanizing China. Renew. Sustain. Energy Rev. 27:464–474.

7

Fuzzy ARTMAP and GARCH-based hybrid model aided with wavelet transform for short-term electricity load forecasting

Swati Takiyar[1], K. G. Upadhyay[1] & Vivek Singh[2]

[1]M.M.M. University of Technology, Gorakhpur, Uttar Pradesh, India
[2]National Hydroelectric Power Corporation Ltd., Faridabad, Haryana, India

Keywords
Artificial neural network, electricity load forecasting, fuzzy ARTMAP, GARCH, wavelet transform

Correspondence
Swati Takiyar, M.M.M. University of Technology, Gorakhpur, Uttar Pradesh, India.
E-mail:takiyar.s@gmail.com

Funding Information
Accepted under 100% fee waiver anniversary edition.

Abstract

With the evolution of the electricity market into a restructured smart version, load forecasting has emerged as an eminent research domain. Many forecasting models have been proposed by researchers for electricity price and load fore-casting. This state of art introduces a load time series modeled with a hybrid technique culminating from the logical amalgamation of GARCH, a conventional hard computing method, Fuzzy ARTMAP, an artificial intelligence- based soft computing technique, and wavelet transform, for treating the load time series. The study investigates into the ability of the proposed hybrid model in tackling the electricity load time series forecasting problems. The work under this study also includes comparisons drawn among models which use either one or two of the mentioned techniques and the model proposed. Results certify the efficacy and effectiveness of the model over others.

Introduction

Forecasting [1, 2] knowingly or unknowingly holds an integral stature in every person's life. We often predict stock market, earthquake and weather in our day to day life. A business manager forecasts product sales [3]. People make future plans based on these forecasts. Here it should be understood that it is an impossible task to make exact forecasts, we can only work incessantly towards attaining higher accuracies. Generally forecasting presumes that future occurrences depend upon past or present observable events; it assumes that some aspects of the past pattern will continue in the future. Through observing and studying past data relationships can be established between the event and the parameters persisting at the then moment [4].

Forecasting with load time series is a challenging application because load time series are inherently nonstationary [5], deterministically chaotic, and highly noisy by nature. Above this by no technique the past information can solely determine the futuristic behavior of the electricity market. Therefore, to maintain global competitiveness, market dependency on advance computer technologies is increasing day by day.

Forecasting can either be spatial or temporal in nature. While spatial forecasts are based on area covered, temporal are based on time horizon utilized. Temporal forecasting can further be divided into three subcategories: Short Term, Medium Term and Long Term [6]. Out of these, short-term forecasting is of most importance because of its utility. They help plan capacity building, estimate load flows and prevent overloading to name a few. Forecasting techniques primarily use either statistical tools [7, 8] or artificial intelligence based algorithms [9].

In this study, we present a new method of forecasting electricity loads using Fuzzy ARTMAP (FA) and GARCH along with wavelet transform (WT) for day-ahead forecasting. The approach implemented aims to develop a sturdy, precise and efficient day-ahead load forecasting tool utilizing data filtering technique, employing WT, fused with a computational model implementing conjointly FA (soft computing model), and GARCH (hard computing model). Comparison results of the proposed models' performance with that utilizing only FA and utilizing both FA and GARCH show a convincing reduction in mean absolute percentage error (MAPE). A simple artificial neural network (ANN) model utilizing the same data has also been implemented simply to draw a wider range comparison and highlight the supremacy of the hybrid model over ANN.

The procedure of forecasting proposed can be summed in three steps:

1. Decomposition of past load series using WT.
2. Using FA model fitted to one approximated and two decomposed series of WT and using GARCH on one decomposed series of WT.
3. Then using inverse WT to reconstruct the forecasted load series.

The literature of the study is arranged as follows:

Section II details WT, FA, and GARCH techniques in brief to enhance reader understanding.
Section III elaborates on the implemented hybrid methodology.
Section IV comprises numerical and graphical results and Section V concluding the study.

Any load data series embodies numerous spikes, nonlinearities, and fluctuations. The 2010 hourly New South Wales electricity load series is no exception to it. As illustrated in Figure 1, it is also characterized by chaotic and random changes. This series of 2148 h has been utilized in this study. Where the hourly data of 92 days or 3 months has been used to train the network and the data of next 24 h has been reserved as validation set data and is also the predicted subset.

Wavelet Transform

Wavelet transform is a mathematical model which transforms the original load series (in time domain) into constituent subseries over time domain of a different scale for processing and analysis. WT is most suitable for the nonstationary data (mean and autocorrelation of series are not constant). It is also well known that most of the load data series are nonstationary, hence the utility of WT [10]. The WT is used to decompose the original load series into several other series with resolution of different levels, which is called multiresolution decomposition [11, 12].

Fourier transform (FT) decomposes the original load series into linear combinations as sine and cosine functions whereas by WT the series is decomposed into a sum of more flexible functions which are localized in both time and frequency [13].

Wavelet transform can be classified into two: continuous wavelet transform (CWT) and discrete wavelet transform (DWT).

The CWT of a continuous time signal $x(t)$ is defined as [4, 6]:

$$\Psi_{a,b}(t) = \int_{-\infty}^{\infty} x(t)\psi_{a,b}^*(t)dt, \qquad (1)$$

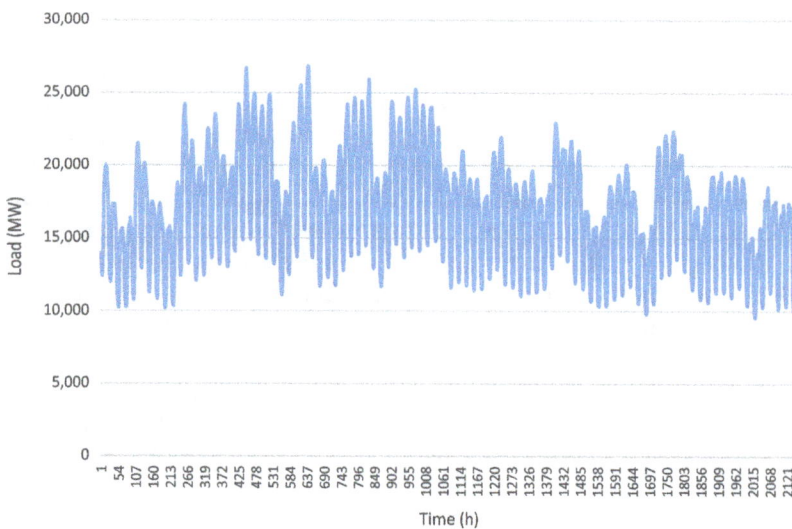

Figure 1. Load data of New South Wales.

where $\psi(t)$ is the mother wavelet, given by eq. (2) where a acts as a scaling parameter and b as a translating parameter.

$$\psi_{a,b}^{*}(t) = \frac{1}{\sqrt{a}}\psi^{*}\left(\frac{t-b}{a}\right). \qquad (2)$$

Each wavelet is formulated by scaling and translating the mother wavelet. The mother wavelet is an oscillatory function characterized by zero average and finite energy.

$$\psi_{c,d} = \sum_{n} x(n)\psi_{c,d}^{*}(n), \qquad (3)$$

where

$$\psi_{c,d}^{*}(n) = \frac{1}{\sqrt{a_{o}^{c}}}\psi^{*}\left(\frac{n - db_{o}a_{o}^{c}}{a_{o}^{c}}\right). \qquad (4)$$

Here, c acts as a scaling coefficient while d acts as a sampling one.

To implement DWT as a filter, Mallat propounded an algorithm called Mallat multiresolution analysis or the Mallat algorithm [12]. It is a two-staged algorithm where decomposition occurs in the first stage followed by reconstruction in the second one. This study implements a three-level decomposition on the original load series yielding three detailed series (D) and one approximated series (A) as illustrated in Figure 2. Decomposing and reconstructing processes both involve filtering for which both high–pass (HPF) and low-pass filters (LPF) are utilized. While down-sampling occurs during wavelet decomposition, up-sampling and filtering is used in wavelet reconstruction. A Daubenchies wavelet function of order 5 (db5) has been utilized in this study as a mother wavelet.

Fuzzy ARTMAP

The FA network is a supervised learning method based on adaptive resonance theory (ART) [14]. FA network carries out learning without forgetting previously learned information [15]. FA is flexible and adaptive to changes in the environment and is self-organizing by nature [16]. FA network is a recent technique that has been utilized in forecasting applications including load forecasting.

Neural network is another popular artificial intelligence technique utilized in forecasting applications. Most neural networks struggle with the plasticity–stability dilemma which probes into ways by which a network can endure adaptive-ness or plasticity toward new inputs while staying aloof of the noisy data inputs, hence stability [17, 18]. A general neural network encounters hindrance in preserving previously learned knowledge while learning newer concepts. The FA confronts this dilemma with a feedback mechanism laid between the competitive and input layers to allow fresh concepts to be absorbed without losing the knowledge attained previously. This results in a firmer learning environment endowed with faster convergence capability compared to traditional soft computing techniques [19]. This is also confirmed by the results of this study. These properties of FA can improve load forecasting performance as load series data are highly stochastic by nature (Figure 1).

The functional layout of FA network is shown in Figure 3. An ARTMAP system embodies twin art modules (ART$_a$ and ART$_b$) to fabricate stable recognition categories corresponding to the arbitrary input patterns. ART$_a$ uses ART-1, a type of ART network which accepts only binary input, while ART$_b$ uses FUZZY ART. This setup enables to switch the binary module notations into a corresponding feature in the fuzzy ART module. For example, the

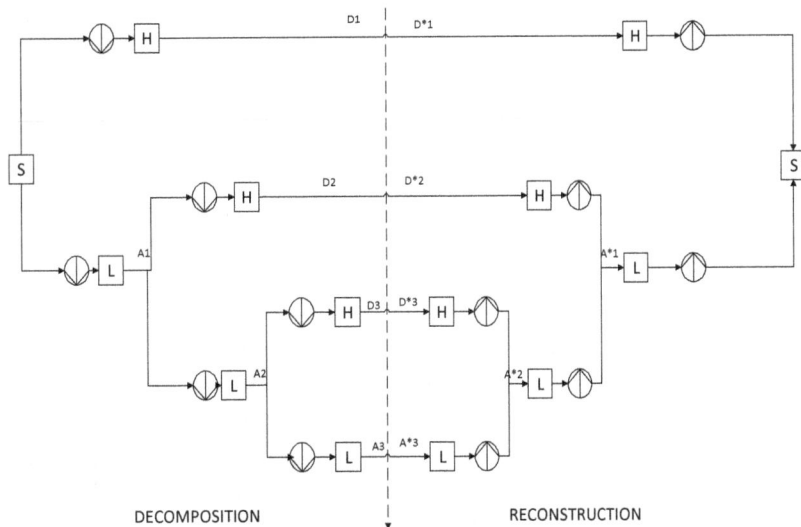

Figure 2. Wavelet transformation (order 3) [23, 28].

DECOMPOSITION   RECONSTRUCTION

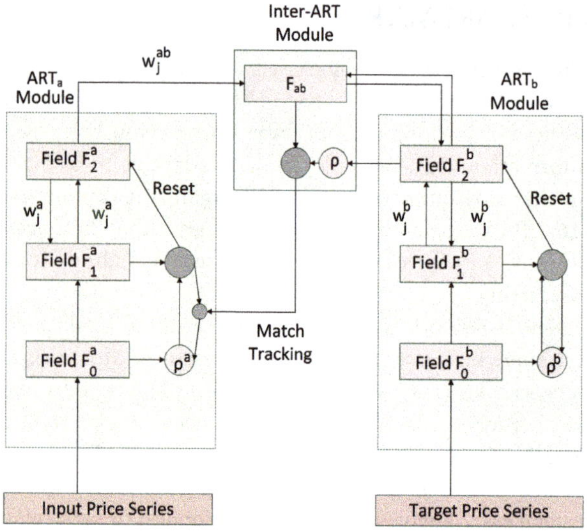

Figure 3. Functional layout of Fuzzy ARTMAP [15, 20, 28].

intersection operator ($\wedge$) of ART_1 is replaced by the operator ($\wedge$) in FUZZY ART. The architecture called FA is achieved by the synthesis of fuzzy logic and ART neural network, employing a close formal similarity between two computations of fuzzy subsets and ART category. Also, FA actualizes a new min–max learning rule that collectively minimizes predictive error and maximizes generalization, or code compression. This is achieved by a match tracking process that increases the ART vigilance parameter by the minimum amount needed to correct a predictive error. As a result, the system automatically learns a minimal number of recognition categories, or "hidden units," to meet the criteria of accuracy. Category proliferation is prevented by normalizing input vectors at a preprocessing stage. A normalization procedure called complement coding [15] leads to a symmetric theory in which the AND operator ($\wedge$) and the OR operator (v) of fuzzy logic plays complementary roles. In training, the best matching category is [19]:

$$J = \arg\max_{0 \leq j \leq N} T_j(I_{tr}), \qquad (5)$$

where

$$T_j(I_{tr}) = \begin{cases} \frac{|I_{tr} \wedge W_j|}{\alpha + |W_j|} & , \text{ If } \frac{|I_{tr} \wedge W_j|}{|I_{tr}|} \geq \rho \\ 0, & \text{Otherwise} \end{cases} \qquad (6)$$

where T_j = choice function, α = choice parameter, $\wedge$ = Fuzzy MIN operator, ρ = vigilance parameter, and $\frac{|I_{tr} \wedge W_j|}{|I_{tr}|} \geq \rho$ is the vigilance criteria. If vigilance criteria satisfy, then resonance occurs. During training, the vigilance criteria vary from baseline vigilance which is the initial value. If vigilance criteria qualify, then category J becomes

representative membership function for time series, and the weighing vector of the winning category W_j is updated as per the following equation:

$$W_j^{(new)} = \beta(I_{tr} \wedge W_j^{(old)}) + (1 - \beta)W_j^{(old)}. \qquad (7)$$

Here β represents the learning rate. If vigilance criteria fail, then category J is deactivated for the present load series by equating choice function equals to zero. If ART_b does not predict the correct output for ART_a, then the vigilance parameter is increased. This is called match tracking, in which the value of the vigilance parameter is slightly increased to a new value [17]:

$$\rho = \frac{|I_{tr} \wedge W_j|}{|I_{tr}|} + \varepsilon, \qquad (8)$$

where ε denotes the learning precision.

The scheme resizes a category on predictive success by amplifying the vigilance parameter ρ by a minimal amount essential to verify the predictive error in ART_b. The parameter ρ holds an inverse relationship with the category size. A lower value leads to a broadly generalized category with higher compressed code. This parameter rates the minimum faith that ART_a should have while accepting a category during hypothesis testing which focuses ART_a on a new cluster. The failures at ART_a increase ρ to that threshold value which in turn triggers ART_a under a process called match tracking. This technique reduces generalization essential to correct a predictive error. The combination of these techniques, i.e. ARTMAP

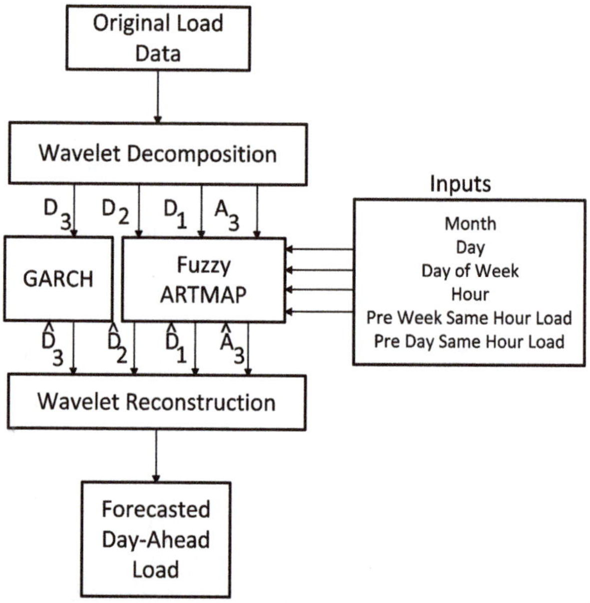

Figure 4. Schematic diagram of the proposed hybrid model.

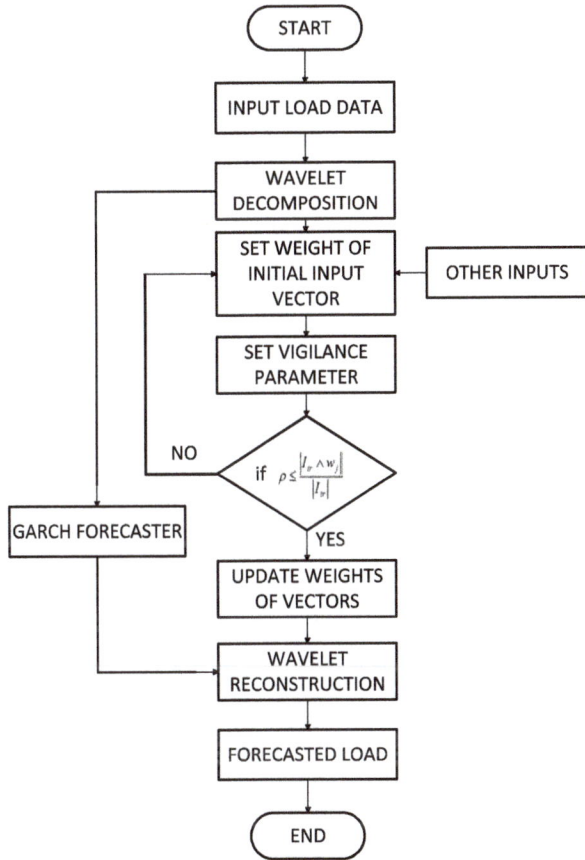

Figure 5. Flowchart for the proposed method for day-ahead load forecasting.

and match tracking leads to a faster learning and erudition from a rare event. The fuzzy ART reduces to ART_1 for a binary input and works as self for a binary input

and works as self for an analog vector. Thus the crisp logics of ART_1 with their fuzzy counterparts form a potent module.

Once the training stage is completed, the FA network is used as a classifier of the input load series which is given to ART_a. ART_b is not used during classification process and the learning capability of the network is deactivated during classifying process (i.e. $\beta = 0$). In this stage we get predicted classified labels in the output of ARTMAP. These classified labels are later de-fuzzified to get the forecasted loads.

GARCH

GARCH stands for Generalized Autoregressive Conditional Hetero-skedasticity which is used to model observed time series. GARCH is effectively implemented to highly volatile time series caused by unexpected random effects [20, 21]. The model GARCH (p, q) is defined as:

$$x_t = \mu + \varepsilon_t, \tag{9}$$

where μ is offset and $\varepsilon_t = \sigma_t z_t$.

Considering a time series x_t with a constant mean offset [4]:

$$\sigma_t^2 = c + \sum_{i=1}^{q} \phi_i \varepsilon_{t-i}^2 + \sum_{i=1}^{p} \psi_i \sigma_{t-i}^2, \tag{10}$$

where p is the order of GARCH terms σ^2 and q is the order of ARCH terms ε^2.

As can be seen in eq. (10), in GARCH (p, q) model is $p = 0$, i.e. a GARCH $(0, q)$ model becomes an ARCH (q) model.

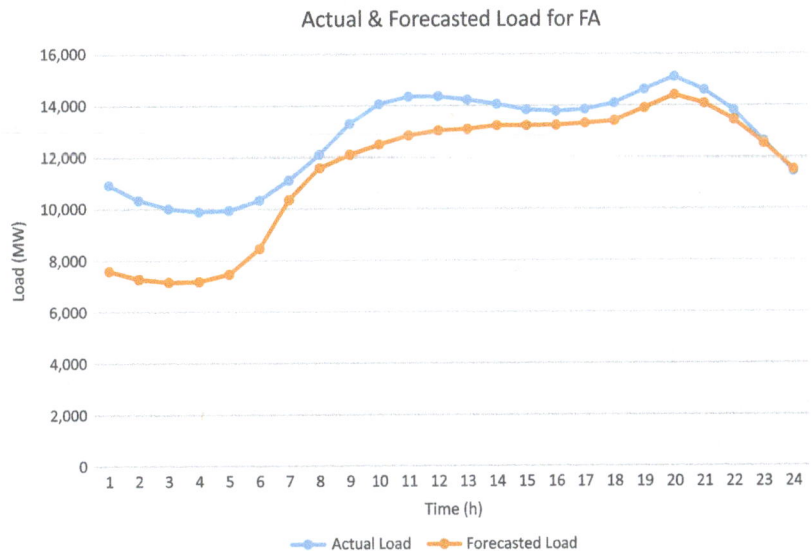

Figure 6. Actual and forecasted load for FA.
FA, Fuzzy ARTMAP.

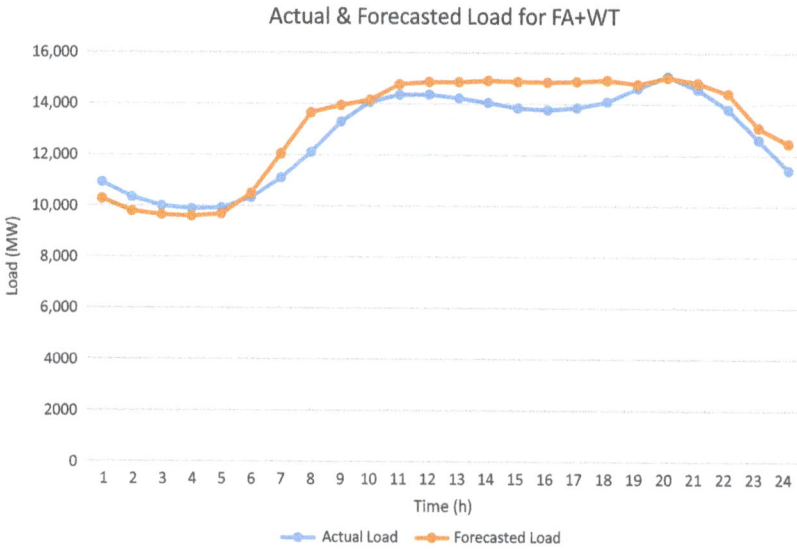

Figure 7. Actual and forecasted load for FA + WT. FA, Fuzzy ARTMAP; WT, wavelet transform.

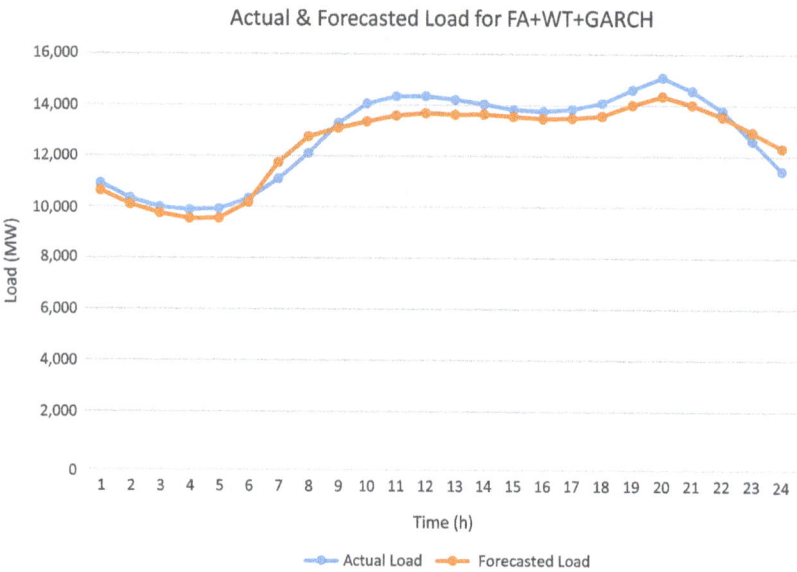

Figure 8. Actual and forecasted load for FA + WT + GARCH. FA, Fuzzy ARTMAP; WT, wavelet transform.

A limitation of the GARCH model is that it can only be specified for stationary time series, hence the below equation must be satisfied for stationary time series only:

$$\sum_{i=1}^{q} \phi_i + \sum_{i=1}^{p} \psi_i < 1. \tag{11}$$

Steps for GARCH modeling [21, 22]:

1. Identify the class of models.
2. Identify the subsets of GARCH models.
3. Parameter estimation.
4. Model validation.
5. Apply forecast.

Proposed Methodology

Figure 4 presents the schematic diagram of the prospective hybrid model for day-ahead electricity load forecasting built on the FA technique combined with GARCH and WT. The procedure for forecasting is as follows:

1. In the proposed hybrid model, input variables are hourly data of electricity load, month, day, day of week, hour, previous week same hour load and previous day same hour load. Only the load series is passed through WT. The load series is decomposed through WT into four components. The decomposed detailed coefficients,

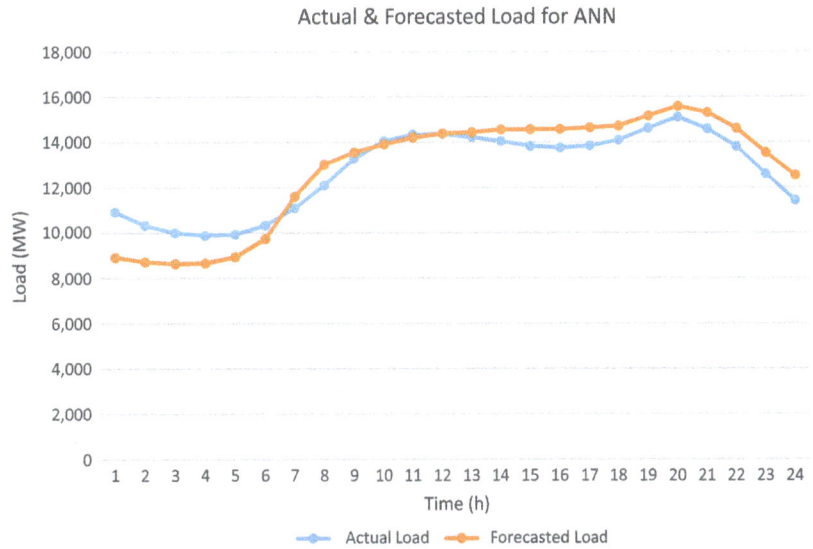

Figure 9. Actual and forecasted load for artificial neural network.

Table 1. Comparative hourly forecasts for respective techniques.

Hours	Actual load	FA		FA + WT		FA + WT + GARCH		ANN	
		Forecasted load	MAPE	Forecasted load	MAPE	Forecasted load	MAPE	Forecasted load	MAPE
1	10940	7599.86	30.53	10304.45	5.81	10654.79	2.61	8945.86	18.23
2	10357	7285.57	29.66	9810.04	5.28	10124.10	2.25	8740.28	15.61
3	10020	7171.58	28.43	9668.12	3.51	9777.74	2.42	8660.86	13.56
4	9906	7190.78	27.41	9607.85	3.01	9566.39	3.43	8691.42	12.26
5	9950	7476.25	24.86	9700.96	2.50	9575.59	3.76	8961.52	9.93
6	10347	8463.64	18.20	10508.24	−1.56	10209.69	1.33	9762.46	5.65
7	11119	10358.68	6.84	12054.61	−8.41	11765.89	−5.82	11636.54	−4.65
8	12130	11593.56	4.42	13668.34	−12.68	12787.44	−5.42	13048.84	−7.57
9	13322	12118.27	9.04	13958.09	−4.77	13122.40	1.50	13583.11	−1.96
10	14081	12509.26	11.16	14160.53	−0.56	13374.91	5.01	13937.94	1.02
11	14369	12851.78	10.56	14778.85	−2.85	13611.55	5.27	14231.49	0.96
12	14376	13049.30	9.23	14868.05	−3.42	13708.11	4.65	14402.53	−0.18
13	14241	13104.27	7.98	14868.00	−4.40	13651.71	4.14	14458.80	−1.53
14	14069	13240.01	5.89	14914.21	−6.01	13668.93	2.84	14580.21	−3.63
15	13852	13243.24	4.39	14887.22	−7.47	13578.73	1.97	14589.42	−5.32
16	13783	13252.21	3.85	14864.27	−7.84	13503.08	2.03	14596.21	−5.90
17	13864	13331.54	3.84	14892.18	−7.42	13519.43	2.49	14663.60	−5.77
18	14107	13415.88	4.90	14943.05	−5.93	13601.57	3.58	14734.08	−4.45
19	14633	13908.10	4.95	14796.38	−1.12	14039.43	4.06	15180.97	−3.74
20	15117	14411.75	4.67	15048.74	0.45	14379.41	4.88	15598.05	−3.18
21	14595	14069.20	3.60	14845.71	−1.72	14042.48	3.79	15325.88	−5.01
22	13813	13441.37	2.69	14420.63	−4.40	13572.74	1.74	14628.50	−5.90
23	12623	12529.19	0.74	13097.21	−3.76	12974.20	−2.78	13556.04	−7.39
24	11446	11518.88	−0.64	12474.47	−8.99	12333.07	−7.75	12557.34	−9.71
Overall MAPE			10.77		4.74	3.56			6.38

ANN, artificial neural network; FA, Fuzzy ARTMAP; MAPE, mean absolute percentage error; WT, wavelet transform.

D1, D2, D3 (high frequency components) and approximated series, A3 (low frequency components) are then obtained by down sampling with HPF and LPF, respectively.

2. The characteristic decomposed load series (D1, D2, and A3) along with other input data are furnished into the FA network and decomposed load series D3 is fed into GARCH.

Table 2. Mean absolute percentage error comparison of various models for load.

MAPE of various models			
ANN	FA	FA + WT	FA + WT + GARCH
6.381%	10.772%	4.745%	3.563%

ANN, artificial neural network; FA, Fuzzy ARTMAP; MAPE, mean absolute percentage error; WT, wavelet transform.

3. The output components of GARCH, FA network, and the decomposed detailed and approximate series are then processed by wavelet reconstruction to produce the day-ahead electricity loads.

This step-by-step summary is shown in Figure 5.

Numerical and Graphical Results

The paper introduces a new hybrid algorithm based on WT, FA, and GARCH, which accounts for the interactions of month, day, day of week, hour, previous week same hour load, and previous day same hour load. The proposed method has worked on the electricity load data of New South Wales. To rank the performance of the proposed model, the results have been compared to other models such as FA, FA + WT, and the most employed artificial intelligence technique, ANN. The summary table of this is illustrated in the Conclusion section.

Above this the outputs of the mentioned models have been tabulated below followed by the respective graphs comparing the forecasted and actual data.

Before one takes a look at these, we have briefed how they have been tabulated and why they serve as an appropriate measure of the efficiency of any forecasting model. Error is defined as the difference between the actual value and the forecasted value for the corresponding period [10, 23–26].

$$\varepsilon_t = A_t - F_t, \qquad (12)$$

where ε_t is the error for the period t, A_t is the actual value for the period t, and F_t is the forecasted value for the period t. MAPE or mean average percentage error is the most widely accepted parameter of forecasting error, which mathematically means:

$$\text{MAPE} = \frac{\sum_{t=1}^{N} \left| \frac{\varepsilon_t}{A_t} \right|}{N}. \qquad (13)$$

In this study, N has been valued 24 for daily electricity load forecasts. N should be valued 168 when we attempt weekly electricity forecasts. The graph shown below uses $N = 24$ as it predicts the day-ahead forecasts. Figures 6–9 represent the actual versus forecasted data for FA, FA + WT, FA + WT + GARCH, and ANN, respectively. Table 1 presents the actual versus forecasted data for all the techniques, namely FA, FA + Wavelet, ANN, and the proposed model (FA + WT + GARCH).

Conclusion

The hybrid model proposed in this paper is for short-term electricity load forecasting. The model is the aftermath of befitting coalition of FA, GARCH, and WT. While WT looks after the ill-behaved load series, FA captures the nonlinear fluctuations by virtue of stability–plasticity dilemma [27]. The attributes of FA renders the proposed hybrid method robustness and higher efficiency enabling forecasting meeting higher accuracy.

The model has also been compared with FA, FA + WT, and ANN. The results certify the efficacy of the proposed load forecasting hybrid model, as can be seen from Table 2.

Acknowledgments

The authors thank the New South Wales System Operator for providing hourly data of electricity load (http://www. asx.com.au/).

Conflict of Interest

None declared.

References

1. X., Zhang, Q. Wu, and J. Zhang. 2010. Crude oil price forecasting using fuzzy time series, 3rd ISKAM, 978-1-4244-8005-0, Pp. 213–216.

2. Georgilakis, P. S. 2006. Market clearing price forecasting in deregulated electricity market using adaptively trained neural networks, SETN 2006. Lecture notes in artificial intelligence LANI 3955. Springer, Berlin, Heidelberg. Pp. 56–66.

3. Gountis, V. P., and A. G. Bakirtzis. 2004. Bidding strategies for electricity producers in a competitive electricity marketplace. IEEE Trans. Power Syst. 19:356–365.

4. Gross, G., and F. D. Galiana. 1987. Short-term load forecasting. Proc. IEEE 75:1558–1573.

5. Binh, P. T. T., N. T. Hung, P. Q. Dung, and L.-H. Hee. 2012. Load forecasting based on wavelet transform and fuzzy logic. Power System Technology (POWERCON), 2012 IEEE International Conference on, pp. 1–6. doi: 10.1109/ PowerCon.2012.6401281.

6. Alfares, H. K., and M. Nazeeruddin. 2002. Electric load forecasting: literature survey and classification of methods. Int. J. Syst. Sci. 33:23–34.

7. Hahn, H., S. Meyer-Nieberg, and S. Pickl. 2009. Electric load forecasting methods: tools for decision making. Eur. J. Operat. Res. 199:902–907.

8. Marin, F. J., and F. Sandoval. 1997. Short-term peak load forecasting: statistical methods versus artificial neural networks. Springer Lect. Notes Comput. Sci. 1240:1334–1343.

9. Sureban, M. S., and S. G. Ankaliki. 2015. Applications of artificial neural networks in various areas of power system: a review. J. Power Sys. Eng. 2:35–44.

10. Al Wadi, M. T. I. S., and M. Tahir Ismaul. 2011. Selecting wavelet transforms model in forecasting financial time series data based on ARIMA Model. Appl. Math. Sci. 5:315–326.

11. Chui, C. K. 1992. An introduction to wavelets. Academic, New York. Pp. 6–18.

12. Mallet, S. 1989. A theory for multiresolution signal decomposition – The wavelet representation. IEEE Trans. Pattern Anal. Mach. Intell. 11:674–693.

13. http://fourier.eng.hmc.edu/e161/lectures/wavelets/node7.html

14. Carpenter, G. A., and S. Grossberg. 1988. The ART of adaptive pattern recognition by a self-organizing neural network. Computer 21:77–88. doi:10.1109/2.33.

15. Carpenter, G., S. Grossberg, N. Markuzon, J. Reynolds, and D. Rosen. 1992. Fuzzy ARTMAP: aneural network architecture for incremental supervised learning of analog multidimensional maps. IEEE Trans. Neural Networks 3:698–713.

16. Lopes, M. L. M., C. R. Minussi, and A. D. P. Lotufo. 2005. Electric load forecasting using a Fuzzy ART and ARTMAP neural network. Appl. Soft. Comput. 5:235–244.

17. Christodoulou, C., and M. Georgiopoulos. 2000. Applications of neural networks in electromagnetics, 1st ed. Artech House, Norwood, MA.

18. Dagher, I., M. Georgiopoulos, G. Heileman, and G. Bebis. 1999. An ordering algorithm for pattern presentation in fuzzy ARTMAP that tends to improve generalization performance. IEEE Trans. Neural Netw. 10:768–778.

19. Zornetzer, S. F., J. L. Davis, and C. Lau, eds. 1990. An introduction to neural and electronic networks. Academic Press Professional Inc, San Diego, CA.

20. Antunes, J. F., de Souza Araujo N. V., and C. R. Miussi. 2013. Multinodal load forecasting using an ART-ARTMAP-Fuzzy neural network and PSO strategy. in PowerTech (POWERTECH), IEEE Grenoble.

21. Garcia, R. C., J. Contreras, M. van Akkeren, and J. B. C. Garcia. 2005. A GARCH forecasting model to predict day-ahead electricity prices. IEEE Trans. Power Syst. 20:867–874.

22. Mathworks 2005. GARCH Toolbox for use with MATLAB. Mathworks User's Guide Version 2.0. Pp. 1–272. Available at http://www.mathworks.com/access/helpdesk/help/pdf_doc/garch/garch.pdf.

23. Mandal, P., A. U. Haque, J. Meng, R. Martinez, and A. K. Srivastava. 2012. A hybrid intelligent algorithm for short-term energy price forecasting in the Ontario market. IEEE 978-1-4673-2729-9.

24. Tan, Z., J. Zhang, J. Wang, and J. Xu. 2010. Day-ahead electricity price forecasting using wavelet transform combined with ARIMA and GARCH models. Appl. Energy 87:3606–3610.

25. Mishra, S., A. Sharma, and G. Panda. 2011. Wind power forecasting model using complex wavelet theory, 978-1-4673-0136-7, IEEE.

26. Mandal, P., A. U. Haque, J. Meng, A. K. Srivastava, and R. Martinez. 2013. A novel hybrid approach using wavelet, firefly algorithm and fuzzy ARTMAP for day-ahead electricity price forecasting. IEEE Trans. Power Syst. 28:1041–1051.

27. Carpenter, G. A., S. Grossberg and D. B. Rosen, 1991. Fuzzy ART: fast stable learning and categorization of analog patterns by an adaptive resonance system. Orig. Res. Art. Neural Netw. 4:759–771.

28. Haque, A. U., P. Mandal, J. Meng, A. K. SrivastavaT.-L. Tseng, and T. Senjyu. 2012. A novel hybrid approach based on wavelet transform and fuzzy ARTMAP network for predicting wind farm power production. Ind. Appl. Soc. Ann. Meet. 1–8.

8

Tracing the energy footprints of Indonesian manufacturing industry

Yales Vivadinar, Widodo W. Purwanto & Asep H. Saputra

Department of Chemical Engineering, Universitas Indonesia, Depok, Indonesia

Keywords

Efficiency energy, energy flows, energy mapping, manufacture, Sankey diagram

Correspondence

Widodo W. Purwanto, Department of Chemical Engineering, Universitas Indonesia, Depok, Jawa Barat 16424 Indonesia.
E-mail: widodo@che.ui.ac.id

Funding Information

No funding information provided.

Abstract

The low energy efficiency is the source of the large energy consumption issue and the rapid growth of energy demand in the manufacturing industry sector in Indonesia. The driving forces behind the low energy efficiency situation in this sector are both economic and efficiency factors. The objective of this study is to generate the map of the energy flow in the manufacturing industry sector to investigate the energy utilization in the industrial process. The data sample that represents 80% of energy consumption in the manufacturing sector is used to generate this map. The generated map shows the heating system is the largest energy consumer among all energy equipment and as the primary source of the energy losses. Additionally, we found the industry groups such as sugar, cement, pulp and paper, and textile use enormous amounts of energy as their source for the heating system; meanwhile, the industries such as basic chemical, metal, and textile are the largest electricity consumers for their motor-driven machinery. The energy flow analysis together with the comparison of the Specific Energy Consumption shows the areas that should be the focus for further energy conservation measures. Recommended measures are also discussed.

Introduction

Improving energy efficiency is the most important first step toward achieving the three goals of efficient energy policy: security of supply, environmental protection, and economic growth [1]. To improve energy efficiency at the national level in Indonesia, the effort should be focusing on the manufacturing and transportation sectors. Both sectors are consuming more than 60% of the total energy supply, and the yearly average demand growth of the manufacturing sector during 2009–2013 is 9.0% [2]. This growth rate is almost double compared with the yearly average growth rate of the total energy demand, which is only at 5.3% during the same period. While at the same time, the contribution of manufacturing sector to gross domestic product (GDP) declined from 21.6% in 2010 down to 20.8% in 2013 [3].

Similar to those in many other countries during the period of 2000–2013, Indonesia's energy intensity also declined during this period. However, compared with other developing countries such as China, the decline rate of Indonesian energy intensity in the last 5-year period

(2009–2013) was much lower. The yearly average decline rate of energy intensity in China was 9%, while Indonesian yearly average decline rates were only 4%. Even though the yearly average demand growth of energy in China during the above-mentioned period was 5.6% or slightly higher than Indonesia's energy growth rates (5.3%), China has successfully decreased the yearly average energy consumption from 10.5% during 2003–2008 to 5.6% during 2009–2013. The government policy on the export tax plays a significant role in this reduction [4]. At the same time, Indonesian yearly average energy demand growth was increasing from 2.6% during 2005–2009 to 5.3% during 2009–2013, where the total energy demand Indonesia in 2013 was 1.2 billion Barrel Oil Equivalent (BOE) [5]. These observations clearly indicate that both energy consumption and energy efficiency are among the main factors that have brought the energy intensity of the manufacturing sector to the higher level.

The Competitive Industrial Performance (CIP) index in 2013 that issued by United Nation Industrial Development Organization (UNIDO) [6] shows Indonesian industry is ranked at 38. This position is much lower compared to the neighboring countries such as Thailand, Malaysia, Singapore, and China that are at rank 6–23. The CIP index is a composite index that measures the ability of the countries to produce and export manufactured goods competitively. The CIP index consists of eight subindicators grouped along three dimensions of the industrial competitiveness in which the level of technological deepening and upgrading of the countries is one of the three dimensions of competitiveness. This report demonstrates the contribution of high technology manufacturers among Indonesian manufacturers is the lowest among all above-mentioned countries. As a result, this technological aspect affects the level of energy consumption as well as the energy efficiency. The typical energy utilization of each industry is so much affected by the various conditions besides the technology aspect such as the scale of the manufacturing plant, and the type of the industry. This study is only focused on the large and medium scale of industries according to the four-digit classification of International Standard of Industry Classification (ISIC) category that assumes the member in each industry group is relatively comparable.

The studies on the energy demand side of the manufacturing sector in Indonesia are very rare. Irawan et al. [7] studied the energy demand of manufacturing sectors between 2002 and 2006. This study used a top-down approach of both Laspeyres index decompositions and Panel Data Analysis that explained the driving force of the changes in energy intensity in 2002 and 2006 was both structural effects and efficiency-related effects. The report from Pambudi [8] on his study of the energy intensity of the medium and large industry has also indicated the similar issue. However, both studies did not explain the driving forces behind the inefficiency and left further questions, such as whether the low efficiency of the industry is related to the type of industry and the energy utilization of energy equipment in the manufacturing plant. Another study of the energy consumption in the manufacturing sector has been reported by Sitompul in 2006 [9]. The report has indicated the industry such as textile, chemical, and paper should be the major priority for energy efficiency improvements by adopting the energy efficient technologies. This study also did not elaborate which part of the production equipment should be improved.

There are also not many studies about the energy flow mapping. The report from Cullen and Allwood [10] is among the very rare studies. This study suggests the investigation of the global improvement of energy efficiency measures that require tracing of the energy flow along energy value chains. Another study on this subject has been reported by the Energy Efficiency and Renewable Energy (EERE) of the US Department of Energy (DOE).

The report of energy flow map was published by EERE in 2012 and called as the energy footprints [11]. This map shows the flows of various types of energy inside an industrial plant consisting of three areas: on-site power generation facility, process purpose, and non-process purpose equipments. The energy footprints produced for industry sector help to see the specific area for potential efficiency improvements and the achievable energy savings. Chen et al. [12] used the energy flow analysis to identify energy use of the pulp and paper industry in Taiwan. This study reported that energy flow analysis has helped to find out the main reasons for energy loss were the internal distribution, boilers, electricity generation, and process equipment inefficiencies. Ferng [13] proposes a calculation framework to estimate energy footprints according to the primary energies embodied in the goods and services consumed by a defined human population and develop scenarios and simulation of policy instruments for reducing energy footprints in Taiwan. Dacombe et al. [14] also showed that the energy footprint approach could identify the major source of energy saving to determine the best options for the waste glass recovery from households.

Another method to identify the potential energy efficiency is by comparing Specific Energy Consumption (SEC). The SEC comparison between the industry sectors with the best practice reference of the similar industry will provide a better understanding of energy efficiency of the above industry.

Among the reports of energy flow study, the study on energy flow mapping of the manufacturing industry and combined with the SEC analysis in the developing countries such as Indonesia has not been reported. To fill in the research gap, the main objective of the study is to generate the map of the energy flow of the manufacturing industry which falls under the category of the medium- and large-scale industry in Indonesia and the SEC analysis for each industry group that considered as the largest energy consumer. These maps can reveal how the industry utilizes and consumes a large amount of energy and guide the measures for improving energy efficiency.

Methodology

In this study, the bottom-up approach was adopted to generate the map of the energy flow in the manufacturing industry of Indonesia. The main challenge in this study is the data availability limitation to construct the energy flow diagram. To overcome this limitation, we collected the site data from 40 industry groups that represent 80% of the total energy consumption in the manufacturing industry. Indonesian manufacturing sector in 2013 consisted of approximately 25,000 industries, which could be divided into 177 industry groups based on ISIC version

4.0. This classification is basically based on the similarity of the industry. The total industry population of these 40 industry groups is approximately 9800 industries where the industry population of each industry group is ranging from 16 to 1485 industries. The target sample of each industry group has been selected randomly using the purposive sampling criteria to ensure that the selected samples will represent the average energy consumption profile of each industry group. The size of the sample is circa 4% of the industry population of each industry group. The total collected samples are 375 industry samples out of 9800 industry populations, representing almost 4% of the target industry populations.

The data collection and site visit have successfully managed to collect the energy consumption data, such as types of energy, demand profiles, energy use, and the historical production data. The collected information shows the manufacturing plant purchases, the various types of energy at the market prices that consist of petroleum fuels, coal, natural gas, biomass, and the grid electricity from outside providers. Bhattachariyya et al. [15] have identified that the bottom-up models are traditionally based on a detailed representation of the end-use energies such as heating, lighting, mechanical energy, and process heat. This study refers to the above classification and combines it with the type of final energy used. The types of energy equipment used in this study are grouped into six categories consisting of heating system, cooling system, motor driven machinery, on-site power plant, Heating, Ventilation and Air Conditioning (HVAC), and the energy source for raw materials.

The above energy utilization data of each industry group only represent the typical energy consumption and utilization of the specific industry group and does not represent all the industry group. This data were then plotted as the energy flow diagram for each industry group. The combination of the energy flow diagram of each industry group will construct the energy flow diagram that explains the energy utilization in the overall manufacturing sector. Finally, the data of the total energy consumption in the manufacturing sector published by the Ministry of Energy and Mineral Resources (MEMR) [5] are applied into this energy flow diagram to obtain the quantity and the utilization of each type of final energy by the energy equipment of the manufacturing industry in 2013.

The energy flow diagram in this study is represented by the Sankey diagram, first introduced by Riall Sankey in 1898 [10]. This diagram shows the chain of the energy flow where the quantity of energy traced through the manufacturing plant facilities was represented by arrows or lines, in which the width of lines or arrows represents the volume of energy flow. The diagram provides the visual comparative scale of the energy flows in which the dominant energy flow can be quickly identified. The energy supply from an outside manufacturing plant is represented by five different colors of arrows where each color represents one type of final energy.

This study also analyzes the typical Specific Energy Consumption (SEC) of the large energy consumer industry. The total energy consumption divided by the total production of the plant provides the SEC ratio that can give the total energy consumed to produce one unit of product. Then, the SEC of the target industry is compared with the best practice of the SEC from other parts of the world. The comparison analysis of SEC only applied to the industry with large energy consumption such as pulp and paper, chemical, textile, basic metal and steel, and sugar processing. This comparison will give the potential energy saving that could be achieved by the industry.

Results and Discussion

Distribution of energy utilization

The overall energy flow in the manufacturing industry in Indonesia is shown in Figure 1. The energy flow diagram shows that coal and gas are the largest final energy supplier in the manufacturing industry. The contribution of coal and gas to the total energy supply is 42% and 29%, respectively. The remaining 29% of the demand is supplied by the other types of energy such as biomass, petroleum fuel, and grid electricity.

The heating system is the largest energy equipment sector in which the main energy suppliers are coal and biomass. Coal and biomass exclusively used by the heating system where the heat output of the heating system utilized by various production equipment such as on-site power plants and the motor-driven machinery.

Motor-driven machinery is the largest energy equipment after the heating system. The motor-driven machinery utilizes almost 78% of the grid electricity supply on the top of the electricity generated by the on-site power plant. The grid electricity supplies 60% of the total final energy demand of motor-driven machinery, while the remaining demand is supplied by petroleum fuel and gas. The motor-driven machinery also gets the energy input from the internal supply such as steam from the heating system and electricity from on-site power plants. The internal energy supply has provided approximately 33% of the total energy demand of this energy equipment. The composition of the final energy utilization shows that the motor-driven machinery relies much on the grid electricity supply.

Energy use for the on-site power plant is sitting on the third highest position among the energy equipment after the heating system, and motor-driven machinery.

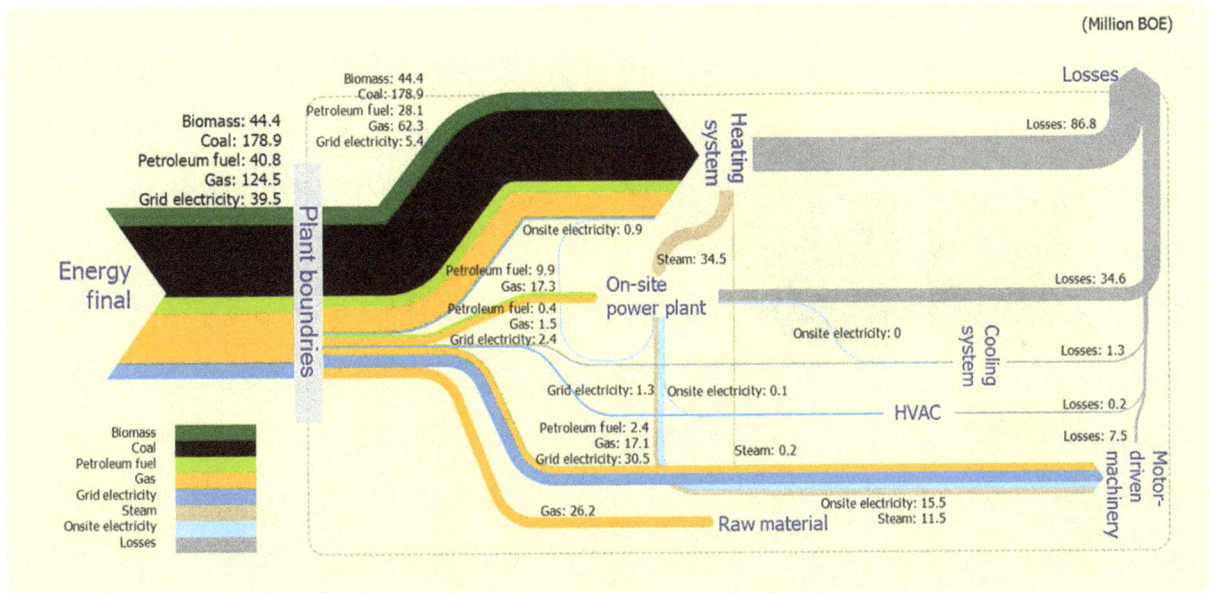

Figure 1. Energy flows of the manufacturing industry sector.

The energy input for the on-site power plant comes from petroleum fuel, natural gas, and the steam from the heating system. The steam supply from heating system is approximately 56% of the total energy demand of on-site power plant and the remaining energy demand is supplied by natural gas and petroleum fuel.

The use of final energy for the feedstock of the production process, cooling system, and HVAC are among the lowest energy equipment in the manufacturing plant. The utilization of the final energy for the raw material purposes was dominated by natural gas for the fertilizer industry. The total volume was only 6% of the total energy supply or equal to 21% of the total gas supply to the manufacturing industry. The energy utilization for the cooling system, and HVAC was even lower at 1% and less than 0.1%, respectively.

Energy utilization of the selected industry group

The top-down analysis by Irawan et al. [7] has indicated that the efficiency-related factor is the key driving force of the energy intensity issue and Sitompul [9] has also reported that the industries such as textile, pulp and paper, and basic chemical are among the priority for adopting energy efficiency technologies. This section will show the typical energy flow and the energy utilization of the large energy users mentioned above.

The cement industry is the largest energy users in the manufacturing sector in Indonesia. Figure 2 shows the energy flows of this industry. The energy input is dominated

by coal that supplies almost 90% of the total energy supply and left only 10% of demand to be shared by other types of energy. The heating system in cement plant only uses coal and biomass as the main energy supplier, while the prime mover of the motor-driven machinery gets the energy supply from the grid electricity. The grid electricity supply is only 18% of the total electricity supply where most of the electricity supply comes from the on-site power plant that used the excess steam from the heating system process. The energy losses of this industry are relatively high or approximately 42% of the total energy input. Most of the energy losses produced by the waste heat of the heating system process where only 23% of the waste heat that used back by the prime mover of the power plant.

The huge coal consumption for cement production in Indonesia basically shows the typical characteristic of cement industry that is dominated by massive kiln technology in their production process in which some of them are still using wet process technology. Unfortunately, the huge amount of waste heat is still underutilized which potentially can be used for further energy conservation.

Pulp and paper industry (see Figs. 3 and 4) consist of two major branches: the pulp industry and paper manufacturing industry. The energy supply for the pulp industry mostly comes from the excess wood of the pulp production process with a small amount of additional energy sources such as coal, oil, natural gas, and grid electricity. The grid electricity only supplies 20% of the total electricity load where 80% of the load is satisfied by the on-site power plant that uses almost half of the excess heat from the heating system process. Reutilization of

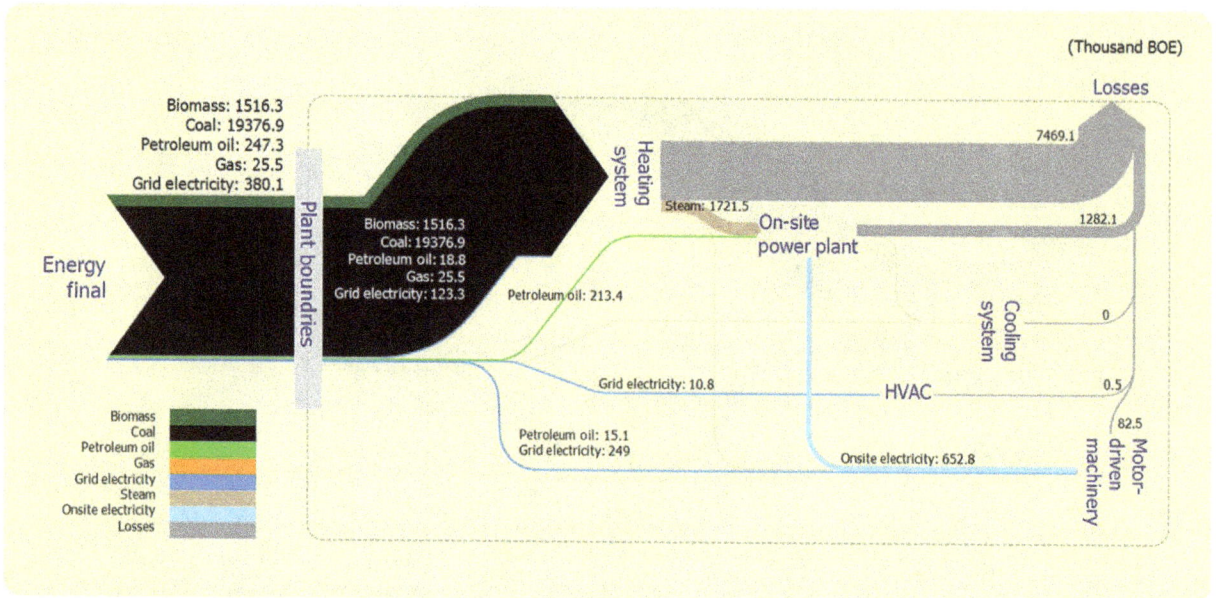

Figure 2. Energy flows of the cement industry.

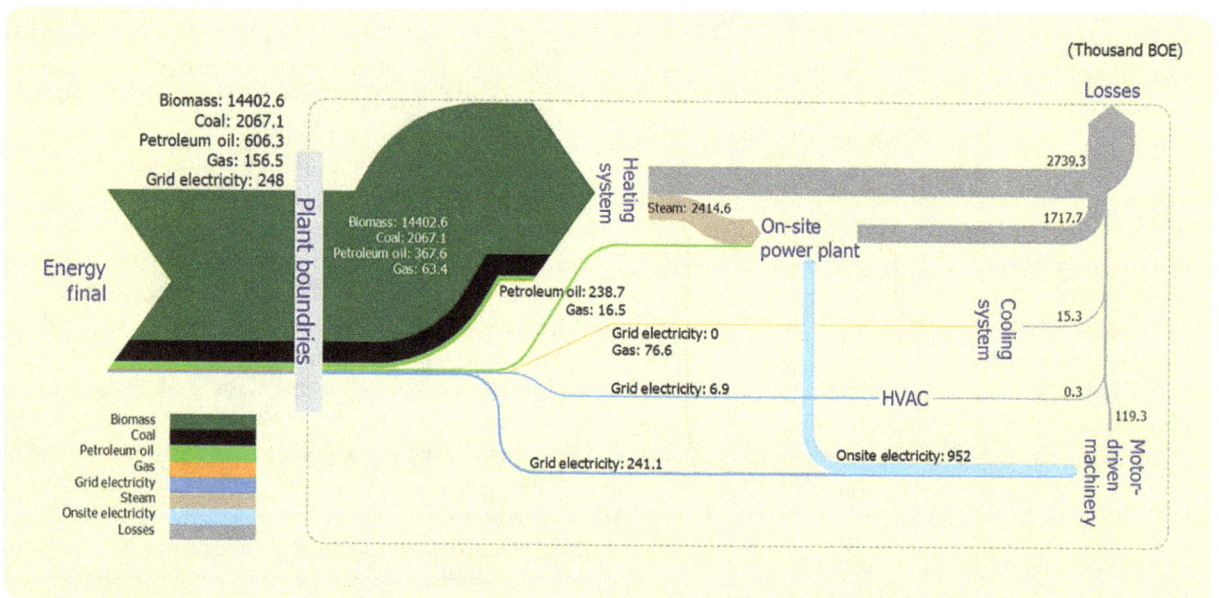

Figure 3. Energy flows of the pulp industry.

excess heat has reduced the total energy loss of this industry group down to 16% of total energy input.

The energy supply for paper industry group is dominated by coal that supplies 60% of the demand and followed by other types of energy. The heating system utilizes 90% of energy inputs and consists of all energy types except grid electricity. The grid electricity mostly supplies the motor-driven machinery beside the cooling system and HVAC. The paper industry group clearly relies much on the grid electricity supply as compared to the pulp industry. Total energy loss from this industry group is approximately 17% and mostly come from the heating system process. The energy loss from the heating system alone is approximately 18% of the total energy input on this process or higher compared with the energy loss from the heating system of the pulp industry.

The large energy consumption for pulp and paper industry is basically dominated by the pulp industry instead of the paper industry. The pulping process is the most energy-consuming phase in the entire production process

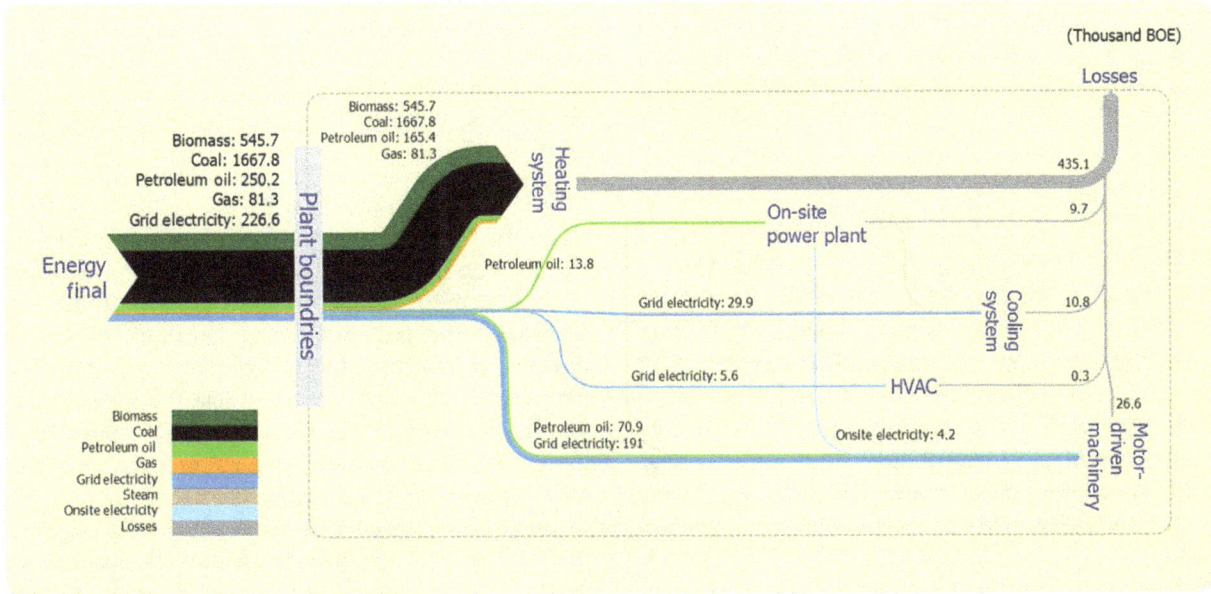

Figure 4. Energy flows of the paper industry.

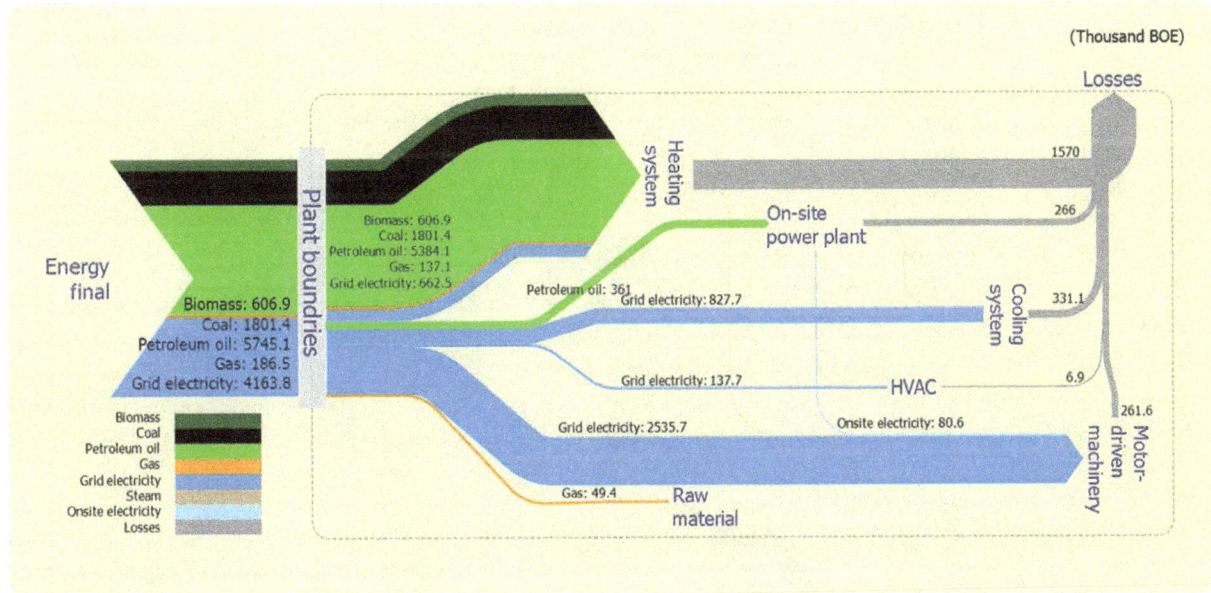

Figure 5. Energy flows of the basic chemical industry.

of the paper manufacture. The availability of the biomass with low handling cost is the solution for this high energy consumption process. The availability of biomass with low handling cost has put less pressure on the energy conservation issue in the pulp industry. The paper industry as the downstream industry uses less biomass compared to the pulp industry except for the paper industry that integrated with the pulp industry. The paper industry that located away from the pulp industry uses more coal instead of biomass and also the significant user of grid

electricity. The availability of the biomass and high handling cost compared to coal are among the main reason why the utilization of biomass is much lesser than the pulp industry.

Energy supply for the basic chemical industry is dominated by petroleum fuel and the grid electricity as shown in Figure 5, where the petroleum fuel supply is approximately 46%, grid electricity supply is 33%, and coal is only 14%. Almost 70% of the energy supplied converted into heat and even 16% of the grid electricity supply also goes to the heating system. Most of the petroleum fuels are used for the heating system, and only <10% goes to the on-site power plant. The on-site power plant in this industry group mostly uses petroleum fuel instead of other types of energy. The utilization of excess steam from the heating system process is also less utilized to produce other forms of energy such as electricity.

The basic chemical industry is the largest grid electricity consumer in the manufacturing industry sector. Its electricity demand is approximately 4% of the total supply of grid electricity devoted to the manufacturing industry sector. The supply of grid electricity is mostly used by the motor-driven machinery and followed by the cooling system. The technical requirement of the production process has forced this industry group to consume a large amount of grid electricity. Unfortunately, the contribution of the on-site power plant to the motor-driven machinery is only less than 3% compared to the supply from grid electricity supply and almost no supply from the on-site power plant to the cooling system. The petroleum fuel supply is approximately 46%, the grid electricity supply is approximately 34%, and coal is only 15%. Almost 68% of the energy supply converted into heat, and 16% of the grid electricity supply goes to the heating system. The estimated energy loss of this industry group is approximately 19% of the total energy input where the heating system contribution is approximately 64%.

The single largest electricity grid user in Indonesia is a petrochemical plant that uses the electricity on its electrolysis process and motor-driven machinery. Some of the members in this industry group also use the grid

electricity supply on the heating process in very substantial quantity for the technical requirement reason.

The textile industry is divided into three major branches: the spinning industry, weaving industry, and textile finishing industry. Figures 6–8 show the energy flow of each industry in the textile industry group. The energy demand behaviors of these three subsectors of the textile industry reflect the production process of each industry group. The weaving industry and finishing industry are the first and the second largest energy users and the consumption of coal and biomass could satisfy approximately 80% of their energy demand. Most of coal and biomass are used for heating system purpose. Meanwhile, the grid electricity supply is dominating the energy supply for spinning industry and the on-site power plant uses the petroleum fuel to generate the electricity.

Although the spinning industry is the smallest energy user compared to the other two industries, this industry is the largest electricity user among of the three textile sub-industry groups. This huge electricity demand is required by the spinning process that need more electricity power to run the machinery instead of heat like on the weaving industry or in the finishing process. The grid electricity supplies for the spinning industry are about three times of the electricity supply to weaving and finishing industry. The energy loss of the spinning industry is much higher compared with two other industries. The total energy loss of the spinning industry is approximately 41% of total energy input while the total energy loss of weaving and finishing industry is only 18% each. The major energy loss contributor of spinning industry comes

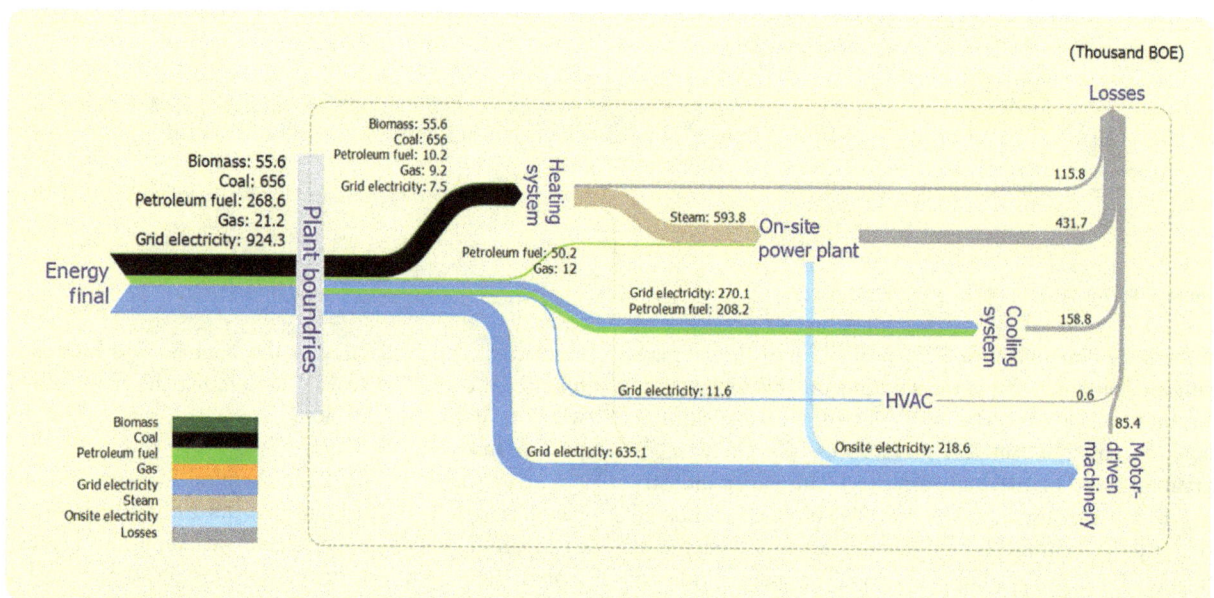

Figure 6. Energy flows of the spinning industry (textile).

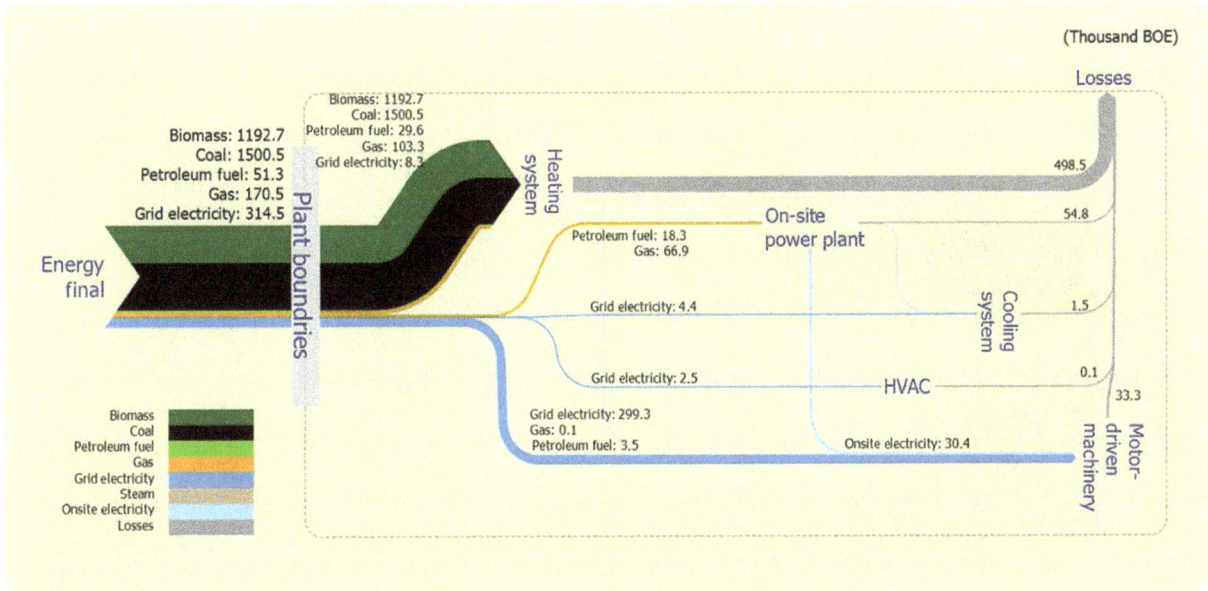

Figure 7. Energy flows of the weaving industry (textile).

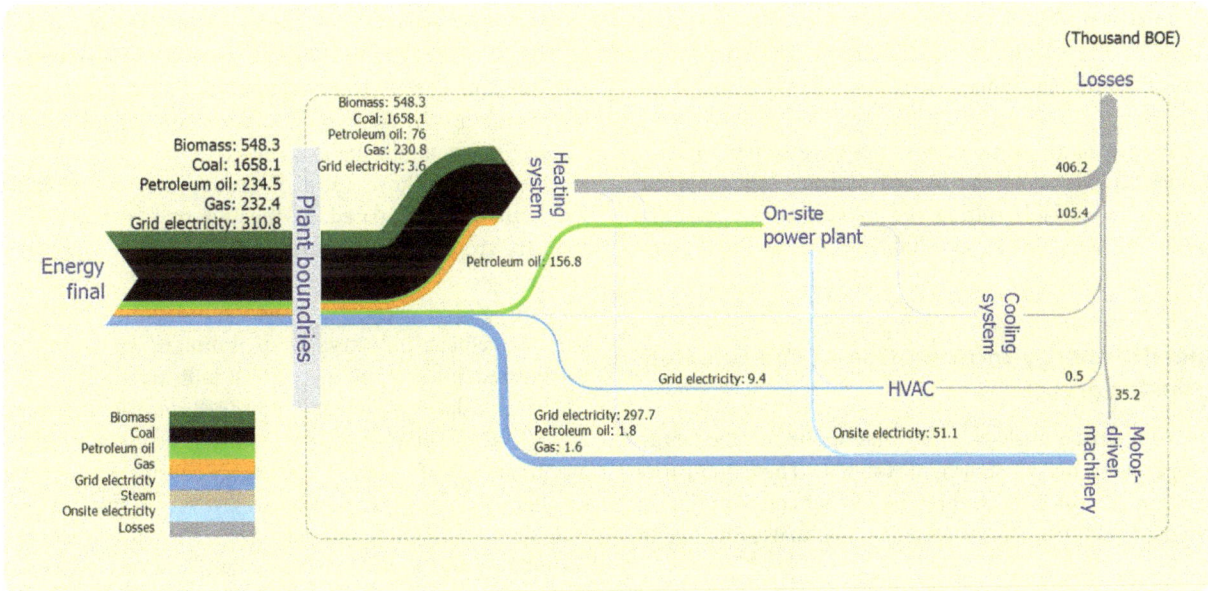

Figure 8. Energy flows of the textile finishing industry.

from electricity generation. The on-site power plant is also consuming petroleum fuel to turn on the engine and then producing a very large amount of energy loss.

The sugar industry is the largest user of biomass (see Fig. 9) where the supply is about 64% of the total energy supply, followed by petroleum oil and coal, which contribute 13% each. Interestingly, the ratio of petroleum oil supply compared to the other types of energy in the sugar industry is relatively higher than the other industries, except for the basic chemical industry and pulp industry.

The petroleum oil mostly used by the on-site power plant and the supply of petroleum oil is about 40% of the total energy demand of the power plant. The remaining energy input for the on-site power plant come from the heating system. The utilization of the on-site power plant is very high, approximately 85% of the total electricity demand.

Just like any other industries, the heating system and the on-site power plant are the largest sources of energy loss. The utilization of heat output from the heating

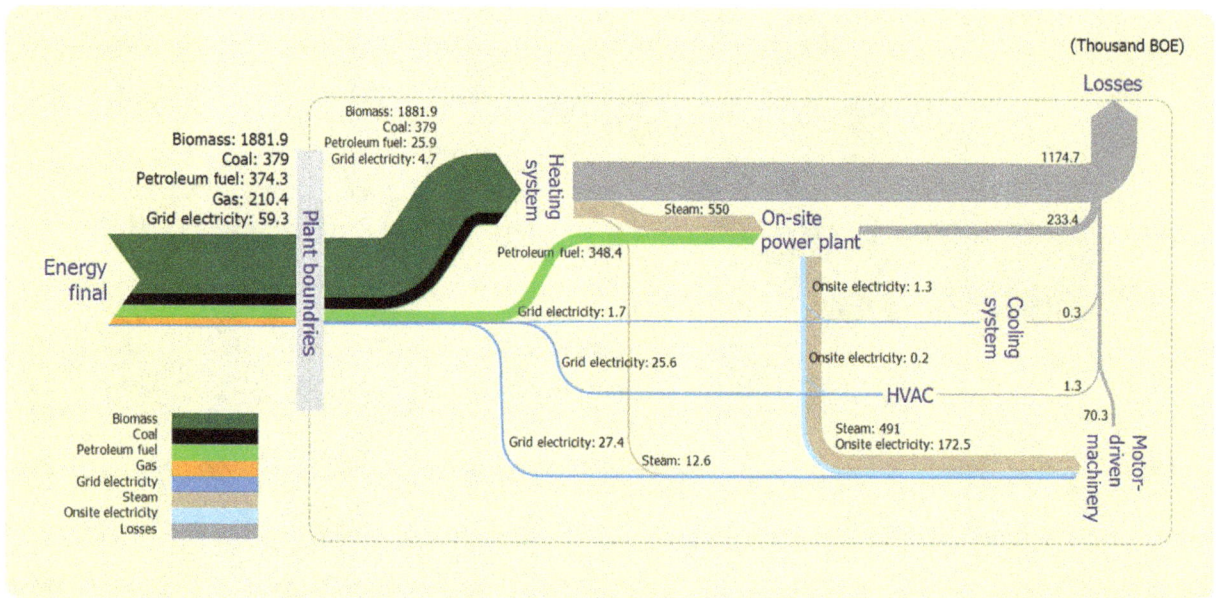

Figure 9. Energy flows of the sugar industry.

system goes to the on-site power plant beside to the prime mover of the motor-driven machinery. The energy loss from the heating system is higher than in the other industries.

The high petroleum oil consumption together with the biomass is related to the harvest season. The supply of biomass is available during the harvest season and the supply of biomass is very limited outside the harvest season.

Specific energy consumption of the selected industry group

The comparison analysis of SEC has applied to the largest energy consumer as shown in Table 1. These industries consume almost 20% of the total energy consumption of the manufacturing sector and the manufacturing sector

Table 1. Specific energy consumption of the selected industry groups.

Industry group	SEC (BOE/Tonne)	SEC reference (BOE/Tonne)	References
Pulp	1.21–2.74	1.13–3.16	[16]
Paper	1.63–2.14	1.23–1.79	
Alcohol	1.11	0.78	[17]
Cement	0.67	0.57	[18]
Spinning	1.48–3.65	0.57–0.59	
Weaving	0.83–8.6	0.82–7.03	
Finishing	8.35	7.8	
Basic Metal and Steel	0.67–0.86	0.46–0.50	
Sugar processing	5.98	4.75	[19]

is the largest energy consumer compared to the others sectors [14]. This large portion of energy consumption has put the energy consumption of the manufacturing sector among the key success factors on the energy efficiency effort at the national level.

This analysis shows most of the industry in Indonesia has high SEC compared to the industry standard. The industry groups such as textile group industry, sugar processing, and alcohol have a relatively high SEC compared to their global peers, but some other industry groups such as cement industry group is not to far compared to the best practice in the similar industry.

The SEC level of cement industry group in Indonesia is 0.67 BOE/Tonne or almost 18% higher compare average industry performance. The energy consumption profile of this industry was dominated by the heating system that derives its energy supply from all types of energy, including grid electricity. The potential energy conservation in the heating system is available on the grid electricity supplies and the utilization of steam from the heating system. Optimizing the excess heat from the heating system will enable the reduction of the utilization of grid electricity.

The SEC of the alcohol and sugar processing are 42% and 26% higher compared with the average SEC figures, while some of the basic metal industries is up to 70% higher to the reference industry. The source of these high figures is mostly from the heating system of the major energy users in the manufacturing plant. The heating system of the sugar industry has low efficiency, while the steam from the heating system of the basic metal industry is underutilized. Most of the steam goes to the heating process only, and very small amounts go to the steam turbine to generate electricity. The improvement of the heating system of these industries should bring these two industries down to the lower energy consumption levels.

Among the textile industry, the SEC level of the spinning industry is the highest SEC level and followed by weaving industry and finishing process industry. The high SEC level of this industry mostly comes from the old production machinery with high energy consumption rate. The similar situation is also happening in the other industry such as sugar industry, where the old production machinery is the main source of the high energy consumption.

The SEC reference has given the idea about the potential energy saving that can possibly be achieved. By comparing the SEC level among the similar industry group and assuming that the high SEC level of the above industry group can be lowered to meet the SEC reference level, then the potential reduction of energy consumption of the industry groups above only could reach up to 10 million BOE or 12% of their total energy consumption in 2013. This figure could be even higher if all high energy intensity industries are taken into account.

Estimated Energy Losses

Understanding the relative importance of energy systems among different industry groups is the key to identifying potential energy efficiency opportunities [20]. This energy flow study has also analyzed the energy losses that were estimated from the typical energy efficiency of the energy equipment as well as the discussion during data collection. The total estimated energy loss is approximately 130 MBOE or 31% of the total energy input including the energy use for raw material. The largest energy losses within the manufacturing industry come from the heating system process; the energy source of the heating system supplied by biomass, coal, gas, petroleum fuel, and even

from the grid electricity with the total energy losses from this equipment is estimated to be 25% of the total energy input. The combined heat and power (CHP) application on this heating system has improved the total energy utilization from this system.

The on-site power generation plant uses the steam output of the heating system to generate the electricity for the internal use. The total energy input from the heating system to the on-site power plant is approximately 11% of the total energy input of the heating system or much higher compared to other final energy supply for the on-site power plant. The steam from the heating system has supplied 56% of the demand in which the gas supply is in the second position (28%) followed by the petroleum fuel with 16% of the energy demand.

The on-site power plant is the second highest source of energy losses in the manufacturing plant. The total loss from this process is 51% out of the total energy input. The energy loss from the cooling system, HVAC, and motor-driven machinery are relatively small compared to the energy loss from the heating system and the on-site power plants.

Figure 10 compares the energy use and losses across the largest energy users while Figure 11 compares the energy losses of each energy equipment across different industry groups. As shown in Figure 10, the largest energy loss comes from the five largest industry groups, and Figure 11 illustrates that 80% of the energy losses of the largest energy users come from the heating system and on-site power plants except for pulp and paper, textile, and basic chemical industries. Both pictures clearly show the largest source of energy losses are the heating system and the on-site power plant which should be the main targets for future energy conservation measures.

Energy Conservation Opportunities

The energy losses in the manufacturing plant represent the immediate target for the improvement of energy

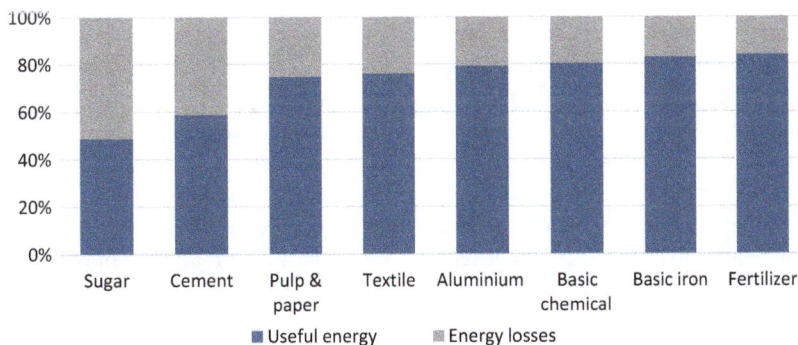

Figure 10. The ratio of energy use and energy losses of the selected industry sectors.

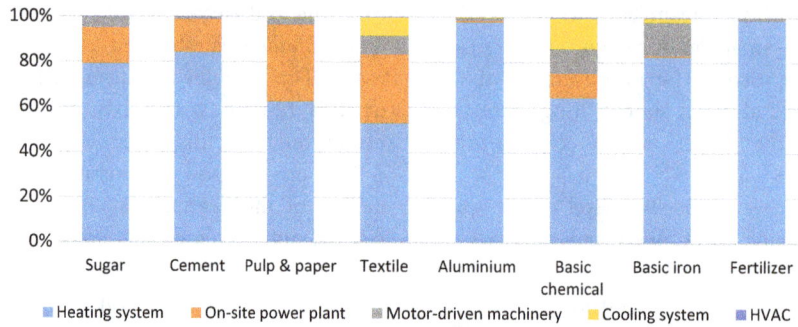

Figure 11. The contribution of each energy equipment to the total energy loss of the selected industry groups.

conservation. The heating systems of the large energy users have wasted substantial amounts of energy in the form of low and high quality of waste liquid and gas. The on-site power plant also has shown an opportunity to reduce energy losses. The substantial load of electricity that has been used to run the motor-driven machinery of the large electricity users has indicated the areas for the potential energy efficiency improvement.

Energy audit reports by the office of MEMR [21] has identified the potential areas for energy efficiency in the selected industry groups such as the cement, basic metal, sugar, textile, basic chemical, and other major energy users. The industry with large energy loss has very wide areas for improvement in the heating system operation. The operation of raw material handling, kiln loading, and the exit temperature of material prior to entering into the preheater process are the most common findings of the energy audit in the cement industry group, besides the improvement in the motor-driven machinery where the low-efficiency motors are still in operation at some cement manufacturing plants [21].

The basic chemical industry groups share the same issue with the cement industry group concerning the heating system efficiency. Optimization of burning systems on the steam boilers as well as the replacement of low-efficiency motor-driven machinery is strongly recommended besides the improvement of the insulation system for the heat transfer.

The improvement area in the sugar industry is quite similar to the above industries. Optimization of the material handling during the heating process and the efficiency of steam boilers are the areas that need more attention as well as the insulation of the heat distribution system. The adjustment of the electricity load factor should also be the potential area to reduce the energy loss besides the replacement of low-efficiency motor-driven machinery.

Industries such as the basic chemical and textile industries are the massive user of electricity, which is mostly used it to energize motor-driven machinery. The main supplier of electricity is the central power plant with a small additional supply from the on-site power plant. The energy efficiency effort should start from the distribution system down to the final users of the electricity supply. The common electricity utilization issues of such industries are the installed capacity that does not match with the average demand profile, the absence of the capacitor bank, effective working voltage, transformers effectiveness, and low-efficiency of the motor-driven machinery. The installation of the capacitor and transformer enhancement to reduce the energy loss along the distribution network and to manage the unbalance current as well as the installation of variable speed drive (VSD) on the motor-driven machinery are highly recommended.

The heavy measures to improve the energy efficiency should prioritize the industry groups with large energy consumption and loss as well as the high SEC level. The SEC level reduction on the above industry groups will give a significant impact to the total energy consumption of the manufacturing sector. This reduction subsequently will reduce the total national energy consumption. By only reducing the SEC level of the industry groups above to the average industry reference, the total national energy demand in 2013 could be reduced by 1–1.5%. The larger reduction will be achieved when the program of energy efficiency improvement is applied to all industry groups.

Conclusions

We successfully constructed the energy flow maps for the specific manufacturing industry groups with large energy consumption in Indonesia. The energy flow map provides the visual comparison of energy used among the industry groups as well as the energy equipment. The energy flow mapping and the SEC comparison analysis not only enhance our understanding of energy utilization in the manufacturing industry sector but also highlight the area for improvement along the energy value chain and provide the framework for energy efficiency measures.

The study result shows that the heating system is not only the largest energy consumer among all energy equipment but also as the primary source of the energy losses. The heating system in the industry such as sugar, cement, pulp and paper, and textile use enormous amounts of energy from coal and biomass; meanwhile, the industries such as basic chemical, metal, and textile are the largest electricity consumers for their motor-driven machinery.

The proposed energy efficiency measures should encompass the potential areas for improvement in heating systems. The installation of energy recovery technologies to capture the waste energy should be a part of the measures in the energy integration. The on-site power plant improvements might include planning with detailed efficiency improvement plans. The roles of the regulators together with the industry players are very important in the effort to improve the energy efficiency and conservation.

Conflict of Interest

None declared.

References

1. IEA. 2007. Tracking industrial energy efficiency and CO_2 emissions. International Energy Agency, Paris, France.
2. Indonesia Statistics. 2013. Manufacturing industrial statistic Indonesia. BPS, Jakarta.
3. MOI. 2014. Industry facts and figures 2014. Ministry of Industry, Jakarta.
4. Li, L., J. Wang, Z. Tan, X. Ge, J. Zhang, and X. Yun. 2014. Policies for eliminating low-efficiency production capacities and improving energy efficiency of energy-intensive industries in China. Renew. Sustainable Energy Rev. 39:312–326.
5. MEMR. 2014. Handbook of energy and economic statistic of Indonesia. Ministry of Energy and Mineral Resources, Jakarta.
6. UNIDO. 2013. The industrial competitiveness of nations, competitive industrial performance report 2012/2013. United Nation Industrial Development Organization, Vienna.
7. Irawan, T., D. Harsono, and N. A. Achsani. 2010. Analysis of energy intensity in Indonesia manufacturing sector. Universitas Padjadjaran, Bandung.
8. Pambudi, H. G. 2009. The analysis of energy intensity determinants on medium and large industry in Indonesia. [Bachelor thesis], Bogor Agricultural University.
9. Sitompul, R. F. 2006. Energy-related CO_2 emissions in the Indonesian manufacturing sector. University of New South Wales, Sydney, Australia.
10. Cullen, J. M., and J. M. Allwood. 2010. The efficient use of energy: tracing the global flow of energy from fuel to service. Energy Pol. 38:75–81.
11. USDOE. 2012. Manufacturing energy use and greenhouse gas emission analysis. US Department of Energy, Washington, DC.
12. Chen, H. W., C. H. Hsu, and G. B. Hong. 2012. The case study of energy flow analysis and strategy in pulp and paper industry. Energy Pol. 43:448–455.
13. Ferng, J. J. 2002. Toward a scenario analysis framework for energy footprints. Ecol. Econ. 40:53–69.
14. Dacombe, P. J., V. Krivtsov, C. J. Banks, and S. Heaven. 2004. Use of energy footprint analysis to determine the best options for management of glass from household waste. Pp. 266–272 In Sustainable Waste Management and Recycling: Glass Waste. Sustainable Management and Recycling: Challenges and Opportunities Conference, Thomas Telford, Southampton, UK.
15. Bhattachariyya, S. C., and G. R. Timilsina. 2009. *Energy demand model for policy formulation – A comparative study of energy demand models WPS4866*. Policy Research Working Paper. World Bank,
16. Worrell, E., L. Price, M. Neelis, C. Galitsky, and Z. Nan. 2008. World best practice energy intensity values for selected industrial sectors. Ernest Orlando Lawrence Berkeley National Laboratory, Berkeley, California.
17. Saygın, D., M. K. Patel, C. Tam, and D. J. Gielen. 2009. Chemical and petrochemical sector, potential of best practice technology and other measures for improving energy efficiency. International Energy Agency, Paris.
18. UNIDO. 2010. Global industrial energy efficiency benchmarking. United Nation Industrial Development Organization, Vienna.
19. Sattari, S., A. Avami, and B. Farahmandpour. 2007. Energy conservation opportunities: sugar industry in Iran. WSEAS, Arcachon, France.
20. EERE; Energetic Inc. 2004. Energy loss reduction and recovery in industrial energy systems. Energy Efficiency and Renewable Energy, Washington, DC.
21. MEMR. 2010. Energi audit pada sektor industri manufaktur. Ministry of Energy and Mineral Resources, Jakarta.

9

A power balancing method of distributed generation and electric vehicle charging for minimizing operation cost of distribution systems with uncertainties

Jiekang Wu[1] (iD), Zhishan Wu[1], Fan Wu[2] & Xiaoming Mao[1]

[1]School of Automation, Guangdong University of Technology, Guangzhou, Guangdong 510006, China
[2]Guangxi Bo Yang Electric Power Survey and Design Co., Ltd., Guangxi Power Grid Co., Ltd., Nanning, Guangxi 530023, China

Keywords
Distributed generation, distribution systems, electric vehicle charging, operation cost, power balancing, quadratic rotated cone programming

Correspondence
Jiekang Wu, Department: School of Automation, Guangdong University of Technology, No. 100 Waihuan Xi Road, Guangzhou Higher Education Mega Center, Panyu Distric, Guangzhou 510006, China.
E-mail: wujiekang@163.com

Funding Information
Guangdong special fund for public welfare study and ability construction (2014A010106026), Natural Science Foundation of Guangdong (2014A030313509, S2013010012431). The National High Technology Research and Development of China (863 Program) (2007AA04Z197). Specialized Research Fund for the Doctoral Program of Higher Education (20094501110002). National Natural Science Foundation of China (50767001).

Abstract

A power balancing method of distributed generation (DG) and electric vehicle charging is presented for minimizing operation costs of distribution systems with uncertainties, which includes the uncertainties in output power of DG and the randomness of charging power of electric vehicles (EV). A probability model is established for the uncertain characteristics of DG and electric vehicle charging. A multi-state optimization coordination method for DG of renewable energy systems and electric vehicle charging in distribution systems based on quadratic rotation cone programming is presented to minimize the expected generation cost of generators in the main power grid, the expected operation cost of DG systems in distribution systems, and the expected social outage loss. An objective function maximizing the operation efficiency of distribution systems with DGs and EVs is proposed. Using quadratic rotated conic programming, the nonlinear objective function and constraint functions are transformed into a linear form. An IEEE 14-node distribution system is used as a study example to illustrate adaptability of the proposed model and the feasibility of the proposed method. The simulation results show that the proposed method simplifies the original problem of the optimization problem and makes its solution faster, more stable, and better.

Introduction

In order to ease the environment and energy crisis, many countries all over the world are studying to apply renewable energy technology in a large scale and efficient way [1], and distributed generation (DG) and electric vehicles (EV) will be more widely installed into distribution systems for some strong motivation of loss reduction, flattering of peak, increasing reliability, and modifying voltage profile [2, 3].

The output power of distributed generation depends on the wind speed, solar light intensity, and other climatic factors, which have intermittent and uncertain characteristics [4]. Because of the difference in energy state of battery, the expected driving distance in a day, the charging power, the expected charging time, and so on, electric vehicle charging also has uncertain characteristics. The uncertainties of distributed generation and electric vehicle charging may bring a great impact on distribution systems. The randomness and uncertainties of distributed generation of renewable energy systems and electric vehicle charging have attracted the great interest of the scholars to study in these topics, and some research results have been obtained. For example, the randomness of electric vehicle charging is considered, and the probability model is established [5–10]; the economic dispatch model for distribution systems with DGs and EVs is constructed, and the influence of DG and EV on the economic performance of the network is analyzed [11]; multiple objective optimization model for DG and EV coordination control is established [12]; a model for optimal siting and sizing of distributed generation in distribution systems with electric vehicles is proposed, and an algorithm based on genetic algorithm is used to solve the proposed optimization problem [13]; and a multi-time-scale-based coordinative dispatch model for distributed generation and electric vehicles charging is proposed to smooth load fluctuation in distribution systems [14].

In view of the uncertainties and the influence of distributed generation and electric vehicle charging in the distribution system, the research work has become an urgent task. These works aim at solving these uncertain problems to enhance security, reliability, and efficiency of distribution systems [15], and they require to carry out necessary risk and security assessment [16], including deterministic evaluation [17], probabilistic evaluation [18], and risk assessment [19]. Deterministic assessment generally gives a set of most credible contingencies, which may result in highly conservative decisions with high operation costs. Probabilistic evaluation uses probability method to evaluate uncertain events in such form as loss of load probability and expected unserved energy. Risk assessment uses risk value calculation method to evaluate the impact of uncertain events and consequences, some of which relate to distributed generation of renewable energy [20]. But less work about the impact on the risk of power grid in the event of the uncertainties related to electric vehicle charging is carried out.

In a distribution system with a large number of distributed generation systems and electric vehicles injected in different nodes, the optimal coordination problem of distributed generation and electric vehicle charging has attracted much attention. Some experts and scholars have carried out a thorough study. Optimal penetration in distribution systems [21] and the charging influence on distribution networks [22] are studied. Using smart metering and demand side management in distribution systems were studied for charging coordination of electric vehicles [23], which may change charging influence on distribution systems. A method for minimizing power loss of distribution systems by coordinating electric vehicle charging [24] is studied. Some scenarios for controlling charging of electric vehicles based on minimizing charging costs [25]

were presented, aiming at putting forward a 1-day charging pattern. However, peak load and uncertainties in charging load demand of electric vehicles are not taken into account in these study works. In fact, uncertainties of electric vehicle charging are significant, and its influence on power loss and nodal voltage is also significant. At the same time, the output power of distributed generation in distributed systems is always uncertain, and its influence on power loss and nodal voltage is also significant. The uncertainties of electric vehicle charging include battery characteristics, state of charge (SOC) levels, arrival times, departure times, charger ratings, and so on.

Coordinated operation and optimal scheduling of distributed generation, distributed energy storage, and electric vehicle can improve the technical and economical performance of power distribution network, and many scholars have carried out in-depth research in this area and achieved gratifying results. A method for the synergy between wind-based distributed generation and plug-in electric vehicles is studied to simulate the impact of wind-based distributed generation on electric energy in distribution systems embedded with electric vehicles, and Monte Carlo method is used to address the uncertainties associated with wind speed variations and charging of plug-in electric vehicles [26]. The impacted electric energy includes the excess in active/reactive power, energy exceeding normal, unserved energy, and energy losses when different DG penetration is up to 35% and different PEV penetration is up to 50%. The coordinated operational problem of a neighborhood of smart households comprising electric vehicles, energy storage, and distributed generation provides a technology for power and energy management optimization problem of distribution systems with renewable energy using mixed-integer linear programming with the objective of minimizing the total energy procurement cost under an hourly varying price tariff scheme [27]. Some studies for seminal insights into the problem of coordinating the smart household activities in a smart grid have carried out, but these studies do not consider the possibility of bidirectional power flow and satisfaction of transformer capacity limitations combined with pricing-based schemes [28–32], or disregard the operational constraints of the DS (Distributed Storage) infrastructure [33–36].

The application of new energy sources in smart household has become a trend, and the operation and control of distributed generation, distributed energy storage, and electric vehicles in the end-user side become more complex and more difficult. With the large-scale penetration of new energy sources on the user side, the problem of power flow diversification becomes more and more prominent, which makes the power and energy management of the distribution network become more complex and more difficult. Distributed energy management in smart home has attracted wide attention, and many scholars have done some research on it. In a smart household with a small-scale distributed renewable energy generation system, electric vehicle capable of operating in vehicle-to-home mode and energy storage system, optimum operating strategy is studied considering the changing load profile imposed by dynamic pricing-based demand response, and the method is based on mixed-integer linear programming modeling framework [37, 38].

In conventional distribution networks, voltage control and reactive power management are performed using on-load/off-load tap-changer of transformers and/or switchable/fixed capacitor banks, while voltage regulation may be carried out by using reactive power management of distributed generation in distribution networks with renewable energy [39–42]. Using electric vehicles for voltage regulation is a technology to be used to mitigate the fluctuation of solar power by smoothing the fluctuation of solar generation. This technology can certainly reduce the difficulties of voltage management [43–47], and this method combining demand response technology may obtain a better result for regulating voltage [48], balancing supply and demand power [48], and automatic generation control [49].

In the new energy environment, the power and energy management of distribution network becomes very urgent. Although there are many scholars involved in this field, some concentrates only on energy management of isolated microgrids [50–54], and some focuses on coordination of operation of separated microgrids [55, 56]. Some scholars involved local energy balancing mechanism in low-voltage networks using the concept of local energy balancing and ancillary services provided by the DG and renewable energy systems [57].

In this article, a power balancing method of distribution systems with distributed generation and electric vehicles is presented by minimizing the expected social cost of the distribution system, the expected interruption cost, the expected generation cost of the generators in the main power grid, and the distributed generators in distribution systems. A probability model is established for the uncertain characteristics of distributed generation and electric vehicle charging. A multi-state optimization coordination method for distributed generation of renewable energy systems and electric vehicle charging in distribution systems based on quadratic rotation cone programming is presented to minimize the expected generation cost of generators in the main power grid, the expected operation cost of distributed generation systems in distribution systems, and the expected social outage loss. An objective function maximizing the operation efficiency of distribution systems with DGs and EVs is proposed, and quadratic rotation cone programming is used to process and transform the nonlinear constraint condition functions. The proposed

method simplifies the original problem of the optimization problem in this article, which makes it faster, more stable, and better.

Probability Model of Distributed Generation

Wind power generation

Probability distribution model of wind speed

The wind speed is generally formulated as Weibull distribution with two parameters [58]:

$$F(v) = 1 - e^{-\left(\frac{v}{c}\right)^k}, \tag{1}$$

$$f(v) = \frac{k}{c}\left(\frac{v}{c}\right)^{k-1} e^{-\left(\frac{v}{c}\right)^k}, \tag{2}$$

where c is the scale parameter, which is dependent to average speed of wind; k is the shape of Weibull distribution, which reflect the slope of Weibull distribution, and its value is usually 1~3.

Different geographic locations and geographic features have different wind speed characteristics, and there are different wind power characteristics. In different geographic position, the scale parameter and shape parameter of wind speed are very different. Using different parameters, the probability distribution of wind speed is also different.

Probabilistic power for wind-power generator

Generally, wind-power-driven generators can work to output electric power into the grid only when the wind speed reaches the requirement of the cut-in wind speed: when the wind speed is below the rated wind speed, the output power of the wind turbine will change with the wind speed. If the wind speed reaches the rated value and above, the output power of the wind turbine will be stable with the rated output power. If the wind speed exceeds the specified speed, it is required to stop the generation of electricity in order to protect the safety of the wind turbine. So it is appropriate to express the relationship between the output power and the wind speed by using the piecewise function.

Using Weibull distribution probability density, wind speed can be sampled at different state. Under different wind speeds with significant uncertainties, the relationship between the output power of the wind turbine and the different states of the wind speed can be expressed as follows [58]:

$$P_e = \begin{cases} 0 & 0 \le v_w \le v_{ci} \\ P_N(A + Bv_w + Cv_w^2) & v_{ci} \le v_w \le v_r \\ P_N & v_r \le v_w \le v_{co} \\ 0 & v_{co} \le v_w \end{cases}, \tag{3}$$

$$A = \frac{1}{(v_{ci}-v_r)^2} \left[v_{ci}(v_{ci}+v_r) - 4v_{ci}v_r \left(\frac{v_{ci}+v_r}{v_r} \right)^3 \right], \quad (4)$$

$$B = \frac{1}{(v_{ci}-v_r)^2} \left[-(3v_{ci}+v_r) + 4v_{ci}v_r \left(\frac{v_{ci}+v_r}{2v_r} \right)^3 \right], \quad (5)$$

$$C = \frac{1}{(v_{ci}-v_r)^2} \left[2 - 4 \left(\frac{v_{ci}+v_r}{2v_r} \right)^3 \right], \quad (6)$$

where A, B, C is coefficient relating to output power of wind-driven generator, P_N, v_r, and v_w is, respectively, the rated power, the rated wind speed, and the actual wind speed of wind-driven generator; v_{ci} and v_{CO} are, respectively, cut-in speed and cut-out speed required for wind turbine.

Photovoltaic power generation

Probability model of solar intensity

Because of the randomness of the solar radiation, the solar intensity can be expressed by Beta distribution [59]:

$$f\left(\frac{r}{r_m}\right) = \frac{\Gamma(\alpha+\beta)}{\Gamma(\alpha)\Gamma(\beta)} \left(\frac{r}{r_m}\right)^{\alpha-1} \left(1-\frac{r}{r_m}\right)^{\beta-1}, \quad (7)$$

where α and β are shape parameter for Beta distribution, and r and r_m are, respectively, the actual intensity and the maximum intensity at the time period.

Probabilistic power for photovoltaic power generation

The output power of a photovoltaic power generation system is not only related to the intensity of solar exposure but also to uncertainties such as temperature, and it may be formulated as follows:

$$P_{PV} = [1 + k(T - T_T)] \frac{r}{r_T} P_T, \quad (8)$$

where T is the actual temperature of the surface of the PV array at this time period, k is temperature coefficient, P_T, r_T, and T_T are, respectively, the maximum output power, solar intensity, and photovoltaic array temperature under the standard test environment.

State probability of power output of distributed generation

The output power of distributed generation is not only affected by the fluctuation of the energy, for example, solar, wind, and so on, but also may be influenced by the fault of distributed generation systems. Therefore, the output power

level of distributed generation can be formulated as the joint probability distribution of output power fluctuations and the operation state of distributed generation systems.

In this article, the probability model with two states is used to reflect the operation state of distributed generation systems, and the forced outage rate can be expressed as follows:

$$P_{FOR} = \frac{\lambda}{\lambda+\mu}, \quad (9)$$

where λ is fault rate of distributed generation systems and μ is repair rate of distributed generation systems.

The probability distribution of the output states of distributed generation systems can be expressed as follows:

$$p_{DG}(i) = \begin{cases} p_{FOR} + (1-p_{FOR}) \times p\left(P_{DG} = \frac{i}{N}\right), i=0 \\ (1-p_{FOR}) \times p\left(P_{DG} = \frac{i}{N}\right), i=1,2,\dots,N \end{cases}, \quad (10)$$

where i/N is the ratio of the output power and the rated power at the ith state, $p\left(P_{DG} = \frac{i}{N}\right)$ is the probability of distributed generation systems at the power output level i/N, P_{DG} is active power of distributed generation of renewable energy system, and N is state number that represents the nonstop of a distributed generation systems.

Probability Distribution of EV Charging State

With a large number of electric vehicles being put into use, uncertainties of electric vehicle charging always cause a certain uncertain impact on distribution systems. In the followings, the charging probability model is established according to the uncertain factors of electric vehicle charging. The electric vehicle charging mode discussed in this article is a conventional charging, which is intent to charge the battery to maximum power for a driving distance of 200 km.

The probability distribution of daily driving distance and battery remaining electricity of electric vehicle

According to the daily driving data of electric vehicles, the probability density function of the driving distance is formulated as follows [60]:

$$f_d(x) = \frac{1}{x\sqrt{2\pi\delta_d^2}} \exp\left[-\frac{(\ln x - \mu_d)^2}{2\delta_d^2}\right], \quad (11)$$

where μ_d and δ_d is, respectively, mean value and standard deviation, generally $\mu_d = 3.20$, $\delta_d = 0.88$, $0 \le x \le 200$.

The ratio of the remaining electricity or SOC and the total electricity amount of the battery of an electric vehicle after driving an expected distance in a day is written as follows:

$$Q = 1 - \frac{x}{D},\qquad(12)$$

where Q is the total electricity amount of the battery, x is the remaining electricity, and D is an expected distance in a day.

The probability density function of the remaining electricity of the battery in an electric vehicle is formulated as follows:

$$\begin{cases} f(Q) = \dfrac{1}{\sqrt{2\pi}(1-Q)D\delta_d}\exp(L) \\ L = \dfrac{\ln(1-Q)+\ln D - \mu_d}{2\delta_d^2} \end{cases},\qquad(13)$$

where L is actual driving distance in a day.

The charging time is

$$T_{EV} = \frac{x \times W}{\eta_{EV}P_{EV}100},\qquad(14)$$

where η_{EV} is charging efficiency, P_{EV} is charging power, and W is the electricity required for a driving distance of 100 km.

Probability distribution of electric vehicle charging state

Assuming that an electric vehicle is charged at the end of the day after the last trip, the probability distribution of electric vehicle charging is approximately submitted as normal distribution [61]:

$$f(t) = \begin{cases} \dfrac{1}{\sqrt{2\pi}\sigma_t}\exp\left[-\dfrac{(t-\mu_t)^2}{2\sigma_t^2}\right], \mu_t - 12 < t \le 24 \\ \dfrac{1}{\sqrt{2\pi}\sigma_t}\exp\left[-\dfrac{(t+24-\mu_t)^2}{2\sigma_t^2}\right], 0 < t \le \mu_t - 12 \end{cases},$$

$$(15)$$

where μ_t and σ_t are, respectively, mean value and standard deviation for charging time of the battery in an electric vehicle, and t is charging time of an electric vehicle for the expected driving distance.

Power Balancing Model and Solving Method

Power balancing model of DGs and EVs

State probability

The state space of the system is supposed to compose of DGs state of renewable energy systems and the charging

state of the electric vehicles, and the state probability of the distribution systems at state k is formulated as follows:

$$p(x_k) = p_{DG}(t) \cdot p_{EV}(t),\qquad(16)$$

where $p_{DG}(t)$ and $p_{EV}(t)$ is, respectively, DG state probability of renewable energy systems and the charging state probability of an electric vehicle at time period t.

The expected operation cost (EOC)

The expected operation cost mainly includes the expected generation cost of the generators in the main power grid and the distributed generators in distribution systems.

The expected generation cost of the generators in the main power grid

The generation cost of conventional generators in the main power grid is formulated as follows:

$$C_{EOC_G}(t) = \sum_{k\in S} p(x_k) \cdot \sum_{n\in N_G} f(P_{Gn}(t)),\qquad(17)$$

where $p(x_k)$ is state probability of system k, $P_{Gn}(t)$ is active power of generator n in the main power grid at time period t, and $f(P_{Gn}(t))$ is cost function of generator n in the main power grid.

The cost function of generator n in the main power grid is as follows:

$$f(P_{Gn}(t)) = a + b \times P_{Gn}(t),\qquad(18)$$

where a and b are, respectively, cost coefficient relating to active power.

The generation cost of DGs in distribution systems

The generation cost of the distributed generators in distribution systems is written as follows:

$$C_{EOC_{DG}}(t) = \sum_{k\in S} p(x_k) \cdot \sum_{m\in N_{DG}} f(P_{DGm}(t)),\qquad(19)$$

where $P_{DGm}(t)$ is active power of the distributed generator m at time period t and $f(P_{DGm}(t))$ is the generation cost function of the distributed generators in distribution systems.

Due to no fuel is needed, when wind, solar, and other renewable energy is used for producing electricity, there is only a very small generation cost. So the operation and maintenance cost of distributed generation is only taken into consideration:

$$f(P_{DGm}(t)) = K \times P_{DGm}(t) \times \Delta t,\qquad(20)$$

where K is coefficient relating to operation cost and Δt is operation time period.

The expected interruption cost (EIC)

The expected interruption cost is formulated as follows:

$$\begin{cases} C_{\text{EIC}}(t) = \sum_{k \in S} p(x_k) \sum_{i \in N_D} h(P_{Ci}(t)) \\ h(P_{Ci}(t)) = P_{Ci}(t) \times VoLL_i \end{cases}, \quad (21)$$

where $h(P_{Ci}(t))$ is outage loss function for the load at node i, $P_{Ci}(t)$ is the active power of the load cut-off from the grid at node i at time period t, $VoLL_i$ [62] is load loss value at node i, and N_D is nodal set of the distribution systems.

Objective function

In this article, a power balancing method of distributed generation and electric vehicle charging is presented for minimizing operation cost of distribution systems with uncertainties, which includes the uncertainties in output power of distributed generation and the randomness of charging power of electric vehicles. In the proposed power balancing model, output power of distributed generation and charging power of electric vehicles is taken for decision-making variables, and its objective is to minimize the expected social cost of the distribution system:

$$\text{Min } C_{\text{ESC}} = C_{\text{EOC}} + C_{\text{EIC}}, \quad (22)$$

Constraint condition

Equal constraint condition for power flow

It is supposed that the injected active and reactive power is equal to the active and reactive power coming from the main power grid and the distributed generation in the distribution systems and active power of cut-off load minus the active power of general load and electric vehicle charging load:

$$\begin{cases} P_i = P_{Gi} + P_{DGi} - P_{EVi} - (P_{LDi} - P_{Ci}) \\ Q_i = Q_{Gi} + Q_{DGi} - Q_{EVi} - (Q_{LDi} - Q_{Ci}) \end{cases}, \quad (23)$$

where P_i and Q_i are, respectively, injected active and reactive power at node i; P_{Gi}, P_{DGi}, P_{EVi}, P_{LDi}, and P_{Ci} are, respectively, active power injected in distribution systems from the main power grid, active power injected in by distributed renewable energy systems in the distribution systems, reactive power consumed by electric vehicles, active power consumed by general load, active power of cut-off load due to uncertainties at node i; Q_{Gi}, Q_{DGi}, Q_{EVi}, Q_{LDi}, and Q_{Ci} are, respectively, reactive power injected in by the main power grid, reactive power injected in by distributed renewable energy systems in the distribution systems, reactive power consumed by electric vehicles, active power consumed by general load, and reactive power of cut-off load due to uncertainties at node i.

According to analysis theory for power flow, the injected power depends on the magnitude and the phase of nodal voltage, as shown in the following equations:

$$\begin{cases} P_i = G_{ii}V_i^2 + \sum_{j=1}^{n} V_i V_j (G_{ij} \cos \theta_{ij} + B_{ij} \sin \theta_{ij}) \\ Q_i = -B_{ii}V_i^2 - \sum_{j=1}^{n} V_i V_j (B_{ij} \cos \theta_{ij} - G_{ij} \sin \theta_{ij}) \end{cases} \quad (24)$$

where V_i and V_j, θ_i and θ_j are, respectively, the amplitude and phase of nodal voltage at node i and j; $\theta_{ij} = \theta_i - \theta_j$; G_{ii} and B_{ii} are, respectively, self conductance and self susceptance at node i; G_{ij} and B_{ij} are, respectively, mutual conductance and mutual susceptance between node i and j, $j \in \partial \Omega(i)$.

Power constraint for generators in the main power grid

The active and reactive power of generator in the main power grid must be not greater than its maximal permitted limit value and be not less than its minimal permitted limit value:

$$P_{Gi}^{\min} \leq P_{Gi} \leq P_{Gi}^{\max}, \quad (25)$$

where $P_{Gi}^{\max}$ and $P_{Gi}^{\min}$ are, respectively, the maximal and minimal permitted limit value for active power injected in by the main power grid at node i.

Constraint condition for nodal voltage

The nodal voltage amplitude must be equal to or less than its maximal permitted limit value, and be not less than its minimal permitted limit value:

$$V_i^{\min} \leq V_i \leq V_i^{\max}, \quad (26)$$

where $V_i^{\min}$, $V_i^{\max}$ are, respectively, the maximal and minimal permitted limit value for voltage at node i.

Constraint condition for line current

The current of the branch must be not greater than its maximal permitted limit value:

$$I_{ij}^2 = (G_{ij}^2 + B_{ij}^2)(V_i^2 + V_j^2 - 2V_i V_j \cos \theta_{ij}) \leq I_{ij\max}^2, \quad (27)$$

where $I_{ij\max}$ is the permitted limit value for branch L_{ij}.

Constraint condition for the cut-off load power

The cut-off load power must be not greater than its actual working value:

$$0 \leq P_{Ci} \leq P_{LDi}, \quad (28)$$

where P_{Ci} is active power of cut-off load at node i, and P_{LDi} is actual active power of the load at node i.

Constraint condition for power output of DGs

The output power of distributed generation must be not greater than its maximal permitted limit value, and be not less than its minimal permitted limit value:

$$0 \leq \sqrt{P_{DGi}^2 + Q_{DGi}^2} \leq S_{DGi}^{max}, \tag{29}$$

where P_{DGi} and Q_{DGi} are active power and reactive power of distributed generation of renewable energy system i, and S_{DGi}^{max} is the maximal capacity of distributed generation system i.

Constraint condition for charging power of electric vehicle

The charging power of electric power must be not greater than its controllable maximal limit value:

$$0 \leq P_{EVi} \leq P_{EV}^{max}, \tag{30}$$

where P_{EV}^{max} is the controllable maximal charging power of an EV.

Solving method

General model for cone programming

Cone optimization is a generalization of linear programming and nonlinear programming. The objective function in cone programming [63] must be linear function, and the feasible region is composed of linear equality or inequality constraints and rotation cone or quadratic cone inequality constraints. Its standard form is as follows:

$$\min c^T x$$
$$s.t.:A_i x + b_i \in K_i \quad \forall i \tag{31}$$

where x is decision-making variables, K_i is a kind of rotating cone which satisfies the following formula:

$$K_i = \{x \in R^{n^i} : 2x_1 x_2 \geq \sum_{j=3}^{n^i} x_j^2, x_1, x_2 \geq 0\}. \tag{32}$$

Cone transformation method for constrained conditions

In this article, the objective function is linear function, which satisfies the cone programming form, so it is not necessary to carry out the transformation of cone form, but the nonlinear constraints must be transformed to the cone form. In order to transform the nonlinear form of

the product of variable V_i, V_j, and θ_{ij} in the constrained conditions into the requirements of the optimal linear conditions, the following variables are introduced to replace the variables related above [64]:

$$\begin{cases} X_i = V_i^2 / \sqrt{2} \\ R_{ij} = V_i V_j \cos \theta_{ij} , \\ Y_{ij} = V_i V_j \sin \theta_{ij} \end{cases} \tag{33}$$

Then, the constraint conditions are transformed to:

$$P_i = \sqrt{2} G_{ii} X_i + \sum_{j=1}^{n} G_{ij} R_{ij} + B_{ij} Y_{ij}, \tag{34}$$

$$Q_i = -\sqrt{2} B_{ii} X_i - \sum_{j=1}^{n} B_{ij} R_{ij} - G_{ij} Y_{ij}, \tag{35}$$

$$V_i^{min} / \sqrt{2} \leq X_i \leq V_i^{max} / \sqrt{2}, \tag{36}$$

$$I_{ij}^2 = (G_{ij}^2 + B_{ij}^2)(\sqrt{2} X_i + \sqrt{2} X_j - 2R_{ij}) \leq V_{ij\,max}^2, \tag{37}$$

$$2 X_i X_j = R_{ij}^2 + Y_{ij}^2, i \in N_D, j \in \Omega(i), \tag{38}$$

When Mosek mathematical tool is used to solve the problem, equation (38) is changed as:

$$2 X_i X_j \geq R_{ij}^2 + Y_{ij}^2 \quad i \in N_D, j \in \Omega(i). \tag{39}$$

The cone transformation of the capacity constraint of distributed generation is formulated as follows:

$$2 \times \frac{S_{DGi}^{max}}{\sqrt{2}} \times \frac{S_{DGi}^{max}}{\sqrt{2}} \geq P_{DGi}^2 + Q_{DGi}^2 \quad i \in N_{DG}. \tag{40}$$

Through the substitution of variables, the formula (34) satisfies the requirements of the linear constraint conditions of cone optimization; equations (39) and (40) constitute the Descartes product form of the quadratic rotating cone, and the search space satisfies the range of the convex cone. When the optimization model is processed into linearization form, the search space of the decision variable is satisfied with the range of the convex cone, which is beneficial to improve the efficiency of the optimization.

The method used in this article is the quadratic rotation cone programming. The proposed algorithm can not only solve the problem of the power flow calculation but also can solve the optimization problem.

Solving steps

The steps for solving the optimization problem are written as follows:

1. DG data initialization: give the initial active and reactive power of distributed generation of renewable energy systems according to different characteristics of renewable energy systems and optimal requirements.

2. EV data initialization: give the initial active and reactive power of electric vehicles according to the expected driving conditions and charging requirements and so on.

3. DG state probability simulation: simulate the state probability of output power of distributed generation of renewable energy systems using Monte Carlo method.

4. EV state probability simulation: simulate the state probability of charging power of electric vehicles in different charging stations using Monte Carlo method.

5. DG and EV power balancing: Search for optimal power balancing solution to the output power of distributed generation and charging power of electric vehicles based on quadratic rotation cone programming model in MOSEK using an objective function minimizing the expected social cost of the distribution system with DGs and EVs.

6. Test for the optimality: test for the optimality, robustness, stability, and convergence of the solution.

7. If the test conditions are satisfied, output optimal power of DGs and charging power of EVs.

8. If the test conditions are not satisfied, go to step e).

Simulation and Analysis for Study Case

Study examples and its data

In this article, the IEEE 14-node distribution system is taken as a study example, in which there are three feeders, the reference capacity is 100 MVA, the reference voltage is 23 kV, the total load is 28.7 MW, as shown in Figure 1.

In this article, the cut-in wind speed of the wind turbine is 3 m/sec, the cut-out wind speed is 25 m/s, the rated wind speed is 13 m/sec, the rated power of the wind-driven generator is 1 MW, and the forced outage rate is 0.05. The maximum irradiance of photovoltaic generation system is 900 W/m^2, and its rated power is 1 MW, the forced outage rate for 0.05. The charging power of an electric vehicle is 5 kW, the electricity consumption for driving 100 km is 14 kWh, and the charging efficiency is 0.95. The optimal scheduling period is 24 h, which is divided into 24 time periods.

The injected power from the main grid at node 1 is supposed to be equivalent to a generator, and its parameters are shown in Table 1 [65].

State probability calculation of distributed generation

State probability of output power of wind-driven generator

According to the probability model of the wind speed and the characteristics of the output power of wind-driven generator, the output power of the wind-driven generator is discretized into 11 states.

If the wind speed at each time period in a day is simulated 10,000 times, the probability distribution function, as shown in Figure 2, and the sampling number of output power of wind-driven generator, as shown in Figure 3, are available for the wind speed simulation of anyone time period in a day.

Table 1. The data of generator.

Cost coefficients of generators		Active power/MW	
a($/h)	b($/MW·h)	Upper limit	Lower limit
213.1	11.669	40	10

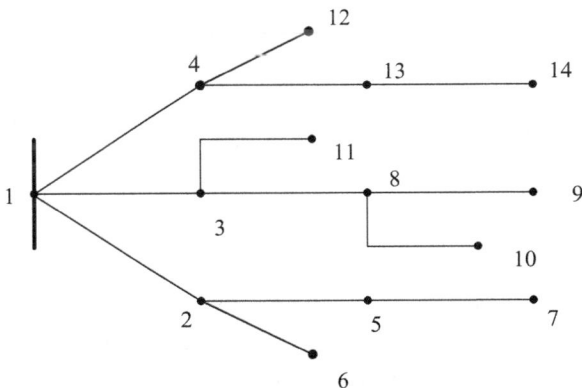

Figure 1. IEEE 14-node distribution system.

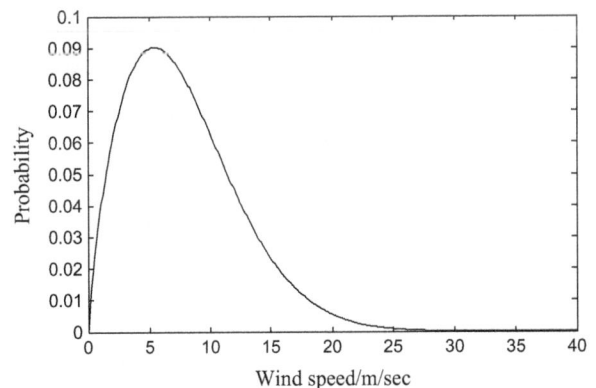

Figure 2. The probability of wind speed.

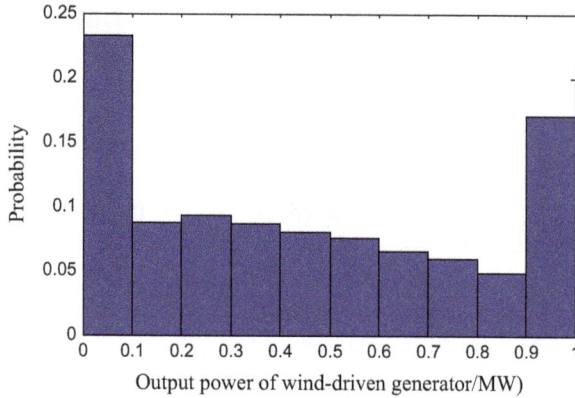

Figure 3. The sampling number of power output of a wind-driven generator.

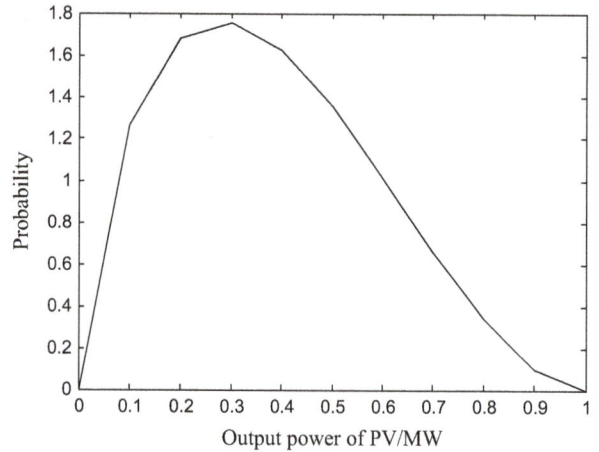

Figure 4. The probability of PV output power.

At each time period, the output power of wind-driven generator is divided into 11 kinds of probability state, as shown in Table 2. According to the formulas (7) and (8), the probability distribution of the output power of photovoltaic power generation at a certain time period can be obtained, as shown in Figure 4. Analysis for state probability of photovoltaic power generation is the same as wind power generation.

State probability calculation of electric vehicle charging

It is assumed that the number of electric vehicle is 10,000. According to formulas (9–13), the probability distribution and the charging power of electric vehicle are shown in Figures 5 and 6, respectively.

For electric vehicles in a day, the charging time is divided into 24-h time period; that is, there are 24 charging states, and the probability of the average charging power

of an electric vehicle at each hour time is shown in Table 3.

Multi-state optimization coordination for distributed generation and electric vehicle charging

In order to coordinate and optimize the operation state of the distributed generation and electric vehicle charging, the optimal operation of the distribution system is optimized, that is, to minimize the expected social cost in this article. The following two cases for optimization calculation are given:

Case 1: when only electric vehicle is installed at some nodes in a distribution system, in which the system has 24 kinds of charging states. In this article, it is assumed that electric vehicle is installed at nodes 7, 10, and 13.

Table 2. The probability of output power of wind generator.

Output power/MW	Probability	
	Not considering the failure of wind turbine	Considering the failure of wind turbine
0	0.1858	0.2265
0.1	0.0914	0.0868
0.2	0.0885	0.0841
0.3	0.0879	0.0835
0.4	0.0793	0.0753
0.5	0.0769	0.0731
0.6	0.0718	0.0682
0.7	0.0618	0.0587
0.8	0.0531	0.0504
0.9	0.0486	0.0462
1	0.1549	0.1472

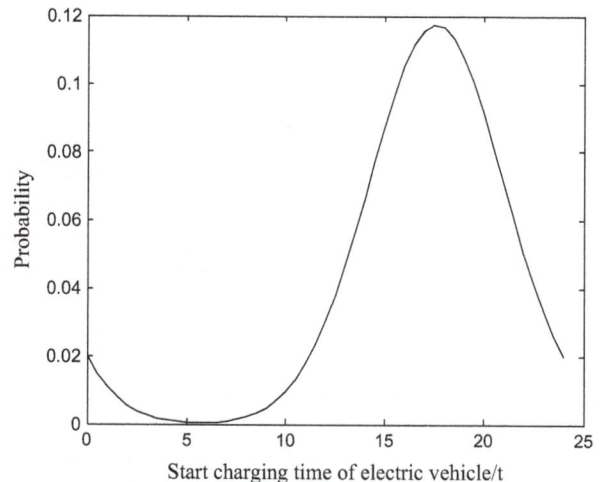

Figure 5. The probability of start charging time of electric vehicle.

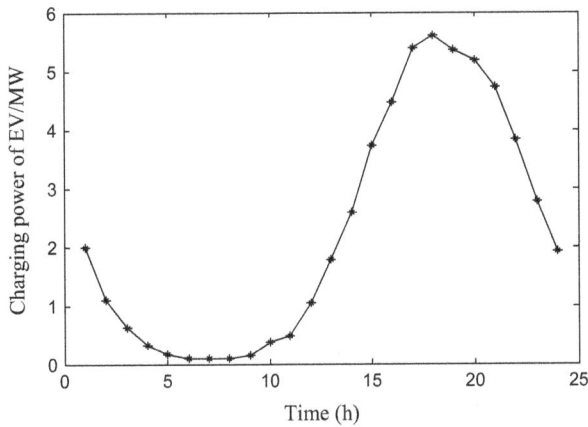

Figure 6. Charging power of an electric vehicle.

Table 3. The probability of electric vehicle charging for 24 h in a day.

Hour period	Charging power/MW	Probability
0–1	1.9675	0.0151
1–2	1.6275	0.0080
2–3	0.9450	0.0039
3–4	0.5475	0.0018
4–5	0.3325	0.0008
5–6	0.2125	0.0005
6–7	0.1275	0.0007
7–8	0.1125	0.0015
8–9	0.1325	0.0033
9–10	0.1775	0.0070
10–11	0.3800	0.0134
11–12	0.8550	0.0237
12–13	1.4700	0.0383
13–14	2.1225	0.0568
14–15	2.9625	0.0774
15–16	3.9225	0.0967
16–17	4.7075	0.1110
17–18	5.4475	0.1169
18–19	5.8000	0.1129
19–20	5.4900	0.1001
20–21	4.9350	0.0815
21–22	4.1250	0.0608
22–23	3.2050	0.0417
23–24	2.3250	0.0262

Case 2: when distributed generation systems and electric vehicle are, respectively, installed at different nodes in distribution system, in which the system has 264 kinds of electric generation states and charging states. In this article, it is assumed that distributed generation system is installed at nodes 3, 4, 7, 9, 10, and 14, and electric vehicle is installed at nodes 7, 10, and 13.

For the optimization calculation of the above scheme, Mosek mathematical tool is used to calculate the cone optimization algorithm introduced above. At the same time, the results are compared with the optimization results using the particle swarm optimization algorithm [66]. The results of the two algorithms are shown in Table 4.

It is seen from comparison of the cost value index of different schemes in Table 4 that the expected operation cost, outage loss, and social cost in the cases when distributed generation system and electric vehicle are, respectively, installed at different nodes in the distribution system is less than that in the cases when electric vehicle is only installed in the distribution system. It illustrates that the operating efficiency of distribution system can be improved when distributed generation system is installed in the distribution systems with a great number of electric vehicles installed at different nodes.

If the fault rate of the distributed generation systems is included, the operation cost of distribution systems may be influenced, as shown in Table 4. It is seen from

Table 4 that the cost value index in the cases considering the fault impact of the distributed generation systems is a little greater than that in the cases without considering the fault impact of the distributed generation systems. It shows that the operation benefit of distribution systems with DGs is reduced, but taking the failure rate of DG into consideration is more practical.

A method-based rotation quadratic cone programming is proposed for modeling optimization coordination of distributed generation and electric vehicle charging in distribution systems and for searching for the effective solution of the proposed optimization problem in this article, and it achieves better results to calculate the cost value index and illustrates the feasibility and superiority of the proposed method in this article. It shows from the results comparison of different optimization algorithms in Table 4 that the cost value index obtained by using rotation quadratic cone programming is smaller than that obtained by using PSO algorithm.

Table 4. Operation cost/worth indices.

Method	Cases		EOC ($/h)	EIC ($/h)	ESC ($/h)
PSO	Case 1		670.97	359.05	1030.02
	Case 2	Without fault	562.65	171.14	733.79
		With DG fault	564.03	176.67	740.70
Cone programming	Case 1		664.44	279.33	943.77
	Case 2	Without DG fault	553.89	152.60	706.49
		With DG fault	561.48	167.27	728.75

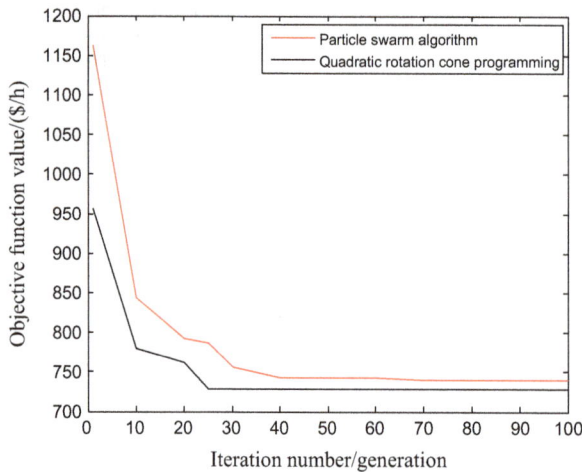

Figure 7. The evolution curve obtained, respectively, by quadratic rotation cone programming and particle swarm algorithm.

It is seen from the results of different optimization algorithms and the number of iterations in Table 4 and Figure 7 that the value of the operation cost obtained by using quadratic rotation cone programming based method is smaller than that of the particle swarm algorithm, and the number of iterations is less and the convergence rate is faster. This proves the feasibility and superiority of the algorithm proposed in this article.

An optimal operation cost value index is obtained by using cone optimization algorithm for optimizing multi-state optimization coordination of distributed generation of renewable energies and electric vehicle charging, and this can provide feasible reference for the operation and control of distribution systems.

Conclusion

1. An optimal power balancing model of distributed systems with distributed generation and electric vehicles is presented based on quadratic rotated conic programming. This method simplifies the original problem of the optimization problem in this article, which makes it faster, more stable, and better. Compared with the results of the particle swarm optimization algorithm, the results of the quadratic rotation cone programming are more excellent, which proves the effectiveness and superiority of the proposed method.
2. Multi-state probability model is used to represent stochastic characteristics of DG and EV, and it shows optimal power balancing results with high economic efficiency.
3. EV charging power is effective controlled by coordinating power output of distributed generation of renewable energy systems with different output power level.

4. The proposed method can improve economic benefits of power systems with DG and EV, in order for minimizing the expected generation cost of generators in the main power grid, the expected operation cost of distributed generation systems in distributed systems, and the expected social outage loss.

Conflict of Interest

None declared.

References

1. Li, X., X. Zhang, L. Wu, P. Lu, and S. Zhang. 2015. Transmission line overload risk assessment for power systems with wind and load-power generation correlation. IEEE Trans. Smart Grid 6:1233–1242.
2. Choudar, A., D. Boukhetala, S. Barkat, and J. M. Brucker. 2015. A local energy management of a hybrid PV-storage based distributed generation for microgrids. Energy Convers. Manage. 90:21–33.
3. Howlader, H. O., H. Matayoshi, and T. Senjyu. 2015. Distributed generation incorporated with the thermal generation for optimum operation of a smart grid considering forecast error. Energy Convers. Manage. 96:303–314.
4. Shaaban, M. F., and E. F. El-Saadany. 2014. Accommodating high penetrations of PEVs and renewable DG considering uncertainties in distribution systems. IEEE Trans. Power Syst. 29:259–270.
5. Yunus, K. 2010. Probabilistic modeling of plug-in electric vehicle charging impacts on power systems. [M.Sc. thesis], Chalmers University Technology, Sweden.
6. Gao, S., K. Chau, C. Liu, D. Wu, and C. Chan. 2014. Integrated energy management of plug-in electric vehicles in power grid with renewables. IEEE Trans. Veh. Technol. 63:3019–3027.
7. Bouallaga, A., A. Merdassi, A. Davigny, B. Robyns, and V. Courtecuisse. 2013. Minimization of energy transmission cost and CO_2 emissions using coordination of electric vehicle and wind power (W2V). Proc. IEEE Grenoble Power Tech., Grenoble, France, pp. 1–6.
8. Sagosen, O., and M. Molinas. 2013. Large scale regional adoption of electric vehicles in Norway and the potential for using wind power as source. Proc. Int. Conf. Clean Elect. Power (ICCEP), Alghero, Italy, pp. 189–196.
9. Cao, Y. et al. 2012. An optimized EV charging model considering TOU price and SOC curve. IEEE Trans. Smart Grid 3:388–393.
10. Wu, D., D. C. Aliprantis, and K. Gkritza. 2011. Electric energy and power consumption by light-duty plug-in electric vehicles. IEEE Trans. Power Syst. 26:738–746.
11. Mwasilu, F., J. J. Justo, E.-K. Kim, T. D. Do, and J.-W. Jung. 2014. Electric vehicles and smart grid interaction

- A review on vehicle to grid and renewable energy sources integration. Renew. Sustain. Energy Rev. 34:501–516.

12. Xu, D. Q., G. Joós, M. Lévesque, and M. Maier. 2013. Integrated V2G, G2V, and renewable energy sources coordination over a converged fiber-wireless broadband access network. IEEE Trans. Smart Grid 4:1381–1390.

13. Das, R., K. Thirugnanam, P. Kumar, R. Lavudiya, and M. Singh. 2014. Mathematical modeling for economic evaluation of electric vehicle to smart grid interaction. IEEE Trans. Smart Grid 5:712–721.

14. Shaaban, M. F., Y. M. Atwa, and E. F. El-Saadany. 2013. DG allocation for benefit maximization in distribution networks. IEEE Trans. Power Syst. 28:639–649.

15. Karatekin, C. Z., and C. Ucak. 2008. Sensitivity analysis based on transmission line susceptances for congestion management. Elect. Power Syst. Res. 78:1485–1493.

16. Jayaweera, D., and S. Islam. 2014. Steady-state security in distribution networks with large wind farms. J. Mod. Power Syst. Clean Energy 2:134–142.

17. McCalley, J., V. Vittal, and N. Abi-Samra. 1999. An overview of risk based security assessment. Proc. IEEE Power Eng. Soc. Sum. Meeting, Edmonton, AB, Canada, pp. 173–178.

18. McCalley, J., S. Asgarpoor, L. Bertling, R. Billinion, H. Chao, J. Chen et al. 2014. Probabilistic security assessment for power system operations. Proc. IEEE Power Eng. Soc. Gen. Meeting (PES), vol. 1. Denver, CO, USA, pp. 212–220.

19. Ni, M., J. D. McCalley, V. Vittal, and T. Tayyib. 2003. Online risk-based security assessment. IEEE Trans. Power Syst. 18:258–265.

20. Dai, Y. J., J. McCalley, N. Abi-Samra, and V. Vittal. 2001. Annual risk assessment for overload security. IEEE Trans. Power Syst. 16:616–623.

21. Hajimiragha, A. H., C. A. Canizares, M. W. Fowler, S. Moazeni, and A. Elkamel. 2011. A robust optimization approach for planning the transition to plug-in hybrid electric vehicles. IEEE Trans. Power Syst. 26:2264–2274.

22. Gomez, J. C., and M. M. Morcos. 2003. Impact of EV battery chargers on the power quality of distribution systems. IEEE Trans. Power Del. 18:975–981.

23. Hadley, S. W., and A. Tsvetkova. 2009. Potential impacts of plug-in hybrid electric vehicles on regional power generation. Electricity J. 22:56–68.

24. Sortomme, E., M. M. Hindi, S. D. J. MacPherson, and S. S. Venkata. 2011. Coordinated charging of plug-in hybrid electric vehicles to minimize distribution system losses. IEEE Trans. Smart Grid 2:198–205.

25. Qian, K., C. Zhou, M. Allan, and Y. Yuan. 2011. Modeling of load demand due to EV battery charging in distribution systems. IEEE Trans. Power Syst. 26:802–810.

26. Abdelsamad, S. F., W. G. Morsi, and T. S. Sidhu. 2015. Impact of wind-based distributed generation on electric energy in distribution systems embedded with electric vehicles. IEEE Trans. Sustain. Energy 6:79–87.

27. Paterakis, N. G., O. Erdinç, I. N. Pappi, A. G. Bakirtzis, and J. P. S. Catalãao. 2016. Coordinated operation of a neighborhood of smart households comprising electric vehicles, energy storage and distributed generation. IEEE Trans. Smart Grid 7:2736–2747.

28. Chang, T.-H., M. Alizadeh, and A. Scaglione. 2013. Real-time power balancing via decentralized coordinated home energy scheduling. IEEE Trans. Smart Grid 4:1490–1504.

29. Brusco, G., A. Burgio, D. Menniti, A. Pinnarelli, and N. Sorrentino. 2014. Energy management system for an energy district with demand response availability. IEEE Trans. Smart Grid 5:2385–2393.

30. Moradzadeh, B., and K. Tomsovic. 2013. Two-stage residential energy management considering network operational constraints. IEEE Trans. Smart Grid 4:2339–2346.

31. Yoon, J. H., R. Baldick, and A. Novoselac. 2014. Dynamic demand response controller based on real-time retail price for residential buildings. IEEE Trans. Smart Grid 5:121–129.

32. Vivekananthan, C., Y. Mishra, G. Ledwich, and L. Fangxing. 2014. Demand response for residential appliances via customer reward scheme. IEEE Trans. Smart Grid 5:809–820.

33. Shi, W., N. Li, X. Xie, C.-C. Chu, and R. Gadh. 2014. Optimal residential demand response in distribution networks. IEEE J. Sel. Areas Commun. 32:1441–1450.

34. Shao, S., M. Pipattanasomporn, and S. Rahman. 2011. Demand response as a load shaping tool in an intelligent grid with electric vehicles. IEEE Trans. Smart Grid 2:624–631.

35. Khamphanchai, W., M. Pipattanasomporn, M. Kuzlu, J. Zhang, and S. Rahman. 2015. An approach for distribution transformer management with a multiagent system. IEEE Trans. Smart Grid 6:1208–1218.

36. Gouveia, C., J. Moreira, C. L. Moreira, and J. A. Pecas Lopes. 2013. Coordinating storage and demand response for microgrid emergency operation. IEEE Trans. Smart Grid 4:1898–1908.

37. Erdinc, O., N. G. Paterakis, I. N. Pappi, A. G. Bakirtzis, and J. P. S. Catalão. 2015. A new perspective for sizing of distributed generation and energy storage for smart households under demand response. Appl. Energy 143:26–37.

38. Kahrobaee, S., S. Asgarpoor, and W. Qiao. 2013. Optimum sizing of distributed generation and storage capacity in smart households. IEEE Trans Smart Grid 4:1791–1801.

39. Ahmadian, A., M. Sedghi, M. Aliakbar-Golkar, M. Fowler, and A. Elkamel. 2016. Two-layer optimization methodology for wind distributed generation planning considering plug-in electric vehicles uncertainty: a

flexible active-reactive power approach. Energy Convers. Manage. 124:231–246.

40. Ullah, N. R., K. Bhattacharya, and T. Thiringer. 2009. Wind farms as reactive power ancillary service providers—technical and economic issues. IEEE Trans. Energy Convers. 24:661–672.

41. Deshmukh, S., B. Natarajan, and A. Pahwa. 2012. Voltage/VAR control in distribution networks via reactive power injection through distributed generators. IEEE Trans Smart Grid 2:1226–1234.

42. Keane, A., L. F. Ochoa, F. Vittal, C. J. Dent, and G. P. Harrison. 2011. Enhanced utilization of voltage control resources with distributed generation. IEEE Trans. Power Syst. 26:252–260.

43. Cheng, L., Y. Chang, and R. Huang. 2015. Mitigating voltage problem in distribution system with distributed solar generation using electric vehicles. IEEE Trans. Sustain. Energy 6:1475–1484.

44. Traube, J., F. Lu, D. Maksimovic, J. Mossoba, M. Kromer, P. Faill et al. 2013. Mitigation of solar irradiance intermittency in photovoltaic power systems with integrated electric-vehicle charging functionality. IEEE Trans. Power Electron. 28:3058–3067.

45. Chukwu, U. C., and S. M. Mahajan. 2014. V2G parking lot with PV rooftop for capacity enhancement of a distribution system. IEEE Trans. Sustain. Energy 5:119–127.

46. Marra, F., G. Y. Yang, Y. T. Fawzy, C. Træholt, E. Larsen, R. Garcia-Valle et al. 2013. Improvement of local voltage in feeders with photovoltaic using electric vehicles. IEEE Trans. Power Syst. 28:3515–3516.

47. Foster, J. M., G. Trevino, M. Kuss, and M. C. Caramanis. 2013. Plug-in electric vehicle and voltage support for distributed solar: theory and application. IEEE Syst. J. 7:881–888.

48. Aghaei, J., A. E. Nezhad, A. Rabiee, and E. Rahimi. 2016. Contribution of plug-in hybrid electric vehicles in power system uncertainty management. Renew. Sustain. Energy Rev. 59:450–458.

49. Battistelli, C., and A. J. Conejo. 2014. Optimal management of the automatic generation control service in smart user grids including electric vehicles and distributed resources. Electric Power Syst. Res. 111:22–31.

50. Chaouachi, A., R. M. Kamel, R. Andoulsi, and K. Nagasaka. 2013. Multiobjective intelligent energy management for a microgrid. IEEE Trans. Ind. Electron. 60:1688–1699.

51. Kanchev, H., L. Di, F. Colas, V. Lazarov, and B. Francois. 2011. Energy management and operational planning of a microgrid with a PV-based active generator for smart grid applications. IEEE Trans. Ind. Electron. 58:4583–4592.

52. Nunna, H. K., and S. Doolla. 2013. Multiagent-based distributed energy-resource management for intelligent microgrids. IEEE Trans. Ind. Electron. 60:1678–1687.

53. Dragicevic, T., J. M. Guerrero, and J. C. Vasquez. 2014. A distributed control strategy for coordination of an autonomous LVDC microgrid based on power-line signaling. IEEE Trans. Ind. Electron. 61:3313–3326.

54. Gburczyk, P., I. Wasiak, R. Miénski, and R. Pawełek. 2011. Energy management system as a mean for the integration of distributed energy sources with low voltage network. Proc. 11th Int. Conf., Elect. Power Quality Utilisation, Lisbon, Spain, Oct. 17–19. pp. 1–5.

55. Guerrero, J. M., J. C. Vasquez, J. Matas, L. G. de Vicuña, and M. Castilla. 2011. Hierarchical control of droop-controlled ac and dc microgrids—A general approach toward standardization. IEEE Trans. Ind. Electron. 58:158–172.

56. Chi, J., W. Peng, X. Jianfang, T. Yi, and C. Fook Hoong. 2014. Implementation of hierarchical control in dc microgrids. IEEE Trans. Ind. Electron. 61:4032–4042.

57. Olek, B., and M. Wierzbowski. 2015. Local energy balancing and ancillary services in low-voltage networks with distributed generation, energy storage, and active loads. IEEE Trans. Industr. Electron. 62:2499–2508.

58. Moharil, R. M., and P. S. Kulkarni. 2010. Reliability analysis of solar photovoltaic system using hourly mean solar radiation data. Sol. Energy 84:691–702.

59. Karki, R., and J. Patel. 2009. Reliability assessment of a wind power delivery system. Proc. Inst. Mech. Eng. Part O J. Risk Reliabil. 223:51–58.

60. Qian, K., C. Zhou, M. Allan, and Y. Yuan. 2010. Load model for prediction of electric vehicle charging demand. 2010 International Conference on Power System Technology, Hangzhou, China. pp. 1–6.

61. Liting, T., S. Shuanglong, and J. Zhuo. 2010. A statistical model for charging power demand of electric vehicles. Power Syst. Technol. 34:126–130.

62. Tollefson, G., R. Billinton, G. Wacker, E. Chan, and J. Aweya. 1991. A Canadian customer survey to assess power system reliability worth. IEEE Trans. Power Syst. 9:443–450.

63. Andersen, E. D., C. Roos, and T. Terlaky. 2003. On implement a primal-dual interior point methods for conic quadratic optimization. Math. Program. 95:249–277.

64. Jabr, R. A. 2006. Radial distribution load flow using conic programming. IEEE Trans. Power Syst. 21:1458–1459.

65. Yuanzhang, S., C. Lin, and H. Jian. 2012. Power system operational reliability theory. Tsinghua University Press, Beijing.

66. Jingjing, Z. et al. 2009. Based on particle swarm optimization (PSO) algorithm for power distribution network reconfiguration and distributed power supply integrated power optimization algorithm. Power Syst. Technol. 33:162–166.

10

Marginal abatement cost curve for wind power in China: a provincial-level analysis

Weiming Xiong, Yuanzhe Yang, Yu Wang & Xiliang Zhang

Institute of Energy, Environment and Economy, Tsinghua University, 100084 Beijing, China

Keywords

China, CO_2 mitigation cost, GIS analysis, wind curtailment, wind power resources

Correspondence

Yu Wang, Institute of Energy, Environment and Economy, Tsinghua University, 100084 Beijing, China. E-mail: y-wang@tsinghua.edu.cn

Funding Information

4th Generation District Heating Technologies and Systems, Ministry of Science and Technology of the People's Republic of China (Grant/Award Number: 2012IM010300) and National Natural Science Foundation of China (Grant/Award Number: 71103111).

Abstract

Wind power has seen a remarkable growth in China since the *Renewable Energy Law* came into force in the beginning of 2006. However, the contribution and the economic cost of wind power development to reduce carbon emission at the provincial level are still unclear. This research combined geographic information system analysis and levelized production cost calculation to provide provincial-level supply curve of carbon dioxide (CO_2) mitigation by wind power in China. The results showed that wind power could be a very competitive mitigation option in China, with a potential to contribute 500 million tons of CO_2 emission mitigation at an average abatement cost of 75 RMB/t of CO_2 mitigation. The emission abatement cost and the potential to reduce carbon emission in different provinces may be used for national planning as well as for the allocation of investments.

Introduction

China has overtaken the United States as the foremost energy consumer in the world since 2011, taking great pressure due to the carbon emission and air pollution caused by its coal-dominated energy system [1]. In 2014, the generation amount of coal-fired power plants, one of the largest carbon emission sectors, was 3951 TWh, accounting for 70.5% of total electricity generation [2]. In order to mitigate climate change and improve air quality, renewable energy, especially wind power, has been regarded as the fundamental solution for sustainable development by China [3]. Wind power has seen significant progress in China, particularly after the *Renewable Energy Law* came into force in 2006. By the end of 2015, the cumulative on-grid wind power capacity reached 129 GW, which is the world's largest wind industry and has contributed

186.3 TWh power, accounting for 3.3% of electricity generation [4]. Meanwhile, the Chinese government has announced its ambitious goal of achieving a wind-installed capacity of 100 GW by 2015 and 200 GW by 2020 [5]. Although the installation of wind turbines is increasing, the annual growth rate of wind capacity has decreased below 33% since 2011 [6]. Besides, the National Energy Administration has announced a decrease in the adjustment of feed-in-tariff for wind power by the end of 2014, which cast a shadow on the rapid development of the new installation in the future [7].

In order to reach the 2015 and 2020 wind power capacity target [5], allocating new installations in different regions with heterogeneous resource and demand characters must be considered as the leading question for future development. From the perspective of carbon emission mitigation, three main types of policies exist to promote the development of wind power among countries: price instrument imposing an external cost of emitted fossil carbon dioxide (CO_2); financial subsidy for wind power to cover the cap between generation cost and market price (e.g., feed-in-tariff); and quantity instrument imposing an emission cap or renewable penetration (e.g., cap-and-trade market). All these policy instruments add an extra cost to the existing power sector in order to reduce CO_2 emission. Therefore, it is critical and essential to investigate the abatement cost and emission mitigation potentials of wind power, which could provide solid supporting information for policy makers. The essential difficulties in estimating appropriate abatement cost and the relevant emission mitigation potential lie in the assessment of regional disparities of wind resource, wind generation cost, and coal power generation cost [8–10]. From the economic perspective, the deployment of wind energy should have the least incremental cost on the existing power system, with the same amount of emission reduction.

A widely used methodology to estimate how costly it would be to achieve the specific emission reduction is the marginal abatement cost curves (MACCs). The MACC can plot the corresponding cost after tightening the emission mitigation target further, which links marginal cost of abating an incremental emission to an emission potential. The purpose of our study was to generate the cost curve of carbon emission for wind power utilization, providing the emission mitigation could be reached by the deployment of wind power and the related emission abatement cost at high spatial resolution. The two important factors influencing carbon emission abatement cost by wind power for each region are the generation cost of wind and the coal-fired power cost, assuming the replacement of traditional coal power to wind power [11]. Thus, the carbon emission abatement cost is defined as the incremental electricity cost between wind power and coal-fired power, divided by the carbon-emitted gap between coal-fired power and wind power.

To investigate the economics of wind deployment, the purpose of our study was to evaluate the reduction in the potential and cost of carbon emission by wind power from the regional perspective. The paper first reviews the existing wind potential assessment in China and describes the wind resource as well as the on-going development plan. Thereafter, the methodology and data are used to calculate the wind power CO_2 mitigation potential as well as the abatement cost. Finally, the results could supply an overall message regarding the cost and amount of CO_2 abatement by wind power, providing some support for both national and local policy makers, investors, and other stakeholders.

Wind Power Potential and Development Plan

The existing assessment on wind potential in China mainly focused on physical capacity or generation of wind resource potential, which represents the upper limit of usable wind electricity with an assumption that all wind generation would be accepted by the power grid. Leading literatures have concluded that China's total wind power capacity ranges from 832 to 2600 GW [12–14], which is several times more than that of China's national targets of 200 GW wind installation by 2020 [5]. The China Meteorological Administration has developed a wind energy numerical simulation and evaluation system, indicating that the theoretical wind resource potential ranges from 2000 to 3400 GW [15]. He and Kammen combined the geographic information system (GIS) and wind hourly profile simulation to make high spatial resolution resource analysis at the provincial level, which indicates technical wind potential varies from 1243 to 2643 TWh due to different assumptions of wind turbine spatial density [14]. For offshore wind, Hong and Möller [16] reported that offshore wind in China could technically contribute 2450 TWh in 2020 and 2758 TWh in 2030. McElroy and Lu concluded that approximately 10% of CO_2 emission in China could be reduced if 0.62 PWh wind electricity is generated per year to replace coal-fired power, below the levelized cost of 0.4 RMB/kWh [17]. In summary, previous studies paid attention to wind resource assessment while the potential to mitigate carbon emission and the related abatement cost were simply calculated using national average coal-fired power generation cost, while the regional heterogeneity of the rest of power system except wind power is overlooked.

Specifically, the majority of existing installed wind turbines are located in provinces with abundant wind resources including high wind density and available land, such as Inner Mongolia, Hebei, Gansu, Jilin, and Xinjiang. By the end of 2014, 16 provinces exceeded 1 GW in terms of cumulative wind installation, and the 10 leading provinces account for more than 80% of the national wind installation, shown in Figure 1 [18]. Along with the existing high level of capacity concentration, the National Energy Administration of China has released development planning for seven 10-GW large-scale wind power installations by the end of 2020. In 2012, China's 12th 5 year plan for renewable energy has announced the 100 and 200 GW wind installation targets respectively by 2015 and 2020, followed by regional targets for wind installations [19]. Most of the announced 10-GW large-scale wind power installations are located in the north and northeast parts of China shown in Figure 2.

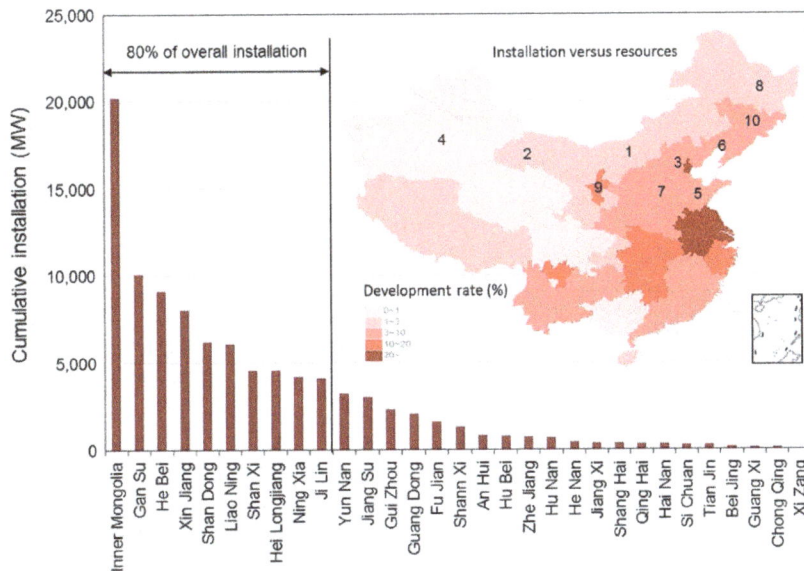

Figure 1. Provincial accumulative on-grid wind capacity by the end of 2014.

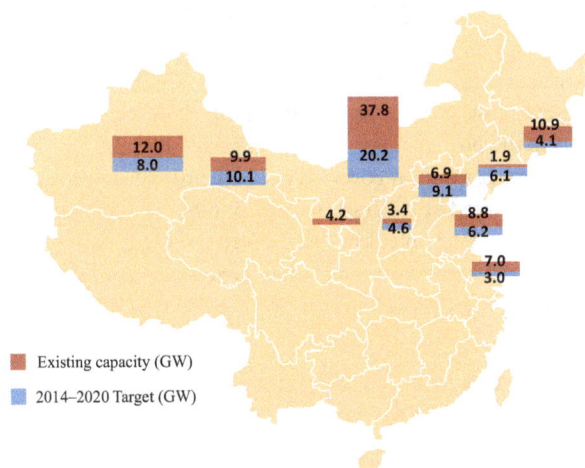

Figure 2. Planned large-scale wind installations toward 2020. For Ningxia province, existing wind capacity has exceeded 2020s target by 0.2 GW.

Methodology and Data

As the 10 leading provinces in wind capacity cover more than 80% of existing wind installation, we focused on the carbon emission mitigation potential of these provinces including Jiangsu,[1] assuming that the provincial capacity target by 2020 is going to be perfectly implemented. In order to generate provincial-level supply cost curve for CO_2 emission mitigation of wind power, annual wind generation and emission abatement cost are necessary. This study combines GIS system analysis and cost–benefit assessment to evaluate the abatement cost

and emission that could be avoided. The purpose of introduction of GIS analysis is to provide detailed description of regional capacity factor, which is inevitable for calculation of abatement cost and emission reduction. The logical framework and data utilization are listed as follows in this section. Particularly, this study shed light on onshore wind while offshore wind is not included.

Wind resource analysis

Wind resource assessment relies heavily on the method of retrospective analysis, or re-analysis. This study uses Modern Era Retrospective-analysis for Research and Applications, which is undertaken by The National Aeronautics and Space Administration's (NASA) Earth Observing System satellites [20]. With high spatial resolution dataset of 0.5° latitude by 0.67° longitude, the annual capacity factor for each grid cell could be calculated by averaging capacity factors from 1979 to 2009, in order to eliminate yearly variation. Based on hourly wind speed at the hub height of 80-m conducted by GEOS-5 Atmospheric Data Assimilation System for each grid cell, hourly wind output is generated assuming that a sole 80-m hub height Sinovel 1.5-MW turbine is fully equipped for all the wind projects. The power curve is shown in Figure 3 [21]. Thus, the annual capacity factor is calculated as the annual wind output divided by the product of nameplate turbine capacity (1.5 MW) and 8760 h (1 year), presented in Figure 4. From the perspective of capacity factor, it is intuitively observed that the high capacity factor areas are highly consistent with planning

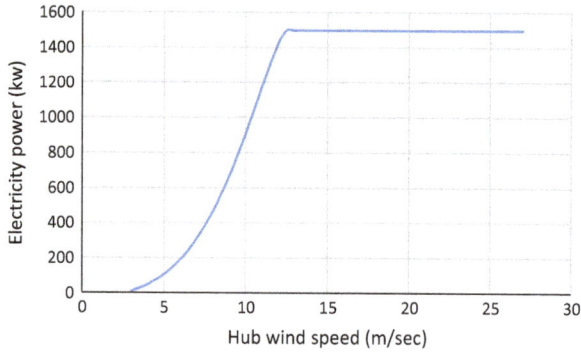

Figure 3. Wind turbine power curve.

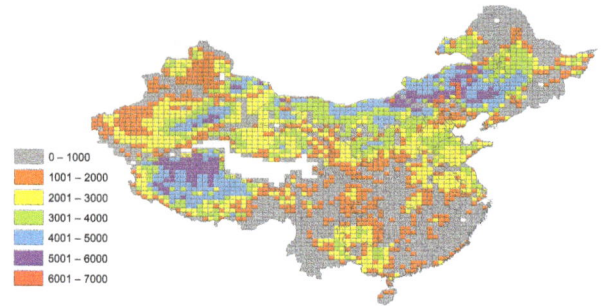

Figure 5. On-shore wind generation potential density per year (MWh/km²) in China. Unavailable grid cells (e.g., natural protection zones) are excluded.

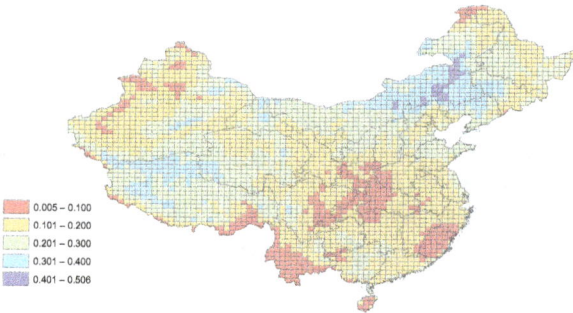

Figure 4. Simulated China wind capacity factor map.

CO$_2$ abatement cost calculation

Levelized production cost

Carbon dioxide abatement cost of wind power is defined as the gap between wind generation cost and coal-power generation, divided by avoided emission per generation unit. Levelized production cost (LPC) was widely used to calculate electricity production cost which reflects the average cost of one production unit (kilowatt or kWh) during the power station's entire expected lifetime [24–26]. The total generation costs over the lifetime of power plant are discounted at the start of operation by a predetermined discount rate, while LPC is derived as the ratio of the discounted total cost and total generation output over lifetime shown as follows:

$$
\begin{cases}
\text{NPV} = \sum_{i=1}^{N} (\text{LPC} \cdot \text{E} - \text{COM}_i - \text{LoanPay}_i - \text{Profit}_i) \cdot (1 + r_\text{d})^{-(i-1)} \\
\text{LoanPay}_i = \dfrac{\text{DebRate} \cdot \text{CapC} \cdot r \cdot (1+r)^{15}}{(1+r)^{15}-1}, i = 1 \sim 15 \\
\text{Profit}_i = \text{EquRate} \cdot \text{CapC} \cdot r_e, i = 1 \sim 20 \\
E = \text{WindCap} \cdot \text{CapacityFactor} \cdot 8760.
\end{cases} \tag{1}
$$

With assumption that capital cost (CapC) is a combination of 80% (DebRate) loan and 20% (EquRate) equity for each wind farm, LocanPay$_i$ is the payment of principal and interest to the loan during 15-year payment period while Profit$_i$ is the required profit for the internal equity. Operation and maintenance cost, COM$_i$, includes cost for regulatory operation fees, local taxes, and insurance. E represents the wind farm annual energy output, which is the product of nameplate capacity, local capacity factor, and 8760 h of 1 year. For a given discount rate r_d (7%), the expected return rate r_e (10%) and loan interest rate r (6.55%), LPC equals to the price to make the net present value (NPV) as zero, assuming a 20-year lifetime for wind project.

wind installations except the Tibet Plateau with harsh natural environments.

We used NASA's Shuttle Radar Topography Mission and land-cover categories of China to remove unavailable areas for turbine siting, such as urban regions, sloping fields with greater sloping factor than 10%, lakes and rivers, cropland, natural protection zones, and major industrial and transportation infrastructures [22]. Wind turbine density is another significant factor to determine the physical upper limit of wind power capacity installation, which ranges from 2 to 5 MW/km² in previous studies [12, 23]. As the other types of unavailable land like natural parks, wetland, and military basements are removed due to deficiency, the lower case of turbine density, 2 MW/km², is selected to calculate the potential wind capacity for each grid cell which equals the product of available area and wind density. Thus, the wind generation potential for each grid equals the product of capacity potential, local capacity factor, and 8760 h. The physical on-shore wind generation potential density for China is shown in Figure 5.

As operation and maintenance cost is relatively smaller from a system perspective, capital cost is the dominated factor to calculate LPC, including wind turbine, transmission, construction, and land use [27, 28]. China Wind Power Outlook [29] showed an obvious decrease in capital cost of wind turbine in from 6000 RMB/KW in 2008 to 4300 RMB/KW in 2014. The capital cost structure of wind power in 2013 and 2012 and related parameters is shown in Table 1.

Estimation of future capital cost

The decreased cost and changing structure of wind power reflects the economies of scale that could be estimated using the learning curve approach [31–33]. Learning curve describes the evolution of new technology resulting in costs reduction by assuming that the change in the costs (C) between two time segments "t_1" and "t_2" is a power function of the change in the capacities (Q) between the time segments, shown as follows:

$$\frac{C_{t_2}}{C_{t_1}} = \left(\frac{Q_{t_2}}{Q_{t_1}}\right)^{-b} \tag{2}$$

where C_{t1} and C_{t2} represent the costs of the examined new technology at the moments t_1 and t_2, respectively, and Q_{t1} and Q_{t2} represent the cumulative installed capacity of the technology at the moment t_1 and t_2. b is the exponent coefficient of the learning curve (usually defined as learning rate), meaning the decrease in new technology cost when the cumulative capacity of this new technology doubles. The cost of the technology at the moment t_2 can be calculated using the next equation.

$$\log(C_{t_2}) = \log(C_{t_1}) - b \cdot \log\left(\frac{Q_{t_2}}{Q_{t_1}}\right) \tag{3}$$

As Q_{t2}, Q_{t1}, and C_{t1} are obtained, the key issue for applying equation (3) to estimate the future cost of wind

systems is the learning rate (b), which is related to the historical variation trend in the cost and the cumulative installed capacity of wind turbines, and the cost reduction potentials of key components. Previous studies focusing on China's wind power have investigated different types of learning curve models, indicating that learning rate ranges from 4.1% to 11% [34–37]. Here, a medium learning rate of 7% is adopted in this paper, while the analysis of learning rate is not our focus. In order to simulate LPC in 2020, we have implemented learning curves analysis on wind power toward 2020. Assuming that the accumulative installation of wind capacity would grow with the same annual increase rate from 2014 to 2020, the capital cost decrease trend in China is simulated under a fixed learning rate in Figure 6, while the averaged capital cost for plain and hills in 2013 listed in Table 1 is set as the benchmark for learning-curve calculation. Despite the variation of capital cost among projects, we utilized projected capital cost of 7252 RMB/kW for overall wind farms in 2020 all over the country.

Wind power CO_2 mitigation cost

The CO_2 abatement cost of wind power means the additional cost of avoiding one unit of CO_2 emission by substituting wind power for baseline units, calculated in

$$\text{AbatementCost}_{r,i} = \frac{(\text{LPC}_{\text{Wind},r,i} - \text{LPC}_{\text{Coal},r}) \cdot \text{Eff}_r}{3.6 \cdot \text{EmFactor}_{\text{Coal}}} \tag{4}$$

where $\text{AbatementCost}_{r,i}$ means the CO_2 abatement cost of wind power in grid cell i of province r, while the gap of levelized cost of wind and coal power is calculated as ($\text{LPC}_{\text{wind},r,i} - \text{LPC}_{\text{Coal},r}$), Eff_r is the average generation efficiency of coal-fired power in each province, and $\text{EmFactor}_{\text{Coal}}$ equals to 0.096 kg/MJ which is the emission

Table 1. Averaged wind power capital cost structure in China from selected projections[30].

Unit (RMB/kW)	2013			2012		
	Plain	Hills	Shoals	Plain	Hills	Shoals
Equipment purchase	5072	5079	5072	5186	5194	5186
Installation	489	627	689	531	676	747
Construction	861	1384	1333	830	1298	1255
Rest	402	461	416	558	565	572
Basic reserve	171	189	188	142	155	155
Interest during construction	189	209	208	196	213	214
Total capital cost	7184	7949	7906	7443	8102	8129

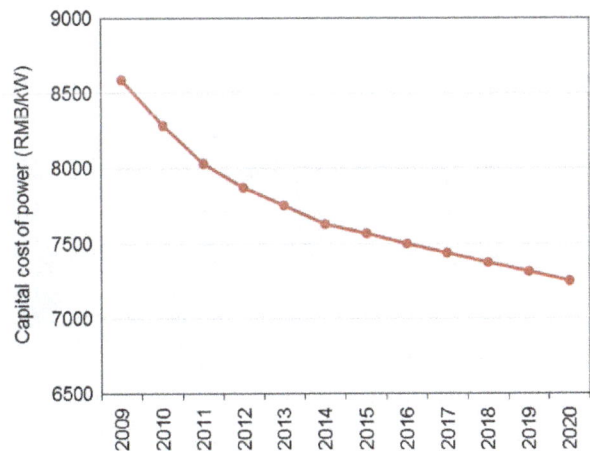

Figure 6. Capital cost reduction pathway toward 2020.

factor of coal from IPCC [38] and 3.6 in formulation (4) is the conversion coefficient which keeps energy unit consistent. The fundamental hypothesis here is that wind power utilization is assumed to replace coal-fired power perfectly in each province, while provincial coal power efficiency disparity is to distinguish the difference in technology of coal-fired power, coal categories, and coal quality in each province. As coal-fired power technology is relatively mature in China, we assume the energy efficiency for coal power will be consistent till 2020. The provincial coal-fired power generation efficiencies for provinces included in this study in 2012 are listed in Figure 7.

Supply curve of CO_2 emission mitigation

Based on the calculation of LPC of wind power for each grid cell in the 11 leading provinces, CO_2 abatement cost could be generated with formulation (4), assuming that wind capital cost remains constant for all provinces and the benchmarking on-grid price of coal fried power excluding taxes for each province is applied to represent the LPC of coal-fired power of each province. The strong projection for coal-power LPC will be affected by coal price, and there is no clear and believable prediction due to various factors. Therefore, emission abatement costs of each grid cell for all provinces are ranked in ascending order of abatement cost, indicating the least-cost order for wind deployment from the perspective of emission mitigation. The amount of carbon emission mitigation for each grid cell is proposed as the local mitigation potential divided by coal power generation efficiency, based on the grid cells of each province. With the assumption that provincial capacity targets would be perfectly finished, the provincial supply curve of carbon emission mitigated by wind power could be generated with aggregating all the grid cells from least to highest emission abatement cost with a certain emission mitigation potential.

Particularly, the average abatement cost of all the grid cells in each province by generation potential would be used to represent the provincial overall cost. Although our emphasis of wind resource evaluation is focusing on the technical available output of wind power in each grid cell, the grid connection issue of wind power to the existing electricity system in China is another vital factor that constrains the technical-feasible wind output from wind farms, which is estimated to be caused by incomplete electricity market, inflexible pricing mechanisms, and inaccurate wind prediction [40–43]. In the following section, we consider two scenarios with different curtailment rate and analyze the underlying influences caused by wind curtailment.

Results and Discussion

Cost curve of carbon emission abatement

Scenario with wind curtailment

Here we assume that the wind curtailment rate by 2020 is the same as that in 2012,[2] which is the share of wind production refused by electricity grid in the total technical-feasible production of wind plants, due to the reasons mentioned above. Each bar in Figure 8 shows the projected abatement cost and CO_2 mitigation potential for each province if the 2020 wind target is perfectly implemented, while the abatement cost is the additional cost between wind LPC and coal-power LPC. Generally, the average abatement costs of grid cells for 10 provinces are positive, which means that the generation cost of wind power is higher than the current electricity price of coal-fired power in most of the provinces. However, the abatement cost of Liaoning

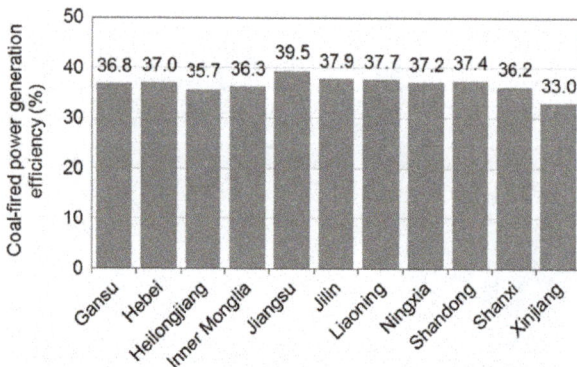

Figure 7. Coal-fired power generation efficiencies for selected provinces [39].

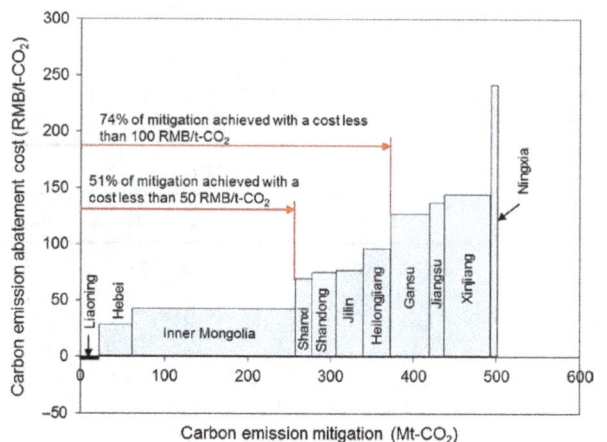

Figure 8. CO_2 mitigation potential and abatement cost for selected provinces.

Table 2. Averaged capacity factor, wind output, and abatement potential by province.

	Wind potential				Emission abatement		
Province	Curtailment rate (%)	Real avg. CF	Planned capacity (GW)	Expected output (TWh)	Coal power efficiency (%)	Abatement potential (Mt-CO_2)	Abatement cost (RMB/t)
Liaoning	6	0.344	8	24.09	37.7	22	−1
Hebei	12	0.304	16	42.66	37.0	40	19
Inner Mongolia	9	0.403	58	204.60	36.3	195	41
Shanxi	0	0.296	8	20.75	36.2	20	69
Shandong	1	0.251	15	33.03	37.4	31	74
Jilin	15	0.275	15	36.09	37.9	33	76
Heilongjiang	12	0.255	15	33.44	35.7	32	95
Gansu	11	0.286	20	50.08	36.8	47	126
Jiangsu	0	0.225	10	19.72	39.5	17	136
Xinjiang	15	0.311	20	54.56	33.0	57	142
Ningxia	0	0.240	4	8.42	37.2	8	240

province is negative, which means the generation cost of wind in 2020 could be cheaper than that of coal-power. We have identified an abatement potential of 500 Mt costing below 250 RMB/t of the CO_2 equivalent. The result indicates that approximately 370 million tons of CO_2 emission (almost 75% of total) could be avoided with an abatement cost lower than 100 RMB/t-CO_2 (16 USD/t-CO_2) in 2020. The overall average abatement cost for the wind installation provinces is 72 RMB/t-CO_2, which is much lower compared to that reported at 38 USD/t-CO_2 (230 RMB/t-CO_2) or 32 Euro/t (210 RMB/t-CO_2) in the existing literature [44, 45].

Table 2 illustrates the average capacity factor simulated by GIS analysis, expected wind output, as well as abatement potential linked with abatement cost for each simulated provinces with large-scale wind capacity. Liaoning, Hebei, and Inner Mongolia have the lowest abatement cost below 50 RMB/t-CO_2, and they are also the leading three provinces in terms of wind resource. Especially, all those make Inner Mongolia the province with the largest abatement potential and relatively low abatement cost. But the abatement costs have different order with capacity factors in the provinces, which indicates that wind resource is not the only factor determining abatement cost while the generation efficiency and generation cost of coal power should be taken into consideration among regions. Inner Mongolia, Xinjiang, and Gansu account for 60% of the total mitigation potential. The highest abatement cost lies in Ningxia province, due to its low capacity factor and low generation cost of coal-power. According to the results, approximately all the mitigation potential could be reached with abatement cost lower than 150 RMB/t. The notable disparity in abatement cost showed a more wider distribution than the capacity factor which varies from 0.24 to 0.40.

Scenario without wind curtailment

The wind curtailment rate of 2012 has been applied in the former calculation, which is based on the assumption that wind integration would still be limited in some provinces by 2020. However, enforcement of flexible generation capacities, market-based pricing system, and regional connection have been planned to improve the integration of large-scale wind power. With the most optimistic prediction, the marginal cost curve can be redraw when wind power is integrated perfectly.

We examined the abatement cost as well as the mitigation potential for no-curtailment assumption, shown in Figure 9. As all available wind generation is 100% absorbed in the electricity grid, the total abatement potential has increased by 10% to 550 Mt-CO_2 compared with the current wind curtailment scenario; meanwhile, the

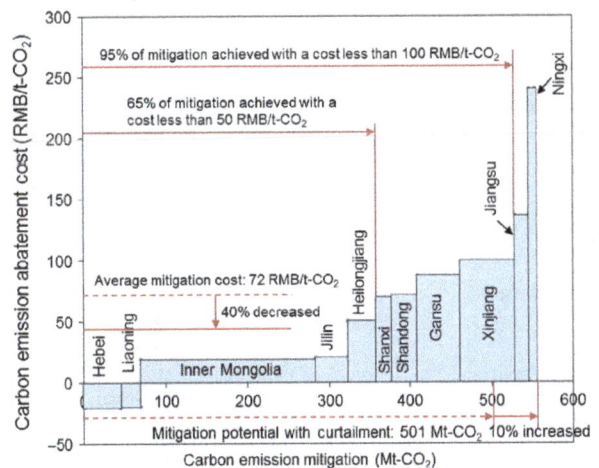

Figure 9. CO_2 mitigation potential and abatement cost without wind curtailments.

averaged abatement cost has decreased by almost 40% to 44 RMB/t-CO_2. The significant change in abatement potential and the related cost indicate the importance of wind integration on economic potential of wind power to replace coal-fired power. Hebei, rather than Liaoning, have become the lowest abatement province. And both provinces have negative abatement cost, suggesting that the expected wind generation is cheaper than coal-fired power. According to the simulation results, approximately 95% of the total emission mitigation potential could be reached with an abatement cost lower than 100 RMB/t. Inner Mongolia, Xinjiang, and Gansu still occupy the leading three places in mitigation potential ranking with 214, 67 and 52 million tons of CO_2. The highest abatement cost reaches the same as the scenario with wind-curtailment scenario in Ningxia province. Compared with the former scenario, here the abatement cost for each province has a similar pattern, but with a slightly different rank, as provinces with severe integration issue such as Hebei, Jilin and Xinjiang would benefit the most from integration improvement.

The provincial differences in abatement potential and the corresponding abatement cost showed that detailed analysis on wind generation cost, integration situation, as well as coal power efficiency and generation cost at the regional level is essential to decarbonize China's power sector by wind power deployment. It is estimated that wind power is a very price-competitive technology to decrease carbon emission; especially, the abatement cost could be negative for some provinces in the near future. However, from a national perspective, financial subsidy or other policy instruments are still needed to achieve the national goal by 2020. The annual wind output could reach 520–580 TWh, representing nearly 500–550 Mt CO_2 emission mitigation. The main abatement potential concentrated in Three-North area in China including North China, Northeast and Northwest, which is the home of

rich wind resource and low-efficiency coal power capacities. In addition, the replacement of existing coal power by wind power could be of mutual benefit to both pollution improvement and the coal-dominated power sector in these provinces.

Potential contribution of wind power development

We compared the provincial potential of abatement in Table 3 with the projected total emission for each province of 2020 and showed the potential share of wind power-contributed abatement in total emission at the provincial level. We use projection for provincial emission of 2020 as the benchmark [46] to predict the potential contribution to provincial emission reduction from the overall system perspective. The total emission for each province is related to many uncertain factors, such as economic development, industry structure, and generation mix. Thus, the share of mitigation contributed by wind power in total emission is the key indicator to show the role of wind power in low-carbon transition. Seen from Table 3, the share of emission mitigation potential contributed by wind in total emission at the provincial level varies at great scale, from 1.5% to approximately 65%, which illustrates the role that wind power could play in decarbonizing the power sector by 2020. In the scenario with wind curtailment, the CO_2 reduction by wind could account for nearly 65% of Inner Mongolia's total CO_2 emitted in 2020. However, for coastal provinces like Jiangsu and Shandong, the potential contribution of on-shore wind power is very limited if they just implement the national planning. Besides, that also makes requests for transmission from wind resources inside other provinces, if wind power is needed to support stringent target of emission reduction. For Inner Mongolia, Xinjiang, and Gansu, wind power could play an important role in

Table 3. Projected share of mitigation potential contributed by wind in provinces in 2020.

Province	Expected emission (Mt)	Abatement potential with curtailment (Mt)	Share (%)	Abatement potential without curtailment (Mt)	Share (%)
Hebei	721.7	39.8	5.5	45.3	6.3
Shanxi	322.5	19.8	6.2	19.8	6.2
Inner Mongolia	302.2	195.0	64.5	214.3	70.9
Liaoning	549.2	22.1	4.0	23.5	4.3
Jilin	284.3	32.9	11.6	38.7	13.6
Heilongjiang	396.0	32.3	8.2	36.7	9.3
Jiangsu	1173.1	17.2	1.5	17.2	1.5
Shandong	1221.4	30.6	2.5	30.9	2.5
Gansu	151.0	47.0	31.1	52.8	35.0
Ningxia	52.3	7.8	14.9	7.8	14.9
Xinjiang	196.2	57.1	29.1	67.1	34.2

reducing local emission, while the high penetration brings great uncertainties on the electricity grid. Therefore, transmission lines to export wind power produces in these provinces, or flexible generation adjusting peak-load like natural gas capacities are strongly needed to balance energy consumption and production, as well as to guarantee the grid security in highly renewable energy penetration situations.

Conclusions

Since the *Renewable Energy Law* came into force in 2006, wind power in China has seen a remarkable growth, reaching 94 GW of installed capacity by the end of 2014. Existing studies have been focusing on wind resource estimation while the abatement potential together with the corresponding abatement cost are disregarded. Due to the lack of marginal abatement cost for each province, former research does not provide the necessary information to estimate the potential contribution and the economic cost of wind power deployment to reduce carbon emission. China's ambitious target has set detailed planning installed capacity for several "wind-installation" provinces by 2020. From the perspective of mitigating carbon emission, knowing the cost curve of wind power is essential to reduce carbon emission at the provincial level, which could help the policy makers on wind development to design policy instruments for wind power promotion.

By combining GIS analysis and LPC calculation, we provide provincial-level supply curve of CO_2 mitigation by wind power in provinces with the mentioned planning capacity target in the 12th Five-year plan. A total of 3783 grid cells are used to represent wind resource spatial disparity for China's on-shore wind, with averaged wind speed data at the hourly level. The available area for wind farms as well as averaged wind capacity factors are then built for each grid cell. Assuming that provincial capacity targets should be perfectly implemented by 2020, the provincial wind capacity factor among grid cells is generated. When learning effect and wind curtailment issues are both considered, provincial abatement cost and abatement potential are estimated by MACCs. Our results indicate significant differences in abatement cost and abatement potential among "wind installations" provinces. The key finding could be used to facilitate or evaluate wind planning by policy makers and academic researchers at both the provincial and national levels.

By 2020, wind power could be a very competitive mitigation option in considering of the current coal-dominated power sector in China. Wind power has the potential to contribute 500 Mt-CO_2 emission reduction, while the overall abatement cost is lower than 75 RMB/t-CO_2. Inner Mongolia, Gansu, and Xinjiang occupied the top three position in

terms of abatement potential. The different order of wind resource and abatement cost have indicated that the local power sector as well as wind curtailment should be regarded as key factors for wind investment in the near future. In addition, wind integration issue is another crucial factor to change the marginal curve order, which should require detailed analysis to actually fulfill carbon emission reduction via wind power installation. Furthermore, the share of abatement potential that could be achieved with 2020 wind target indicated that import/export of wind power could also have an important impact on the achievement of wind power.

In summary, this study has extended existing resources by a detailed investigation into the abatement potential and abatement cost in China at high spatial resolution. Economic access of wind power should be carefully considered rather than following the ambitious national planning. However, in this study, we assumed that wind power generation has caused the replacement of coal-fired power but wind power deployment and utilization are much more complex to perfectly simulate. We would like to implement power sector modeling which captures the rest of the power sector, for future studies of wind power and its linked carbon emission reduction.

Acknowledgments

This study was supported by the National Natural Science Foundation of China (Nos. 71103111, 713111048, 71573152) and Ministry of Science and Technology (2012IM010300). The authors would like to acknowledge the Life Academy, Sweden, who provided an opportunity for participating in training on wind power systems. The Chinese Wind Energy Association provided useful data on wind power resource and operation. The analyses in this article are elaborated as part of the 4DH (4th Generation District Heating Technologies and Systems) project supported by The Danish Council for Strategic Research. Finally, the authors wish to give special thanks to Michael Davidson from MIT Joint Program on the Science and Policy of Global Change for kindly providing the GIS database.

Conflict of Interest

None declared.

Notes

[1] Although Jiangsu is not in the list of top 10 provinces with the biggest wind capacity, it is still selected in our analysis as Jiangsu is one of the major wind installations in the *12th Five-year* Renewable Energy Development Plan.

2 Wind curtailment rate is collected from the national renewable energy information management center, which is available from: http://news.bjx.com.cn/html/20140904/543677.shtml.

References

1. Lin, B. Q., and X. L. Ouyang. 2014. Energy demand in China: comparison of characteristics between the US and China in rapid urbanization stage. Energy Convers. Manage. 79:128–139.
2. China Electricity Council. 2015. China's electricity basic statistical data in 2014. Available at http://www.cec.org.cn/guihuayutongji/tongjxinxi/niandushuju/2015-11-30/146012.html (accessed 22 June 2016).
3. Dai, H., X. Xie, Y. Xie, J. Liu, and T. Masui. 2016. Green growth: the economic impacts of large-scale renewable energy development in China. Appl. Energy 162:435–449.
4. National Energy Administration, China. 2016. China's wind power industry development situation in 2015. Available at http://www.nea.gov.cn/2016-02/02/c_135066586.htm (accessed 22 Jun 2016).
5. National Energy Administration, China. 2012. 12th Five-Year Plan for Wind Power Development. Available at http://news.bjx.com.cn/html/20120914/388348.shtml (accessed 20 November 2015).
6. Chinese Wind Energy Association. 2014. Chinese wind power installation capacity statistic. Available at http://www.cwea.org.cn/upload/2014%E5%B9%B4%E4%B8%AD%E5%9B%BD%E9%A3%8E%E7%94%B5%E8%A3%85%E6%9C%BA%E5%AE%B9%E9%87%8F%E7%BB%9F%E8%AE%A1.pdf (accessed 20 November 2015).
7. National Development and Reform Commission, China. 2014. Notice of the national development and Reform Commission on the adjustment of onshore wind power electricity price benchmark. Available at http://www.sdpc.gov.cn/zcfb/zcfbtz/201501/t20150109_659876.html (accessed 1 June 2015).
8. Yu, D., J. Liang, X. Han, and J. Zhao. 2011. Profiling the regional wind power fluctuation in China. Energy Pol. 39:299–306.
9. Zhang, D., M. Davidson, B. Gunturu, X. Zhang, and V. J. Karplus. An integrated assessment of China's wind energy, in MIT JPSPGC Report 2612014. Available at http://hdl.handle.net/1721.1/88605 (accessed 18 November 2015).
10. Shu, Z. R., Q. S. Li, and P. W. Chan. 2015. Investigation of offshore wind energy potential in Hong Kong based on Weibull distribution function. Appl. Energy 156:362–373.
11. Brouwer, A. S., van den Broek, M., Zappa, W., Turkenburg, W. C., and Faaij, A. 2016. Least-cost options for integrating intermittent renewables in low-carbon power systems. Appl. Energy 161:48–74.
12. Energy Research Institute. China Wind Energy Development Roadmap 2050. Available at http://www.cnrec.org.cn/yjcg/fn/2012-02-10-54.html. (accessed 1 June 2015).
13. Energy Research Institute, China. 2010. China wind power development towards 2030 – feasibility study on wind power contribution to 10% of power demand in China. Energy Foundation, Beijing.
14. He, G., and D. M. Kammen. 2014. Where, when and how much wind is available? A provincial-scale wind resource assessment for China. Energy Pol. 74:116–122.
15. China Meteorological Administration. 2010. China wind power resources evaluation (2009). China Meteorological Press, Beijing.
16. Hong, L. X., and B. Moller. 2011. Offshore wind energy potential in China: under technical, spatial and economic constraints. Energy 36:4482–4491.
17. McElroy, M. B., X. Lu, C. P. Nielsen, and Y. Wang. 2009. Potential for wind-generated electricity in China. Science 325:1378–1380.
18. Chinese Wind Energy Association. Wind capacity statistic report 2014. Available at http://www.cwea.org.cn/upload/2014%E5%B9%B4%E4%B8%AD%E5%9B%BD%E9%A3%8E%E7%94%B5%E8%A3%85%E6%9C%BA%E5%AE%B9%E9%87%8F%E7%BB%9F%E8%AE%A1.pdf (accessed 1 June 2015).
19. National Energy Administration. 2012. China 12th five-year plan for renewable energy development. Available at http://www.fjdpc.gov.cn/Upload/File/2013/2013122794149729.pdf (accessed 1 June 2015).
20. Rienecker, M. M., M. J. Suarez, R. Gelaro, R. Todling, J. Bacmeister, and E. Liu, et al. 2011. MERRA: NASA's modern-era retrospective analysis for research and applications. J. Clim. 24:3624–3648.
21. Sinovel. Technical Instruction Q/GW 2CP1500.9-2011. Available at http://wenku.baidu.com/view/bc570b0452ea551810a68756.html?from=search (accessed 1 June 2015).
22. Liu, J. Y., D. F. Zhuang, D. Luo, and X. M. Xiao. 2003. Land-cover classification of China: integrated analysis of AVHRR imagery and geophysical data. Int. J. Remote Sens. 24:2485–2500.
23. Wang, Z. Y., D. M. Ren, and G. Hu. 2012. China non-fossil energy road. Economic Press China, Beijing.
24. Walraven, D., B. Laenen, and W. D'haeseleer. 2015. Minimizing the levelized cost of electricity production from low-temperature geothermal heat sources with ORCs: water or air cooled? Appl. Energy 142:144–153.
25. Drechsler, M., C. Ohl, J. Meyerhoff, M. Eichhorn, and J. Monsees. 2011. Combining spatial modeling and choice experiments for the optimal spatial allocation of wind turbines. Energy Pol. 39:3845–3854.
26. Mari, C. 2014. Hedging electricity price volatility using nuclear power. Appl. Energy 113:615–621.

27. Jaramillo, O. A., R. Saldana, and U. Miranda. 2004. Wind power potential of Baja California Sur, Mexico. Renewable Energy 29:2087–2100.

28. Sliz-Szkliniarz, B., and J. Vogt. 2011. GIS-based approach for the evaluation of wind energy potential: a case study for the Kujawsko-Pomorskie Voivodeship. Renew. Sustain. Energy Rev. 15:1696–1707.

29. Chinese Renewable Energy Industries Association. 2014 China wind power review and outlook. Available at www.gwec.net/2014_China_Wind_Power_Review_and_Outlook (accessed 1 June 2015).

30. China National Renewable Energy Centre. 2014. Beijing. Renewable energy data manual 2014.

31. Li, S., Zhang, X., Gao, L., and Jin, H. 2012. Learning rates and future cost curves for fossil fuel energy systems with CO_2 capture: methodology and case studies. Appl. Energy 93:348–356.

32. Rochedo, P. R. R., and A. Szklo. 2013. Designing learning curves for carbon capture based on chemical absorption according to the minimum work of separation. Appl. Energy 108:383–391.

33. Neij, L. 2008. Cost development of future technologies for power generation – a study based on experience curves and complementary bottom-up assessments. Energy Pol. 36:2200–2211.

34. Qiu, Y., and L. D. Anadon. 2012. The price of wind power in China during its expansion: technology adoption, learning-by-doing, economies of scale, and manufacturing localization. Energ. Econ. 34:772–785.

35. Xiang Yin, W.C.. 2012. Cost of carbon capture and storage and renewable energy generation based on the learning curve method. J. Tsinghua Univ. Sci. Technol. 52:243–248.

36. Zhu, Y.-C., et al. 2012. Analysis of wind power cost based on learning curve. Dianli Xuqiuce Guanli (Power Demand Side Manage.) 14:11–13.

37. Yao, X., Y. Liu, and S. Qu. 2015. When will wind energy achieve grid parity in China? – Connecting technological learning and climate finance. Appl. Energy 160:697–704.

38. IPCC. 2006. 2006 IPCC guidelines for national greenhouse gas inventories. Available at http://www.ipcc-nggip.iges.or.jp/public/2006gl/index.html (accessed at 1 June 2015).

39. China Electricity Council. 2012. Statistic report on power sector 2012.

40. Zhao, X., Zhang, S., Yang, R., and Wang, M. 2012. Constraints on the effective utilization of wind power in China: an illustration from the northeast China grid. Renew. Sustain. Energy Rev. 16:4508–4514.

41. Zhao, X. L., F. Wang, and M. Wang. 2012. Large-scale utilization of wind power in China: obstacles of conflict between market and planning. Energy Pol. 48:222–232.

42. Vilim, M., and A. Botterud. 2014. Wind power bidding in electricity markets with high wind penetration. Appl. Energ. 118:141–155.

43. Soman, S. S., Zareipour, H., Malik, O., and Mandal, P. 2010. A review of wind power and wind speed forecasting methods with different time horizons.North American Power Symposium (NAPS), 2010. IEEE.

44. Cai, W., Wang, C., Wang, K., Zhang, Y., and Chen, J. 2007. Scenario analysis on CO 2 emissions reduction potential in China's electricity sector. Energ. Pol. 35:6445–6456.

45. Mckinsey & Company. 2009. China's green revolution: prioritizing technologies to achieve energy and environmental sustainability. Available at http://www.mckinsey.com/~/media/mckinsey/dotcom/client_service/sustainability/cost%20curve%20pdfs/china_green_revolution.ashx (accessed at 1 June 2015).

46. Wang, K., Zhang, X., Wei, Y. M., and Yu, S. 2013. Regional allocation of CO_2 emissions allowance over provinces in China by 2020. Energ. Pol. 54:214–229.

11

Study of a new hydraulic pumping unit based on the offshore platform

Yanqun Yu[1,2], Zongyu Chang[1], Yaoguang Qi[2], Xin Xue[2] & Jiannan Zhao[3]

[1]College of Engineering, Ocean University of China, Qingdao, Shandong Province 266100, China
[2]College of Mechanical and Electronic Engineering, China University of Petroleum, Qingdao, Shandong Province 266580, China
[3]School of Petroleum Engineering, China University of Petroleum, Qingdao, Shandong Province 266580, China

Keywords
Characteristic of polished rod, composite hydraulic cylinder, hydraulic pumping unit, model prototype, offshore platform, oil recovery with rod pumping

Correspondence
Yanqun Yu, 66 Changjiang West Road, Qingdao Economic Development Zone, Qingdao Shandong 266580, China.
E-mail: yuyq_hdpu@126.com

Funding Information
This study was supported by the National Natural Science Foundation of China (51174224), the National Science and Technology major projects of oil and gas (2016ZX05017004, 2016ZX05042003-001), and the Natural Science Fund Project of Shandong Province (ZR2014El015).

Abstract

This article introduces a new technology about a rod pumping in the offshore platform according to the demand of offshore heavy oil thermal recovery and the production of stripper well, analyzes the research status of hydraulic pumping unit at home and abroad, and designs a new kind of miniature hydraulic pumping unit with long-stroke, low pumping speed and compact structure to resolve the problem of space limitation. The article also describes the whole structure and the working principle of this pumping unit, determines the choice of stroke and rate of the pumping unit, and establishes mathematical models based on the polished rod loads. A new composite hydraulic cylinder with a special structure was designed by combining the hydraulic cylinder with the energy accumulator. This composite hydraulic cylinder is applied on land, and the model prototype runs smoothly, which indicates that the whole structure design of the pumping unit is reasonable and the control strategy is correct.

Introduction

Energy shortage is a major issue in the world at present, with the continuous exploitation of conventional crude oil, whose reserves and production gradually decline each year. Available light oil resources for mining leave only 1700×10^8 t, while the proven heavy crude oil resources throughout the world are more than 3000×10^8 t [1, 2]. Thanks to the continuous progress of Chinese crude oil production technology; production of CNOOC, Sinopec, and PetroChina offshore oil fields has been improved greatly in recent years. In 2013, annual crude oil production of CNOOC was 66.84 million tons and natural gas was 19.6 billion cubic meters; CNOOC has become the second largest crude oil production enterprise in China. The main producing area of domestic offshore oilfields diverted gradually from the northern South China sea to Bohai Bay oilfield during the last century. At the beginning of this century, Bohai Bay oil production rose rapidly, for example, offshore oil production of Shengli Oilfield Company, Sinopec, reached 5 million tons, and Tianjin Branch of CNOOC accounted for 68.5% of production of CNOOC offshore oil production and became the main oil field of CNOOC. The geological conditions of the Bohai Bay oil region have the following characteristics: marginal, small, and medium size oilfields that account for a relatively high proportion. Large (more than 100 million tons of geological reserves) oilfields mainly consist of heavy crude oil, whose reserves account for 85% of total proven oil reserves. High-quality blocks gradually go into the late stage of development, which brings tremendous difficulty to offshore crude oil production in

Bohai Bay oil region. Therefore, studying the development of heavy crude oil occupy a pivotal position in reserves discovery and application, capacity construction, and oilfield development of the Bohai Bay.

Currently, conventional offshore oilfields lifting technology mainly includes spray oil production and mechanical artificial lift oil production. With the gradually deep development of the crude oil exploitation, the formation of the original pressure falling, artificial lift oil production has gradually become a major offshore oil recovery scheme. Affected by the offshore platform size limit and drainage technology, offshore oilfields use submersible pumps as the main lifting equipment [3, 4]. However, problems emerged as a result of using submersible pumps to develop heavy crude oil and marginal oil fields: the electric submersible pump restricts the process and temperature of heavy crude oil thermal recovery options. For more stable production of offshore oil wells, their fluid production has fallen below the reasonable range of electric submersible pump operation, and the efficiency of development and cost dropped substantially, increasing the operating and development costs.

Therefore, it has become an urgent need to study new offshore oilfield artificial lift way to meet the need of efficient development of heavy crude oil thermal and low-field wells. An artificial lift pumping mode consisting of pumping machine–rod–pumps is the most widely used among the oil exploration methods in the world. According to statistics, 80% of onshore oil wells using this kind of the lift mode produced more than 75% of the total crude oil [5, 6]. So we can use three mature artificial lift pumping modes onshore on the offshore oil platform, which can solve many problems in offshore field exploitation. Among them, three pumping lifting equipment R&D is the premise and key.

Overall Structural Design

The pumping unit is the ground driver device of three pumping artificial lift systems. The main type of the current pumping unit has a beam pumping unit and a non-beam pumping unit. The beam pumping unit includes a conventional beam pumping unit, a front-mounted beam pumping unit, a variable parameters mechanism beam pumping unit, and a multiple rod mechanism beam pumping unit. Although the conventional beam pumping unit is a mature technology in using and manufacturing experience and the main rod pumping equipment, it has larger overall weight, especially the size and space required exceeding the limit of the offshore oil platform. The non-beam pumping unit includes a rotary motor-reversing drum pumping unit, a linear motor-reversing drum pump-

ing unit, a mechanical reversing drum pumping unit, a chain reversing pumping unit, a gear and rack reversing pumping unit, spiral reversing pumping unit, and other reversing pumping units [7]. The pumping unit with a long stroke and a low speed of nonbeam pumping unit mainly has a chain pumping unit, a motor-reversing type pumping unit, a gear and rack pumping unit, and a hydraulic pumping unit, etc. A chain pumping unit has a big occupied area, the motor-reversing type pumping unit uses the balanced weight box and has big weight, the gear and rack pumping unit has a complex structure and high weight, and all of them cannot be used on the off-shore platform.

The hydraulic pumping unit can adapt well to change in the well, with a compact construction and light weight, and it transfers energy intensively and adapts to a wide range of operating conditions to make a long and low stroke, being adjustable in times and frequency of strokes [8].

Currently, the hydraulic pumping unit can be divided into two kinds of hydraulic pumping units with a single counterweight and hydraulic pumping unit with no single counterweight through the different way of balance [9]. The hydraulic pumping unit with no single counterweight including a hydraulic pumping unit adopted the tubing string to balance, hydraulic pumping unit adopted the accumulator to balance, and hydraulic pumping unit adopted the tubing string and the accumulator to balance.

The MaPe-mode drum-type long-stroke hydraulic pumping unit produced by France MaPe companies is a kind of nonbeam hydraulic pumping unit with long stroke [10]. The application of a high-raised frame structure enables this kind of pumping units to fulfill the task of minor workover taking the place of service machines. The stable reliable reversing lowers the load of the polished rod, and the application of vertical construction ensures a less occupied area, which makes it more suitable for cluster wells and the recovery of heavy oil in marine areas. But counterbalance weight features a bigger total weight of this kind of pumping unit, which makes it hard for the offshore platform.

The hydraulic pumping unit made by Western Gear Corp (Harrison, Ohio, United States). America has the following three operation modes: manual operation, inching operation, and automatic operation [11]. The application of monolithic low-rise structure makes sure of a smaller overall dimension and convenient transportation. Low additional load enables the pumping unit to run smoothly, which makes for a longer service life. This kind of pumping unit can reduce the friction between the pumping rod and the tubing, as well as the abrasion of the pump. But during the workover operation, the pumping unit must be moved away from the wellhead and repositioned after completing the workover operation, which increases the difficulty and the workload.

The ATH-mode (Awang Tengah Hasip) hydraulic pumping unit was developed by the Soviet Union; by making tubing as counterweight, it avoids using additional balancing weight or other balancing devices and increases the actual stroke length of the pumping unit, which remarkably improves the effectiveness of the rod pumping [12]. During an operation cycle, the suspension point has a longer steady motion segment, which makes for the operation of the rod string and the downhole pump. But the pumping unit is right installed at the wellhead, causing a higher chance of accidents for the wellhead which can be broken easily in this way. Hence, the situation above should be avoided in the design of a pumping unit if possible.

There are two types of hydraulic pumping units made by Company Curtis Hoover, (Fort St John, British Columbia, Canada): high-raised frame and low-raised frame. The application of bladder accumulators in the balancing system and the use of the solid-state circuit as well as the electromagnetic limit sensor eliminate the wear on moving parts. The pumping unit has the advantages of simple device, small size, light weight, convenient installation, and low manufacturing cost. The frame of this pumping unit is also installed at the wellhead, which not only brings about inconvenience during the workover operation, but also easily causes breakage of the wellhead. At this point, it is not advisable to adopt this design.

The offshore platform hydraulic pumping unit jointly developed by Zhejiang University and Shengli Oilfield Company is suitable for the offshore platform [13]. It adopts a fully hydraulic reversing control loop and accumulator-type closed circuit. With a compact structure and a less occupied area, it is quite suitable for the platform whose area is greatly limited. But this kind of pumping unit has a complex hydraulic circuit. Besides, high manufacturing cost also makes it difficult to be applied widely.

The combined hydraulic cylinder energy-saving hydraulic pumping unit was jointly developed by the Yantai University and the Shandong Kangda Oil Pump Co., Ltd. It adopts a special structure with a combined hydraulic cylinder and accumulator which recover the gravitational potential energy released during the downward movement of the sucker rod and reuse it during the upward movement, contributing a remarkable energy-conservation effect [14]. However, the force analysis indicates asymmetric loading and eccentric wear of the hydraulic cylinder and the piston.

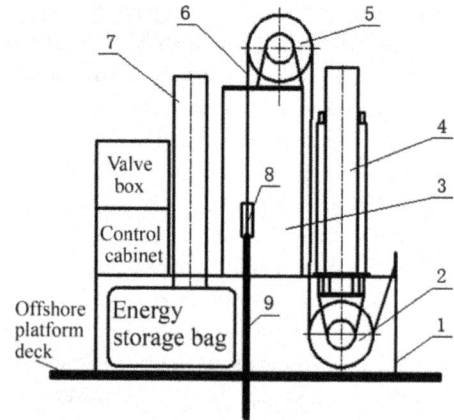

Figure 1. Hydraulic pumping offshore platform structure schematic diagram. 1, Base; 2, driven pulley; 3, towers; 4, cylinder combination; 5, crane wheel; 6, lifting strip; 7, energy storage cylinder; 8, hanging rope; 9, polished rod.

Due to the limited space of wellhead and workover needs, a new R&D hydraulic pumping unit was applied to the offshore oil platform as shown in Figure 1, which mainly consists of a pumping base, a combination of hydraulic cylinders, a pulley group, a energy storage tank group, a hydraulic control system, and power systems. The size of the whole machine was 1.0 m × 0.7 m × 4.0 m.

The pumping unit is mounted above the middle platform of the offshore oil platform and the combination cylinder rod extends during upstroke, lifting the rod by strengthening the connection between the lifting belt and the polished rod via the hanging rope. The rod drops during downstroke and presses the hydraulic oil in the combination cylinder fluid chamber back to energy storage cylinder when the lifting belt returns pressure to the piston rod, which realizes the gravitational potential energy storage of the rod during downstroke. The energy that the energy storage device stored in downstroke releases in upstroke, which achieves the load balancing between upstroke and downstroke and cylinder protection. The detailed operating principle is as shown in Figure 2.

The hydraulic structural schematic diagram of the pumping unit is shown in Figure 2. During the upstroke operation, the electrohydraulic proportional valve (3) is in the lower position. The hydraulic oil flows through the hydraulic lock (4), gets pressed into the center bore of the combined cylinder (5), balances the load of the upstroke with the help of the hydraulic oil of the balance hydraulic cylinder (6) and the accumulator (7) collectively, and pushes the load end to move upward. Then, the hydraulic oil of the side bore flows back into the oil tank through the hydraulic lock (4), indicating the end of the upstroke operation. During the downstroke operation, the electrohydraulic proportional valve (3) is in the upper position, the hydraulic oil is pressed into the side bore of the combined cylinder through the hydraulic lock, pushing the load end to move upward with the help of the load of the downstroke. The hydraulic oil in the mandrel of the combined cylinder is pressed into the energy storage device.

Design Parameters

Pumping stroke and stroke rate

Daily pump displacement is:

$$V = 24 \times 60 \times \frac{1}{4} \pi d^2 S \cdot N \cdot f_p. \quad (1)$$

In this formula, d is the pump diameter (m^2), S the pumping stroke (m), N the pumping stroke times (min^{-1}), and f_p the pump efficiency (0.8).

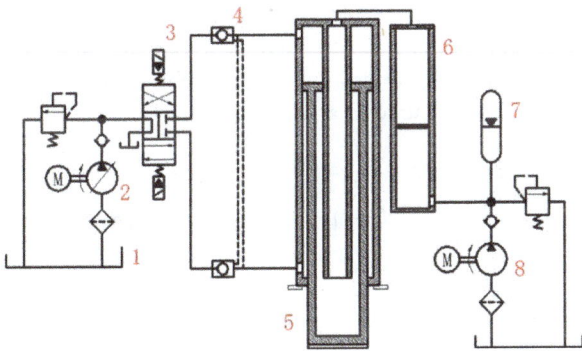

Figure 2. Hydraulic structural schematic diagram. 1, Oil tank; 2, variable pump; 3, electrohydraulic proportional valve; 4, hydraulic lock; 5, combined cylinder; 6, balance hydraulic cylinder; 7, accumulator; 8, slippage pump.

Single well production fluid volume, pump diameter, and pumping stroke times are shown in Figure 3.

Heavy crude oil production generally uses a principle of a large pump diameter, long stroke, and low stroke times. Based on Figure 3, we can determine: Single well liquid production is less than 100 m^3/day, configure "70 pump + 5 meters pump stroke pump unit." Single well liquid production is between 100 and 150 m^3/day, configure "80 pump +6 meters pump stroke pump unit." If single well liquid production is 150 m^3/day, the hydraulic pump unit stroke is 6 m, the stroke times is 4.32 min^{-1}, design and calculation according to the 5 min^{-1}.

Characteristics of suspended point

Rod load calculation

Based on the polished rod force balance, the mechanical model for rod during the upstroke is as follows:

$$F_u + F_f + F_y = G_g + G_o + F_p + F_t + F_g + F_o. \quad (2)$$

In this formula, F_u is the rod load during upstroke (N), F_f the rod buoyancy (N), F_y the force of the lower surface of the pump plunger (N), G_g the total gravity load of sucker rod (N), G_o the total gravity load of the liquid column (N), F_p the friction load between the cylinder and the plunger (N), F_t the friction load between liquid column and tubing (N), F_g the inertial load of sucker rod string (N), and F_o the inertial load of liquid column (N).

Among them,

$$F_y = \frac{1}{4} \pi d^2 (P_h + \rho_o g h). \quad (3)$$

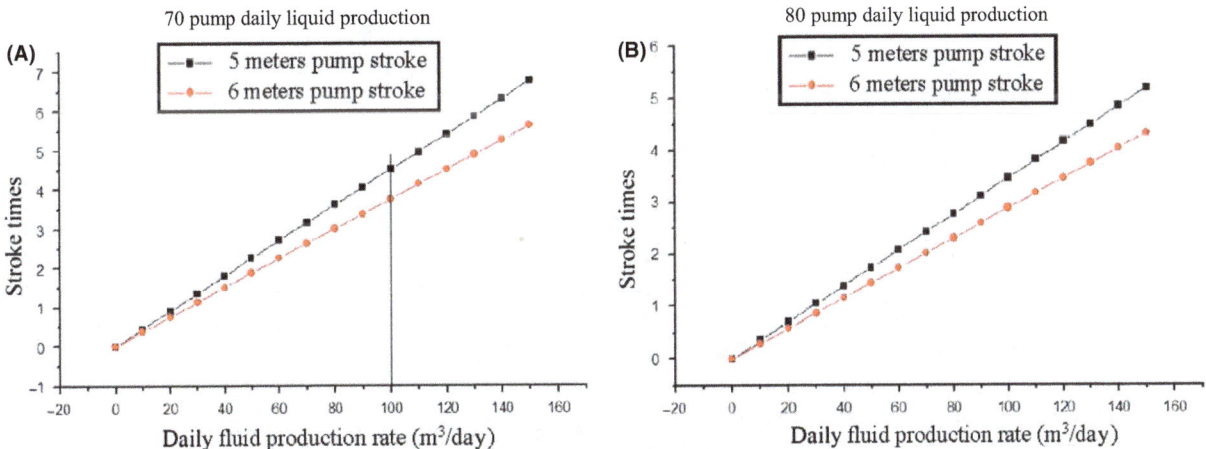

Figure 3. Relationship between single well liquid production and pumping parameter. (A) 70 pump daily liquid production. (B) 80 pump daily liquid production.

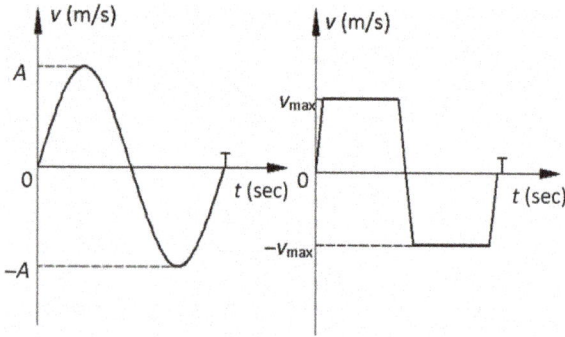

Figure 4. Sinusoidal and trapezoidal profiles of suspension center.

In this formula, P_h is the wellhead back pressure (Pa), h the pump submergence depth (m), d the pump piston diameter (m), and ρ_o the density of heavy crude oil (density of heavy crude oil in Bohai Bay region [15] is 0.94–0.98 × 10^3 kg/m³).

Therefore, the load of suspension center during the upstroke is:

$$F_\mu = G_g + G_o + F_p + F_t + F_g + F_o - F_f - F_y. \tag{4}$$

And the load of downstroke is:

$$F_d = G_g + F_g - F_f. \tag{5}$$

Velocity curve design of suspension center

The velocity curve of pump suspension center can be designed for the sinusoidal or trapezoidal profile, wherein the trapezoidal profile is divided into three stages: acceleration, uniform speed, and deceleration; the acceleration and deceleration phases account for 1/10 of the stroke and the uniform phase accounts for 8/10 [16, 17].

In Figure 4, we assume the equation of motion for sine harmonic curve as follow:

$$V = A \sin(\theta t) \tag{6}$$

We already know that the pump stroke times is 5 min^{-1} and the stroke is 6 m. Hence, we calculate that in Formula (6) $A = 1.57$, $\theta = \varpi/6$; thus, the maximum speed of suspension point in the sinusoidal velocity curve is $V_{max} = 1.57$ m/sec.

For trapezoidal velocity profile, $V_{max} = 1.11$ m/sec, which is equal to 70% the maximum in sinusoidal velocity curve. To decrease the flow rate of the hydraulic system, reduce the installed power, and the suspension center velocity should be designed according to the trapezoid curve mode. However, the acceleration curve of the trapezoidal velocity profile is not continuous, resulting in generating periodic shock loads when deflected, causing pumping vibration, and shortening the operating life of

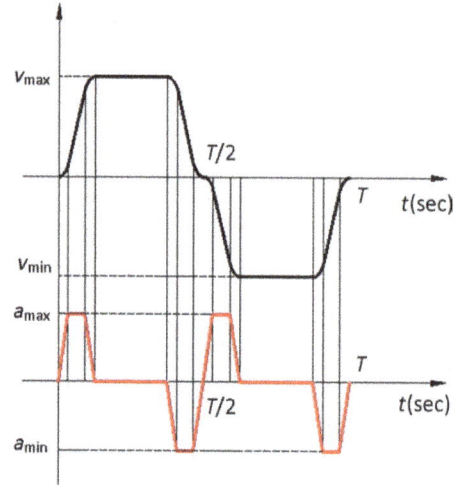

Figure 5. Velocity curve of the pump suspension center.

the pump unit. Combined with advantages of the sinusoidal and trapezoidal curve, the design of the pump suspension center velocity curve is shown in Figure 5.

The analytical expression of Figure 5 for the suspension center velocity curve is as follows:

$$\begin{cases} v = 2.12 \left(1 - \cos\left(\dfrac{\pi}{3}t\right)\right), t \in \text{rest} \\ v = 1.26, t \in \left(\begin{array}{c} 12n + 1.5 \le t \le 12n + 4.5 \\ 12n + 7.5 \le t \le 12n + 10.5 \end{array}\right) \end{cases} \tag{7}$$

n is a natural number in the formula.

Design of the main system

As the combined hydraulic cylinder is the predominant actuator of this pump, its structure and performance determine the control strategies and the operating mode of the system. The structure of the combined hydraulic cylinder is shown in Figure 6.

The outer diameter of the combined hydraulic cylinder is Φ_1, the inner diameter is Φ_2, the outer diameter of the mandrel II is D_1, the inner diameter is D_2, the outer diameter of the piston rod III is d_1, and the inner diameter is d_2.

When the depth of the heavy crude oil is 1000 m, the pump submergence depth is 300 m, the wellhead back pressure is 0 MPa, the maximum rod load of upstroke is 6.0 × 10^4 N, and the minimum rod load of downstroke is 2.0 × 10^4 N.

Piston rod size

The equivalent force of the load act on the piston rod is $F_h = 1.2 \times 10^5$ N, rod stroke is $L = 3$ m, so it is a

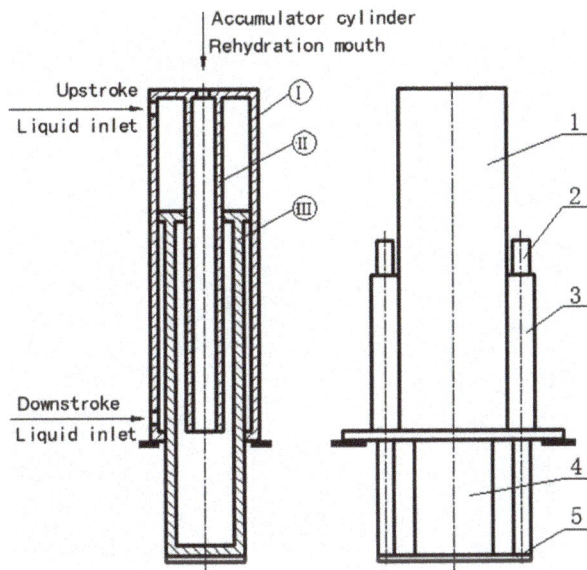

Figure 6. Structure of combined hydraulic cylinder. 1, Cylinder; 2, guide shaft; 3, guide sleeve; 4, piston rod; 5, pulley mounting plate; I, cylinder; II, mandrel; III, piston rod.

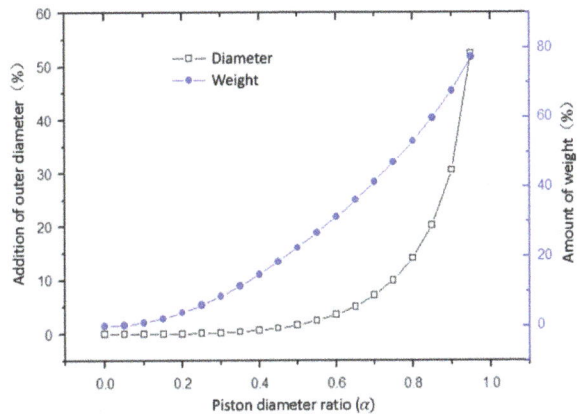

Figure 7. Relationship between piston rod diameter and weight.

typical elongated bar; Euler's formula should be used to calculate critical load F_k.

A fix end and a free end of the column stability calculated by the Euler formula [18] is as follows:

$$F_k = \frac{\pi^2 EI}{(2L)^2}. \qquad (8)$$

In the formula, I is the moment of inertia of the piston rod, E the elastic modulus (210 GPa), and L the length of the piston rod (m).

According to the hydraulic cylinder manual [19], the stability coefficient is around 2–4. Because of the complex offshore platform work environment, we should increase the strength and the durability of hydraulic cylinders, so the safety factor takes $n_k = 4$, thus:

$$\frac{F_k}{n_k} \geq F_h. \qquad (9)$$

According to equation (9), to meet the stability of sucker rod, we should decrease the overall size and weight of the offshore platform by reducing the outer diameter and the weight of piston rod. Here, α is defined as the ratio of the inner and the outer diameters of the piston rod. The curve of the increasing amount of the piston rod's outer diameter and the decreasing amount of piston rod's weight with α is shown in Figure 7.

According to Figure 7, the curve analysis shows that when $\alpha = 0.75$, the piston rod's outer diameter increases by 10%, its weight can reduce by 47%, which achieves

the optimal results. According to equations (8) and (9), we can calculate the outer diameter $d_1 \geq 125.5$ mm; here the piston rod's outer diameter $d_1 = 126$ mm, inner diameter $d_2 = 94$ mm, and wall thickness is 16 mm.

Mandrel size

The load of suspension center is 6.0×10^4 N and 2.0×10^4 N in upstroke and downstroke, respectively, and can be converted to the cylinder load $F_{max} = 1.2 \times 10^5$ N and $F_{min} = 0.4 \times 10^5$ N; thus, the balancing load of the accumulator is:

$$F_p = F_{min} + \frac{F_{max} - F_{min}}{2}. \qquad (10)$$

Inner pressure of the piston rod:

$$P_3 = \frac{F_p}{1/4\pi d_2^2}. \qquad (11)$$

Spindle end force:

$$F_3 = \frac{1}{4\pi D_1^2 - D_2^2 P_3}. \qquad (12)$$

According to the pressure bar stability Euler's formula $\frac{F_k}{n_k} \geq F_3$ and equation (8), we can get:

$$D_1 \geq \sqrt{\frac{256 F_p n_k L^2}{d_2^2 \pi^3 E (1 + \beta^2)}}. \qquad (13)$$

In the formula, β is the ratio of inner and outer diameter of mandrel, $\beta = \frac{D_2}{D_1}$.

The mandrel is mainly used for fluid passage, so in order to decrease the weight of the whole cylinder, select $\beta = 0.8$. After the calculation and optimization, the outer diameter of mandrel D_1 = 88 mm, inner diameter D_2 = 70 mm, and wall thickness is 9 mm.

Cylinder size

Considering the extreme conditions such as storage system failing, the hydraulic cylinder can still work properly, the thrust of the hydraulic cylinder is:

$$F_t = \frac{1}{4}\pi \left(\emptyset_2^2 - D_1^2\right) P. \qquad (14)$$

In this formula, F_t is the thrust of the cylinder (N) and P the working pressure of the selected system (MPa).

According to the national standard of hydraulic transmission's recommendation about working pressure and velocity ratio of the hydraulic cylinder [20], here we determine working pressure by 16 MPa of medium pressure, the thrust of the hydraulic cylinder F_t > 120 kN, thus Φ_2 > 131.5 mm. At the same time, considering the flow rate of the upstroke, the downstroke, and the overall size control of the hydraulic cylinder, we can determine the speed ratio ψ = 1.61.

$$\psi = \frac{\frac{1}{4}\pi\left(\emptyset_2^2 - D_1^2\right)}{\frac{1}{4}\pi\left(\emptyset_2^2 - d_1^2\right)}. \qquad (15)$$

According to equation (15), Φ_2 = 159.1 mm, so the inner diameter of cylinder is Φ_2 = 160 mm.

Wall thickness of the cylinder block:

$$\delta \geq \frac{p_{max} D}{2[\sigma]}. \qquad (16)$$

In this formula, $[\sigma]$ is the cylinder allowable stress, the allowable stress of 45# is 210 MPa.

According to equation (16), $\delta \geq 6.1$ mm. Considering the length of the cylinder and the incretion of the strength and stability, we can determine δ = 10 mm and the outer diameter, Φ_1 = 180 mm.

The main structure size of the combined cylinder is shown in Table 1.

Accumulator size

Accumulator selection is mainly determined by the volume and the allowable working pressure. According to the

Table 1. Main structure parameters of combined cylinder (/mm).

Part	Outer diameter	Inner diameter	Wall thickness
Cylinder	180	160	10
Mandrel	88	70	9
Piston rod	126	94	16

Figure 8. Laboratory model prototype and combined hydraulic cylinder onshore test.

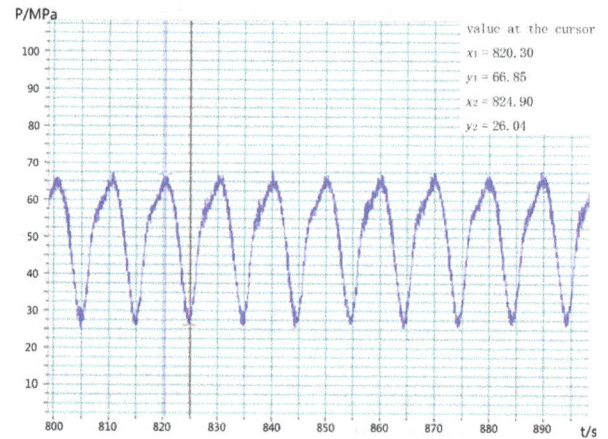

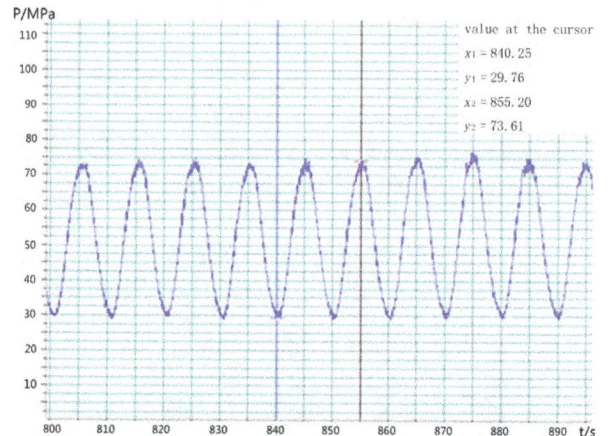

Figure 9. Stress curve of support structure.

characteristics of the accumulator working circuits, the normal operating pressure is:

$$P_x = \frac{F_p}{\frac{1}{4}\pi d_2^2} = 11.53 \ (\text{MPa}). \tag{17}$$

When the pump unit is running, the pressure fluctuation provided by the accumulator should be controlled within 10% [19], thus, the minimum operating pressure of the accumulator circuit is $P_{\min} = 10.38$ MPa, the maximum is $P_{\max} = 12.68$ MPa, and the charge pressure is $P_0 = 0.85$, $P_{\min} = 8.82$ MPa.

The hydraulic stroke $L = 3$ m, the effective volume of the accumulator is:

$$\Delta V = \frac{1}{4}\pi d_2^2 L = 20.82 \ (\text{l}). \tag{18}$$

The working process of the accumulator is calculated as isothermal process, the required volume of the accumulator is:

$$V = \frac{\Delta V}{P_0/P_{\min} - P_0/P_{\max}} = 134.83 \ (\text{l}). \tag{19}$$

Based on the above mathematical models and the designing calculation results, an equal-sized laboratory prototype model is made and shown in Figure 8.

Safety requirement for offshore platform equipment is extremely high. Ground testing is needed for the utmost application on offshore platform over a long-lasting time, combined hydraulic cylinder onshore test is showed as Figure 8. The ground testing has monitored the state of stress for the two main support structure. The state of stress is demonstrated in Figure 9.

According to the Figure 9, bearing of both sides is adjacent, showing the overall structure is adequate and of no partial loading. And the good periodicity of stress curve illustrates the stable operating of the unit. The onshore wellhead prototype has been performed for 6 months, which can meet the oil well's running requirements and lays a foundation for offshore platform application.

Conclusion

In response to the heavy oil thermal recovery of offshore oil fields and to achieve high efficiency recovery in stripper wells, a new technology of the pumping unit–sucker rod–pump combining the artificial lift in the offshore platform is first proposed, and a new type of the hydraulic pumping unit with a long stroke and low stroke times is specially designed for the offshore platform.

This article studies the characteristics of the suspension point during the operation of the pumping unit and designs the structure of the combined hydraulic cylinder in detail. An equal-sized laboratory prototype is made based on mathematical models and the designing calculation results. The prototype runs smoothly and meets the laboratory requirements, which verifies the rationality and effectiveness of the overall designing scheme and lays foundations for the field test.

The pumping unit is designed for offshore platform. Currently, offshore production utilizes electric submersible pump, and the operating process is as follows: produce stopping, pump lifting, steam injection, huff and puff flowing production, well killing, pump lowering then production. The designing of suitable structured pumping unit can realize the replacement from electric submersible pump to rod pump, then the integration of injection and production. Thus, the operating process can be reduced to steam injection, huff and puff flowing production, and production, which will greatly improve the operation efficiency and reduce the cost.

Conflict of Interest

None declared.

References

1. Duan, Y. 2010. Study on multisystem combined with chemical profile control and displacement for heavy crude oil thermal recovery. China University of Petroleum.

2. Rahnema, H., and D. Mamora. 2010. Combustion assisted gravity drainage (CAGD) appears promising Society of Petroleum Engineers – Canadian Unconventional Resources and International Petroleum Conference 2010, v 1, p 206–216,

3. Anderson, G., B. Liang, and B. Liang. 2001. The successful application of new technology of oil production in offshore. Foreign Oilfield Engineering, 17:28–30.

4. Dong, Z., M. Zhang, X. Zhang, and X. Pang. 2008. Study on reasonable choice of electric submersible pump. Acta Petrolei Sinica, 29:128–131.

5. Wang, S., H. Zhao, Q. Shang, and X. Zhou. 2010. Technology of linear submersible motor and its application. Petrol. Drill. Tech., 38:95–97.

6. Liu, C.. 2010. Application of three methods for lifting. Oil-Gas Field Surf. Eng., 29:45–46.

7. Qi, Y., L. Shi, and F. Gao. 1995. Introduction to the type and development trend of domestic pumping unit. Drill. Prod. Technol., 18:46–49.

8. Xu, B., F. Huang, and B. Zhang. 2006. Design of a new energy-saving hydraulic pumping unitwith variable speed control and closed oil circuits. Mach. Tool Hydraul., 34:72–74.

9. Xue, X., Y. Qi, Y. Yu, Z. Li, J. Sun, and J. Mao. 2016. The adaptability analysis and application status of hydraulic pumping units. Mach. Tool Hydraul., 44:151–154.

10. Lea, J. F., PL Tech LLC and H. W. Winkler. 2011. What's new in artificial lift. World Oil 232:231–243.

11. Bo, T.. 2002. The development and state of art of the hydraulic pumping unit in China. Drill. Prod. Technol. 25:60–62.

12. Xun, H.. 2001. The development of hydraulic pumping unit with foreign technology level. Xinjiang Petrol. Sci. Technol., 1:62–64.

13. Ma, H.. 2005. A hydraulic pumping unit for offshore oil exploitation. China Petrol. Machin., 33:54–56.

14. Zhang, L., and H. Xie. 2008. Study on energy-saving hydraulic sucker rig with compound cylinder. Chin. Hydraul. Pneumat. 9:63–64.

15. Guo, Y., X. Zhou, J. Li, Y. Ling, and J. Yang. 2010. Crude features and origins of the Neogene heavy crude oil reservoirs in the Bohai bay. Oil Gas Geol. 31:375–380.

16. Guan, D., X. Wang, C. Li, K. Gao, and Z. Wei. 2012. Analysis of heavy crude oil characteristics and densification factors of PLA oil-bearing structures in the water area of Bohai. Petrol. Geol. Eng. 26:69–72.

17. Wei, X., J. Sun, H. Zhou, and B. Xu. 2009. Design of new hydraulic pumping unit with electro-hydraulic proportional control. Mach. Tool Hydraul., 37:39–41.

18. Liu, H.. 2011. Mechanics of materials. Higher Education Press, Beijing, China.

19. Lu, Y.. 2002. Hydraulic and pneumatic technology handbook. China Machine Press, Beijing, China.

20. Zang, K.. 2010. Hydraulic cylinder. Chemical Industry Press, Beijing, China.

12

The remarkable impact of renewable energy generation in Sicily onto electricity price formation in Italy

Francesco Meneguzzo[1], Rosaria Ciriminna[2], Lorenzo Albanese[1] & Mario Pagliaro[2]

[1]Institute of Biometeorology, CNR, via Caproni 8, 50145 Firenze, Italy
[2]Institute for the Study of Nanostructured Materials, CNR, via U. La Malfa 153, 90146 Palermo, Italy

Keywords
Energy end use, merit order effect, photovoltaics, Sicily, wind energy

Correspondence
Francesco Meneguzzo, Institute of Biometerology, CNR, via Caproni 8, 50145 Firenze FI, Italy. E-mail: f.meneguzzo@ibimet.cnr.it
and
Mario Pagliaro, Institute for the Study of Nanostructured Materials, CNR, via U. La Malfa 153, 90146 Palermo PA, Italy. E-mail: mario.pagliaro@cnr.it

Funding Information
No funding information provided.

Abstract

During the first half of 2015, for the first time, the zonal electricity price in Sicily decreased below than the national wholesale price in Italy. Showing the unique pattern of electricity consumption in Italy's largest region at different time scales, we identify the effectiveness of the impact of renewable power generation on utility-scale in Sicily upon the whole of Italy's electricity market. Increasing the electrification of the energy end uses, as it is happening despite prolonged reduction in electricity demand, will lead to further benefits for power consumers throughout the whole country.

Introduction

Sicily is the largest and the most solar irradiated among Italy's regions. For comparison, the global horizontal irradiation yearly total of the second largest region (Lombardia) is 1100 kWh/m^2 whereas Sicily has an average value of 1800 kWh/m^2 [1].

While photovoltaic (PV) power generation was negligible till 2008–2009, the deployment of the Feed-In-Tariff (FiT) incentive scheme in Italy between 2006 and 2013 caused an impressive surge in the PV installed power nationwide as well as in Sicily, jumping from few tenths of MW to 1270 MW during 3 years (2011–2013) [2].

Even without the FiT subsidies (practically terminated on June 2013), in 2014 Sicily saw the installation of another 130 MW bringing the overall power to 1400 MW [3], with owners of the new PV systems benefiting of net metering, namely self-consuming the energy produced by their PV installations. This accelerated deployment of solar electricity took place almost concomitantly to massive adoption of wind energy as Sicily is also the most ventilated region of Italy, so that the installed wind power rose from zero in 2005 to about 2000 MW at the end of 2014 [3].

A preliminary estimate suggests that the PV power deployed in Sicily rose further up to more than 1400 MW during 2015, while the combined PV and wind power generation exceeded 2 TWh [4].

This rapid change, along with significant economic interests behind the incentives that reward only the electricity produced and actually fed into the grid, pushed the State owner of the Italian grid to deploy massive investments to widen and improve Sicily's high voltage grid [5]. The new "Sorgente-Rizziconi" connection between Sicily and the Italian peninsula [6], now almost completed, is a 380 kV line that by the end of 2016 will bring the interconnection capacity from the existing 1000–3000 MW, relieving the bottleneck that historically caused higher electricity prices in Sicily compared to the rest of the country.

From the viewpoint of electric interconnection, indeed, Sicily is an isolated island. For example, in 2014 the existing 1000 MW line was congested for 90% of the year's time [3], limiting, for example, the possibility to export the surplus of renewable energy during holydays and weekends.

We have recently shown why and how the growing penetration of PV renewable power generation in Italy (18 GW as of late 2014) has caused a substantial fall in the price of electricity in the Italian wholesale electricity market (IPEX) [7]. Now we show how the significant PV and wind power generation has changed the electricity price formation in Sicily affecting the national price (PUN) to such an extent to cause a significant decrease in the national electricity price.

In other words, the Sicily zone is substantially a one-way feeder into the Italian electricity market, not only on the side of locally generated power but also as a price steering element, since the zonal price affects the PUN in the weighted average leading to its formation. We also develop a model linking the daily zonal power price in Sicily to the demand not met by PV and wind energy.

The study is concluded suggesting a few practical ways to achieve further economic (and environmental) benefits for electricity consumers not only in Sicily, but also throughout the whole country.

Italy's and Sicily's Electricity Markets

As mentioned above, until recently wholesale electricity prices in Sicily have been constantly higher than in the rest of the country. The monthly average was more than 40% higher than the average national prices in Italy in 2008 and 2009, exceeding the PUN by more than €20/MWh for more than 50% of the months between January 2008 and December 2014 (Fig. 1). Following the spikes in 2008–2009, the Italian energy regulator (AEEG) launched an investigation into Sicilian prices whose outcomes, on August 2009, were transferred to the antitrust regulator (ICA) [8].

According to two following ICA investigations into Sicily wholesale electricity prices launched on January 2010, capacity withholding (mainly economically) was used to raise Sicily's zonal price, in turn affecting the national wholesale purchasing price. Eventually, the two major utilities agreed, the first, to limit the price of power produced in Sicily (a bid cap) to €190 per MWh in 2011, and the other to set prices based on market [9].

Nevertheless, it was only in February 2015 that the Sicily's zonal price fell below €10/MWh over the PUN, never exceeding that threshold again during the subsequent 13-months (until February 2016), even falling *below* the PUN by €1.51/MWh on July 2015 (Fig. 1).

The trends of absolute monthly power prices, PUN and Sicily's zonal price share few key features, such as the spikes in 2008 and the sustained downward trend after 2012 (Fig. 2). Again, the closing gap between Sicily's and Italy's prices becomes apparent since early 2015.

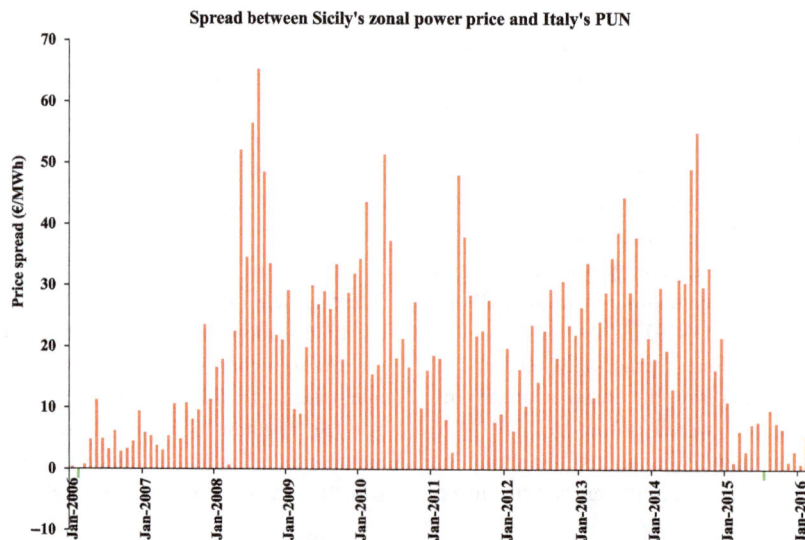

Figure 1. Spread between monthly average Sicily's zonal power price and Italy's PUN. Positive spread in red, negative in green. Source: Energy Markets Authority (GME).

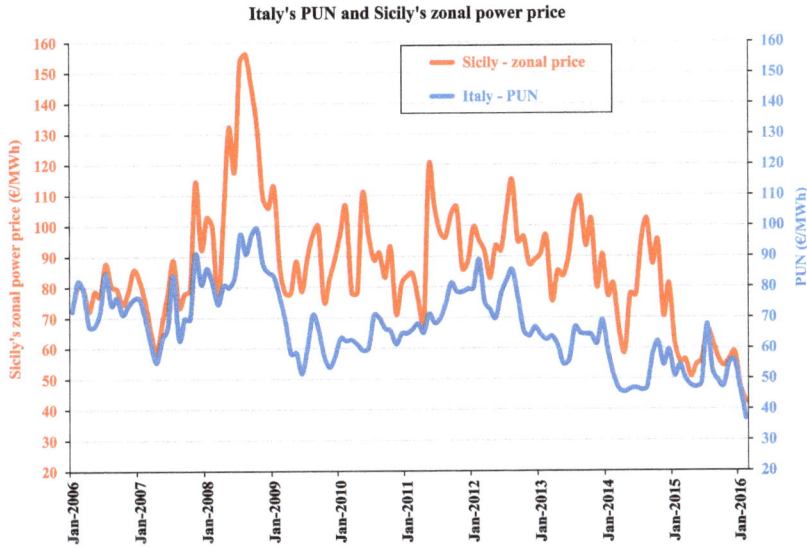

Figure 2. Yearly average prices on the day ahead market in Italy (PUN) and in the Sicily's zone. Source: Energy Markets Authority (GME).

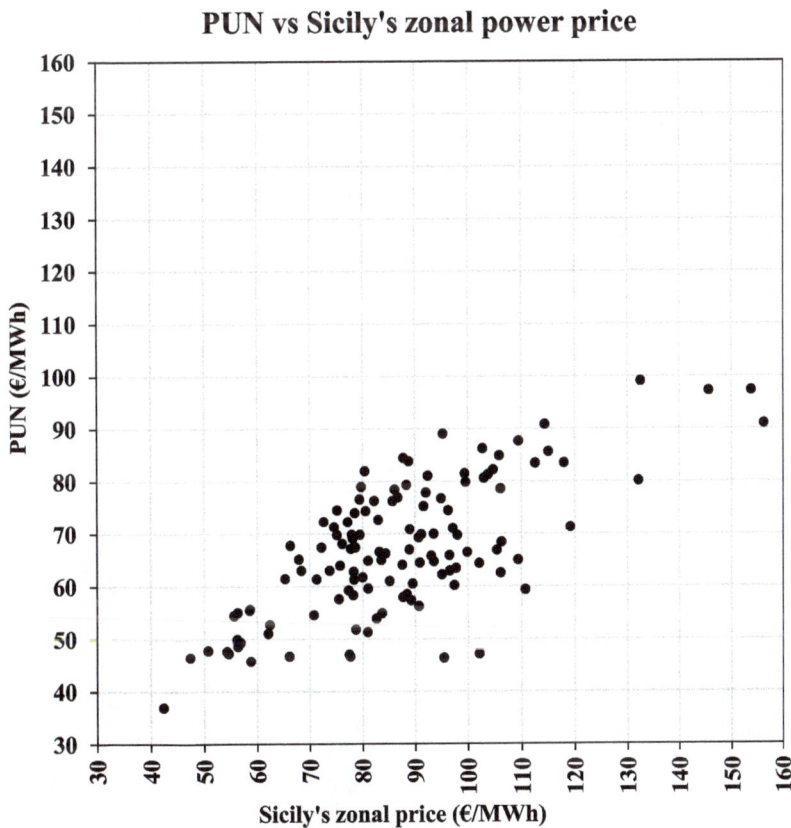

Figure 3. Sicily's zonal power price explains 50% of the variance of the national wholesale price (PUN).

The linear correlation between the PUN and the Sicily's zonal price is especially noticeable, since as much as 50% of the variance of the former is explained by the latter, as shown in Figure 3, a value which is not so distant from the correlation figure found for the peak PUN (wholesale power price during peak hours) model using an autoregressive term, without which the correlation strength fell below 60% [7].

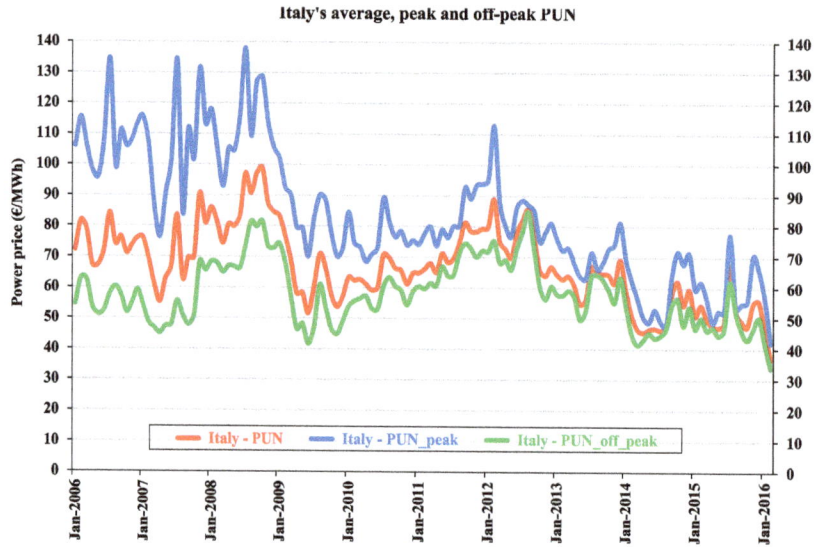

Figure 4. Italy's average, peak and off-peak wholesale power price (PUN). Source: GME.

Power prices formed in the Sicily's zone obviously reflect the impact of the changing prices in raw fuels such as oil and natural gas needed to feed conventional generation plants, and changes in the regional power demand. The closing gap between peak PUN (when most of the demand occurs) and off-peak national prices shown in Figure 4 may be attributed to the growing national generation from renewable sources, mainly PV due to its straightforward impact upon the peak diurnal demand in the merit-order price formation scheme [7].

In the year 2015, while the overall contribution of the renewable energy sources (hydroelectric, wind and PV) to the domestic power generation fell to 33.2% from 37.9% in 2014, mainly due mainly to significantly lower (−24.9%) hydropower generation, the PV production increased by as much as 13%, totaling almost 25 TWh, that is more than 55% of the hydropower and exceeding 9% of the domestic generation (8% in 2014) [10].

It was previously shown that the PV generation affects the peak PUN more than the other renewable sources [7], therefore contributing to close the gap between peak and off-peak PUN. Figure 5 shows the yearly series of PV and wind power generation in Italy where, after the sharp increase from 2010 to 2013, a slower yet sustained upward trend persisted until early 2016.

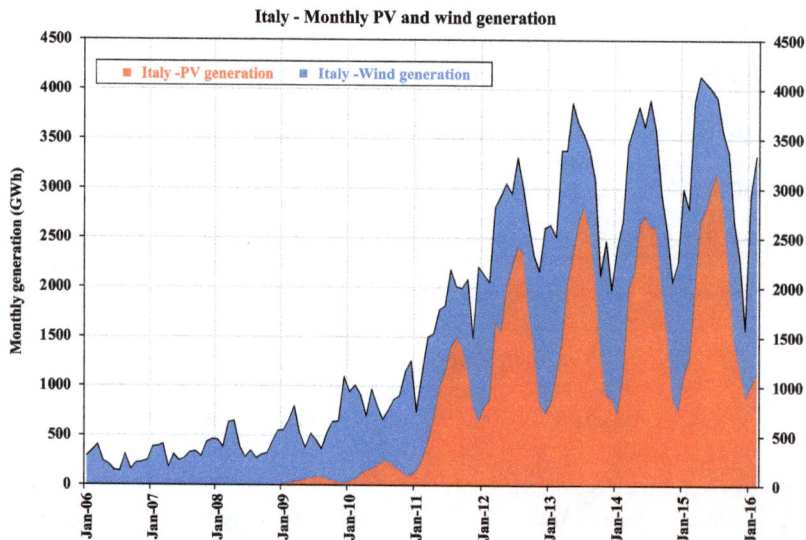

Figure 5. PV and wind power generation in Italy. Source: Terna.

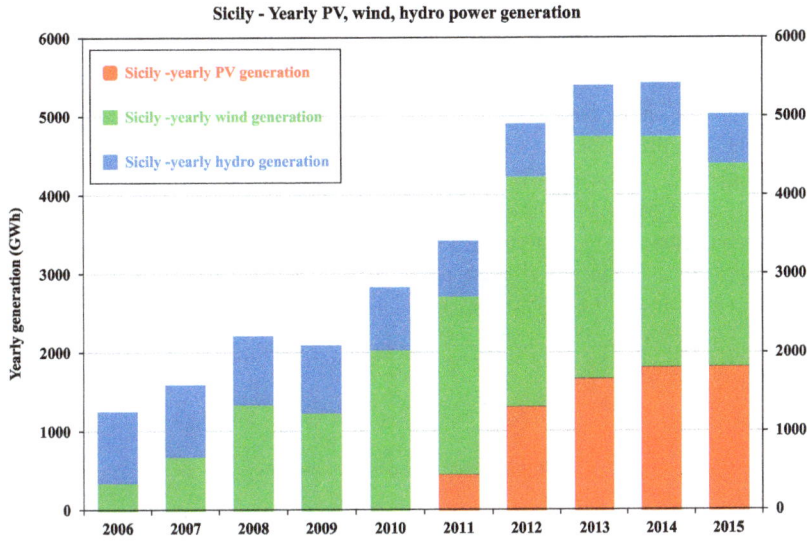

Figure 6. PV, wind, and hydro power generation in Sicily. Source: Terna.

A reputed market research company concluded that, thanks to the renewable energy boom, Italy's electricity market now faces a dramatic overcapacity, with the largest utility having announced closures of several plants for at least 11 GW of capacity, and the fifth largest producer exiting the market with sale of its thermo-electric assets [11].

In this context, in 2013 – following a sharp increase the year before – the amount of renewable energy produced in Sicily exceeded for the first time the threshold of 5 TWh, namely almost 25% of the about 23 TWh produced and fed into the grid [2]. Such a threshold was exceeded in each of the subsequent 2 years, despite a noticeable climatically-forced drop of wind and hydro contributions in 2015, which were partially offset by the slowly increasing PV generation (Fig. 6) [3, 4].

Following the early drop occurring as a consequence of the economic crisis during 2008–2009, both Italy's and Sicily's power demands underwent sustained fall starting in 2012. Yet, while the former halted in 2015, demand of electricity in Sicily nosedived, hitting the 11-years lowest point in February 2016 (Fig. 7). No wonder, therefore, that the historically lowest price was observed in the same month, in its turn helping to explain the partly unexpected record low PUN as a result of the steering effect of Sicily's zonal power price (Fig. 2).

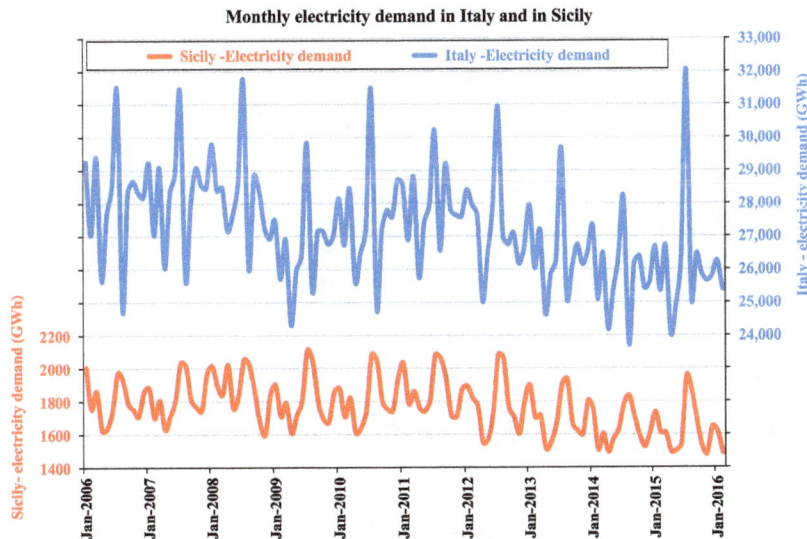

Figure 7. Monthly power demand in Italy and in Sicily. Source: Terna.

In view of the above-mentioned drastic regulatory intervention upon the formation of the zonal power price in Sicily, defining a statistical model linking the monthly Sicily's regional power price with local demand, renewable generation and conventional fuel price, as was done for the Italian market as a whole [7], de facto is impracticable. Nevertheless, there is room for a modeling effort on a daily basis over periods encompassing different seasons, that is, different compositions of the power generation mix, with particular reference to the PV share of the overall generation mix. This is the subject of the next Section.

Impact of Solar and Wind Power Generation

Aiming to show the remarkable impact of solar and wind electric generation in Sicily (and in Italy), we focus on the zonal and national prices of power in the course of the first 6 months of 2015. Prices on the day-ahead market for both Sicily and Italy were collected from GME [12], the manager of the Italian energy markets.

As expected, during the first half of 2015, different climatic conditions affected Sicily, leading to strong changes of solar PV and wind power generation superimposed to

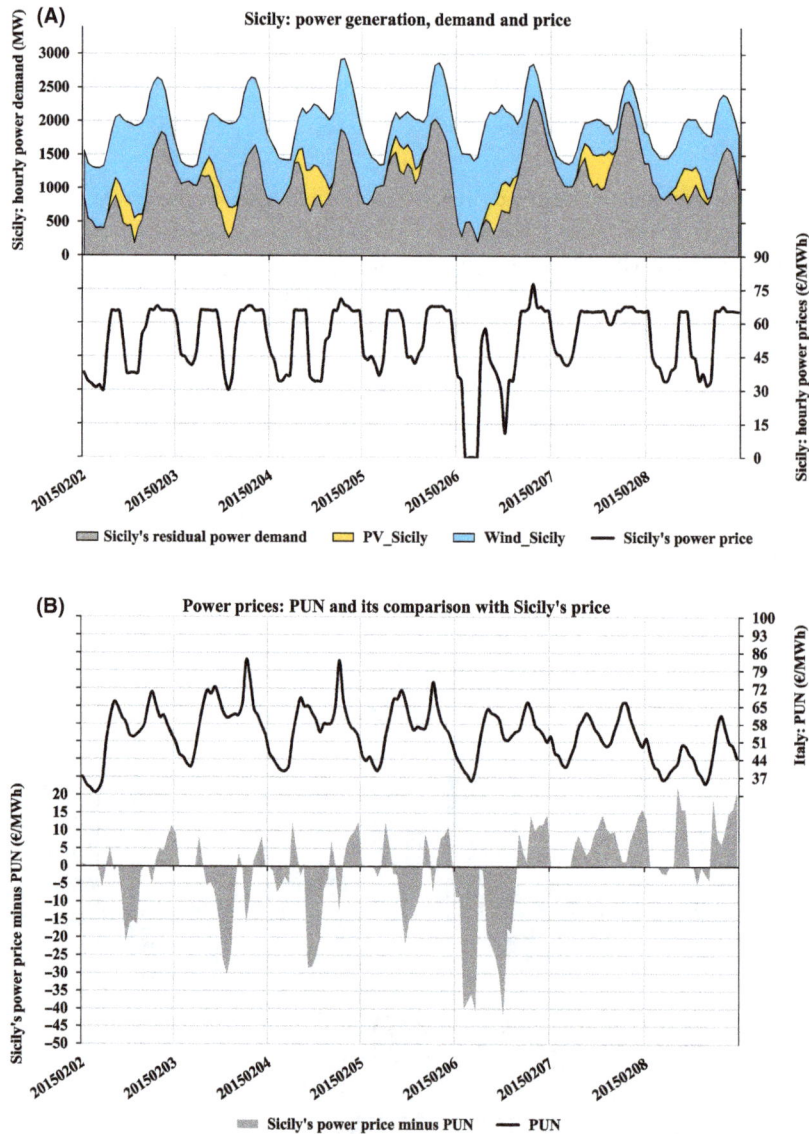

Figure 8. (A) Solar PV and wind hourly power generation superimposed to the residual power demand in Sicily, along with the respective power zonal price in the first week of February 2015; (B) national wholesale power price (PUN) along with the difference between Sicily's power price and PUN in the first week of February 2015. (C) Same as (A) but for the last week of June 2015; (D) same as (B) but for the last week of June 2015.

the daily cycle. Looking at the first week of February (Monday to Sunday) when wind generation was far higher than solar PV generation, Figure 8A shows that the residual electricity demand, that is, demand resulting after subtracting from the total the renewable (solar PV and wind) power generation, is correlated with the zonal power prices on an hourly basis. In other words, price grows along with residual demand, both during day and night, not only during working days but also during the weekend.

A further comparison between the values of Sicily's zonal power price and the national wholesale price during the same week (Fig. 8B) shows that low residual demand in Sicily leads to zonal prices lower than PUN, thereby lowering the national electricity bill.

In Italy, the national power price is formed in the day-ahead market, which works as a periodic multi-unit price auction [13]. The GME uses a merit-order principle to construct and aggregate the supply and demand curves. If electricity flows through the grid violate the transmission constraints (as it happens every day), the market is split into zonal areas: North, Center North, Center South, South, Sardinia and Sicily. Power-generating companies receive the zonal prices, whereas buyers pay the *Prezzo Unico Nazionale* (PUN) which is the average of the zonal prices, weighted by the zonal electricity consumption levels.

Similar results concerning residual electricity demand and zonal hourly power prices were observed during the

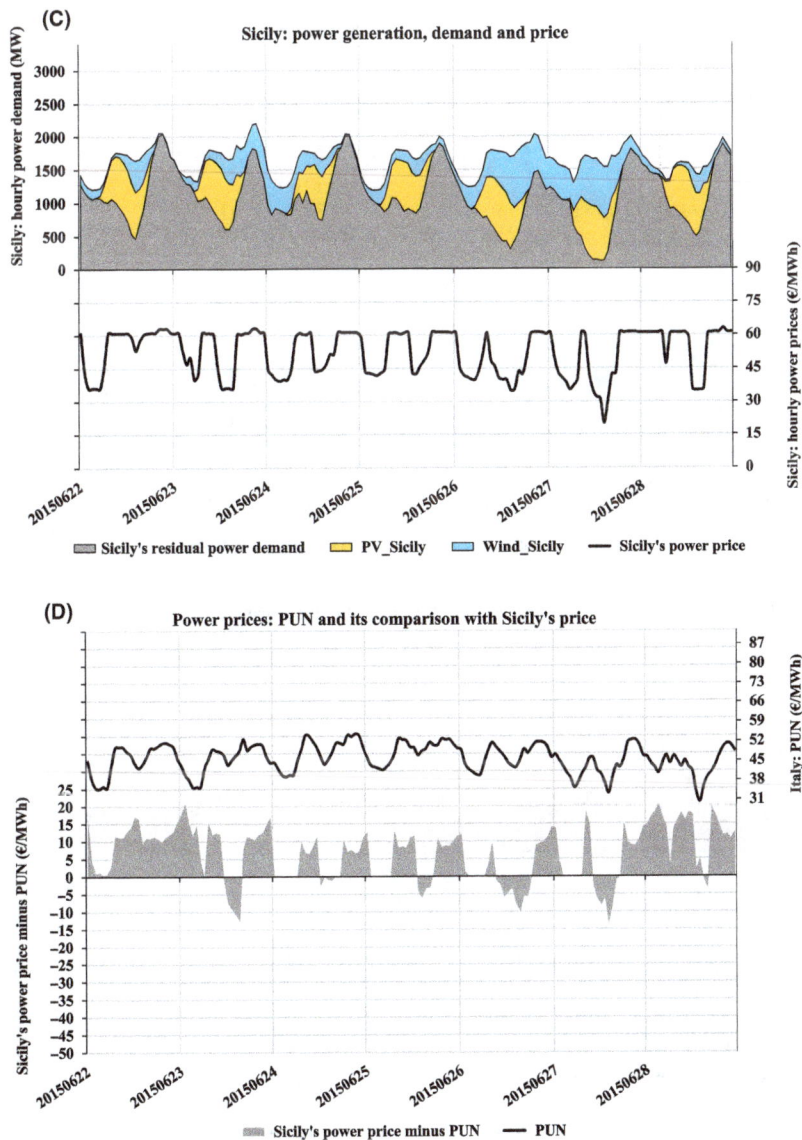

Figure 8. (*Continued*)

last week of June 2015 (Monday to Sunday), when solar PV generation was generally higher than wind-derived generation (Fig. 8C).

However, the substantially lower PUN values due to widespread solar PV generation across Italy, led to negative values of the difference between Sicily's zonal price and PUN only when the zonal residual demand fell below about 500 MW (Fig. 8D).

With reference to the same 2 weeks of 2015 in February and June, comparison between the hourly overall electricity demand in Sicily and Italy shows striking differences (Fig. 9). While the morning and evening peak demands are indeed comparable at the national level, in Sicily the

evening peak is much larger both in February (Fig. 9A) and in June (Fig. 9B).

Moreover, the decrease in the electricity demand in Sicily during the weekend is much smaller than in Italy. Both these outcomes show evidence of the general lack of widespread significant industrial activities in Sicily, where three large, energy-intensive oil refineries and two petrochemical parks are installed in northern (Milazzo) and southern Sicily (Gela, Augusta and Priolo Gargallo), working in the continuous mode, independently of regular or weekend days of the week.

Widening the analysis to the whole period 1 January 2015 through 30 June 2015, the relationship between the

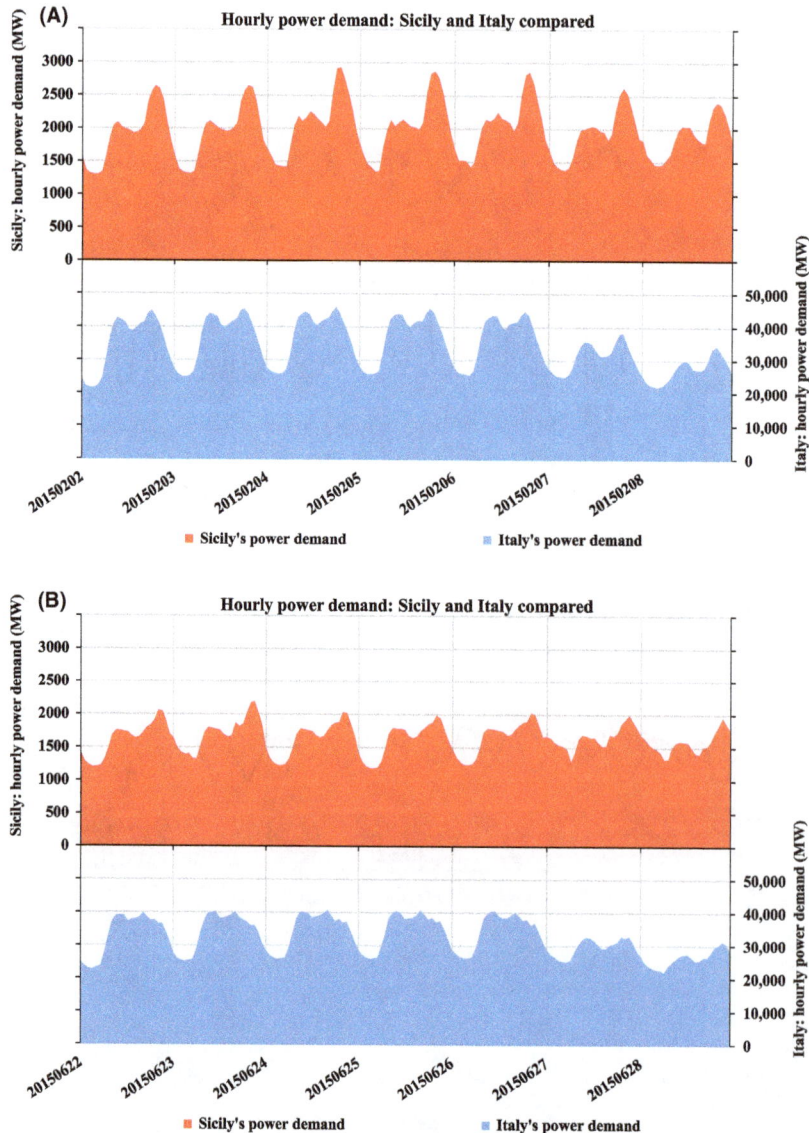

Figure 9. (A) Hourly series of power demand in Sicily and Italy in the first week of February, 2015. (B) Same as (A) but for the last week of June, 2015.

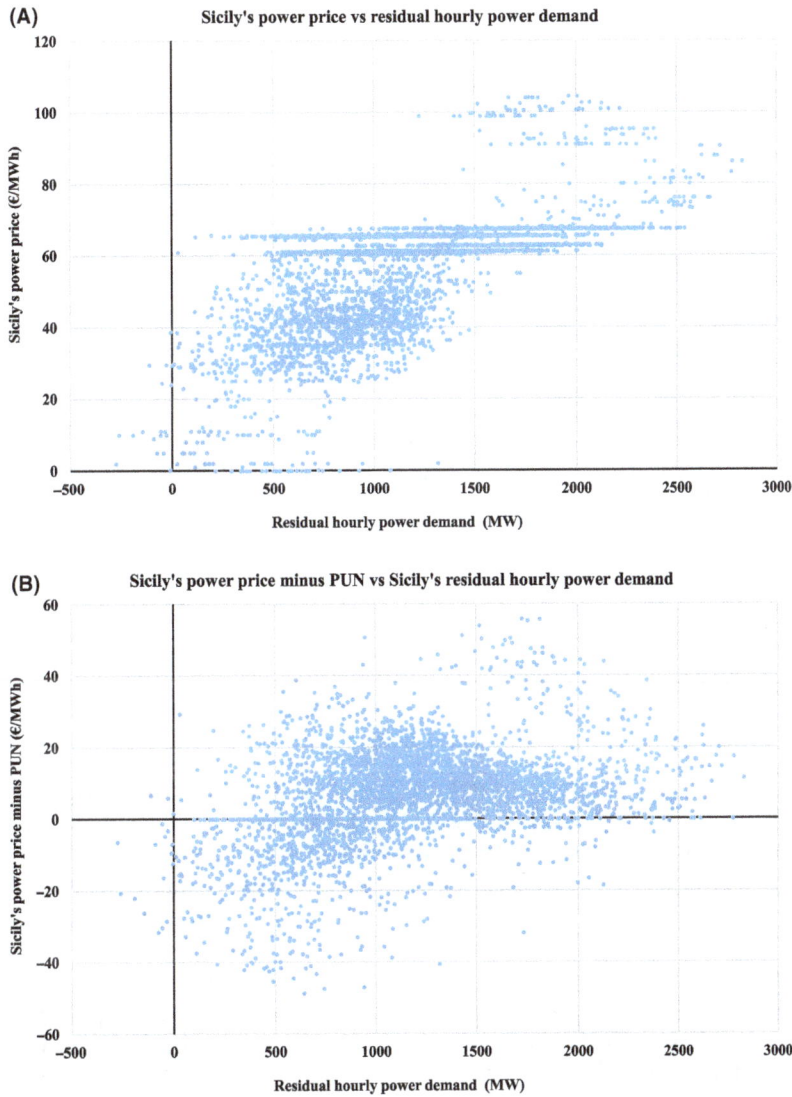

Figure 10. (A) Power price in the Sicily's zone against zonal residual electricity demand; (B) difference between Sicily's zone power price and national wholesale power price (PUN) against zonal residual electricity demand.

Sicily's zonal price and the zonal residual electricity demand, that is the remaining demand after subtraction of solar PV and wind generation, becomes apparent from the plots shown in Figure 10.

Most Sicily's zonal prices fall below the €40/MWh threshold when the residual demand lies below 500 MW (Fig. 10a). Similarly, during the hours in which the residual demand of electricity in Sicily is not generated by renewable sources lies below the same 500 MW threshold, the zonal prices lie below the PUN (Fig. 10B).

In order to outline a more homogeneous statistics, the power prices in the Sicily's zone were averaged across the values observed in six classes of the residual demand.

Figure 11 shows the results along with the standard deviation associated to the average price in each class.

Best fit to this curve is a monotonic, logarithmic increase in the zonal price with the residual demand (eq. 1, where price is in €/MWh and the residual hourly demand is in units of MWh) explaining about 98.7% of the sample variance:

$$\text{Sicily's hourly zonal price} = \tag{1}$$
$$21.6 \ln(\text{Residual hourly demand}) - 93.1.$$

Electrification of Energy End Uses

It is now well understood that the beneficial effects of PV generation onto wholesale electricity price formation rapidly increase with electricity demand [7]. Hence, to allow further reduction in the cost of electricity in both

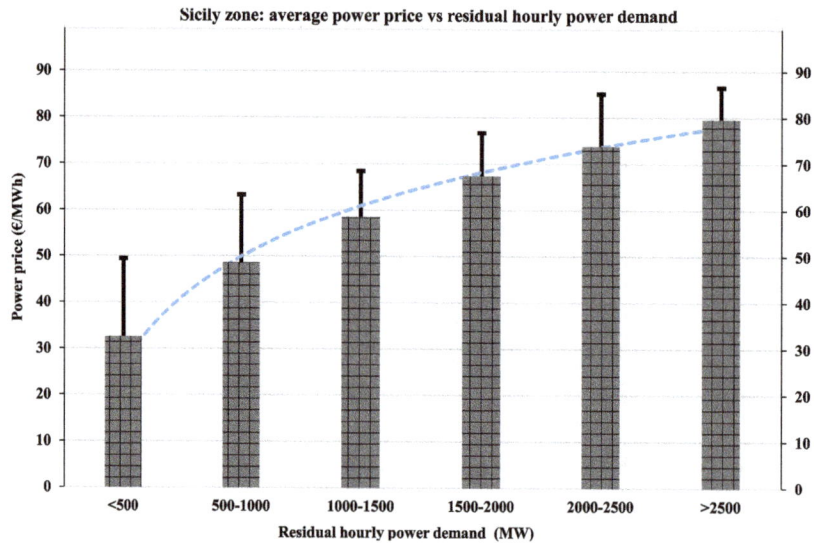

Figure 11. Average power price in the Sicily's zone in classes of the residual electricity demand.

Sicily and in Italy, it is important that the power demand in Sicily returns to grow.

Nevertheless, the contrary occurred both in Sicily and in Italy since 2007, as shown in Figure 7, down to unprecedented low consumption, which weakens the efficiency of solar PV generation concerning its ability to lower the power prices according to the merit-order price formation scheme.

In brief, policy makers in Sicily should boost the electrification of the energy end uses [14], as it happened recently with the new train connection linking the two main cities of the island (Palermo and Catania), as well as replacing obsolete heating systems based on burning natural gas, with much more efficient heat pumps in both residential, commercial, industrial, and public buildings.

Furthermore, new legislation should be passed and aimed to promote the use of pumping storage as an ideally suited option to store the excess of renewable PV and wind energy, using the associated significant (600 MW) hydroelectric capacity installed in Sicily [15].

Of course, electrification means also the adoption of storage systems on a large scale: because of its poor interconnection with Italy's grid, Sicily cannot rely on neighboring regions to balance short-term fluctuations in its grid.

For that reason, the owner of Italy's grid (Terna) has built the new "Sorgente-Rizziconi" and Sicily–Malta cable connection lines, and is adding new storage capacity in Sicily (using lithium-ion batteries) to make the existing grid more flexible, and thus more stable [5].

Conclusions and Perspectives

The analysis of the wholesale electricity prices in Sicily and in Italy reported in this study shows large, positive changes in Sicily's and Italy's electricity markets, due to significant penetration of new renewable energy sources, namely solar PV and wind energy. In 2015, the historical spread between Sicily's electricity zonal price and the national price (the PUN) eventually vanished, occasionally assuming negative values.

Furthermore, the zonal price has been shown to increase with the electric demand not met by renewable energy, according, as we find out, to a log-normal equation.

The forthcoming inauguration of the new 3000 MW high voltage connection between Sicily and the rest of Italy will only enhance such impact that so far has been restricted by the easily saturated 1000 MW existing connection.

Consistently lower values of the electricity national price (the PUN) will be achieved, especially during the sunniest months of the year when the electricity peak demand that has recently surpassed winter's demand (56,883 MW as of 8 July 2015) is met in large part by concomitant PV generation (40% of the above-mentioned peak of July 2015) [16].

Electrification of energy uses is slowly, but inevitably, taking place along with the global solar energy revolution [17].

Railway transportation in Italy, including Sicily, is knowing a renaissance with new high speed trains, new tram and new underground lines. The use of heat pumps in place of conventional heaters to produce low temperature heat is similarly undergoing a slow yet sustained growth [18], starting from large commercial buildings. The ongoing boom of electric bikes and motorbikes is prodromal to the forthcoming diffusion of electric cars [19].

Under these conditions, the reshaping of Sicily's and Italy's electricity market will continue. Investing in the electrification of energy user needs will translate into further economic (and environmental) benefits not only for Sicily's but also for all Italy's electricity consumers.

Acknowledgments

This article is dedicated to Daniele Tringali, eminent engineer at AMG Energia, whose daily engineering and management work does citizen's good in the city of Palermo.

Conflict of Interest

None declared.

References

1. European Commission, Joint Research Centre, Institute for Energy and Transport. Photovoltaic Geographical Information System (PVGIS). Available at http://re.jrc.ec.europa.eu/pvgis/ (accessed 20 April 2016).

2. Palmisano, G., M. Pagliaro, F. Meneguzzo, and R. Ciriminna. 2014. Sicily's solar report 2014. Simplicissimus Book Farm, Catania, Italy. ISBN 9788868859985.

3. Pagliaro, M., R. Ciriminna, F. Meneguzzo, and L. Albanese. 2015. Sicily's solar report 2015, Simplicissimus Book Farm, Catania, Italy. ISBN 9788869094231.

4. Meneguzzo, F., M. Pecoraino, L. Albanese, and M. Pagliaro. 2016. Sicily's solar report 2016. Simplicissimus Book Farm, Catania, Italy. ISBN 9788892500037.

5. Bastioli, C., M. Dal Fante, and P. Cristofori. Terna strategic plan 2015–2019. Available at http://download.terna.it/terna/0000/0086/99.pdf (accessed 20 April 2016).

6. Ries, J., L. Gaudard, and F. Romerio. 2016. Interconnecting an isolated electricity system to the European market: The case of Malta. Utilities Policy (In Press).

7. Meneguzzo, F., F. Zabini, R. Ciriminna, and M. Pagliaro. 2014. Assessment of the minimum value of photovoltaic electricity in Italy. Energy Science and Engineering 2:94–105.

8. Noce, A. Abuse of dominant position by energy incumbents. The Italian experience, Energy Community Competition Network Meeting, Athens. Available at https://www.energy-community.org/pls/portal/docs/2106181.PDF (accessed 4 June 2013).

9. Reuters. 2010. Enel to cap Sicily power price – antitrust watchdog. Available at http://fr.reuters.com/article/idUKLDE6781GW20100809 (accessed 20 April 2016).

10. Terna. Available at https://www.terna.it/it-it/sistemaelettrico/dispacciamento/datiesercizio/rapportomensile.aspx (accessed 20 April 2016).

11. Rossetto, N. An oversized electricity system for Italy. Available at http://www.ispionline.it/it/energy-watch/oversized-electricity-system-italy-12135#sthash.exajQhg8.dpuf (accessed 22 January 2015).

12. Energy Markets Authority (GME). Available at http://www.mercatoelettrico.org/en/Statistiche/ME/DatiSintesi.aspx (accessed 20 April 2016).

13. Petrella, A., and S. Sapio. 2010. No PUN intended: a time series analysis of the Italian day-ahead electricity prices, EUI RSCAS, 2010/03; Loyola de Palacio Programme on Energy Policy, Bruxelles.

14. Edmonds, J., T. Wilson, M. Wise, and J. Weyant. 2006. Electrification of the economy and CO_2 emissions mitigation. Environmental Economics and Policy Studies 7:175–203.

15. Artizzu, G. 2013. Mercato elettrico: il PUN in altalena e lo zampino delle fonti rinnovabili. Available at http://www.qualenergia.it/articoli/20130414-mercato-elettrico-il-PUN-in-altalena-e-lo-zampino-delle-energie-rinnovabili (accessed 28 March 2016).

16. Terna. Flegetonte spinge le rinnovabili e il fabbisogno energetico. 56.883 MW: nuovo record assoluto dei consumi elettrici in Italia, press release, Rome, July 8, 2015. Available at https://www.terna.it/ViewDocumenti/tabid/1095/docid/33932/docType/TCAT-CS/language/it-IT/Default.aspx (accessed 28 March 2016).

17. Meneguzzo, F., R. Ciriminna, L. Albanese, and M. Pagliaro. 2015. The great solar boom: a global perspective into the far reaching impact of an unexpected energy revolution. Energy Science and Engineering 3:300–309.

18. Franci (REF-E), T. 2014. Efficienza energetica: evoluzione della domanda e tendenze di sviluppo tecnologico. Presentazione dei risultati della ricerca di mercato, Nuova etichettatura energetica per gli impianti, Congresso nazionale Domotecnica, Torino, 19 September 2014.

19. Albanese, L., R. Ciriminna, F. Meneguzzo, and M. Pagliaro. 2015. The impact of electric vehicles on the power market. Energy Science and Engineering 3:300–309.

13

Exergy analysis and economic evaluation of the steam superheat utilization using regenerative turbine in ultra-supercritical power plants under design/off-design conditions

Zhou Luyao[1,2] (iD), Xu Cheng[1], Xu Gang[1], Bai Pu[1] & Yang Yongping[1]

[1]National Thermal Power Engineering and Technology Research Center, North China Electric Power University, Beijing 102206, China
[2]Jiangsu Maritime Institute, Nanjing, 210000, China

Keywords
Design/off-design conditions, economic analysis, exergy destruction, regenerative turbine, steam superheat utilization, thermodynamic analysis

Correspondence
Cheng Xu, National Thermal Power Engineering and Technology Research Center, North China Electric Power University, Beijing 102206, China. E-mail: ncepu_xucheng@126.com

Funding Information
This study is supported by National Basic Research Program of China (Grant No. 2015CB251504), the National Nature Science Fund of China (No. 51476053).

Abstract

In order to effectively utilize the steam superheat of ultra-supercritical power plants, a regenerative turbine (RT) system was proposed. In the RT system, a portion of the high-pressure turbine exhaust steam is sent to an extra turbine termed as RT, and the extracted steam of several regenerative heaters (RHs) would be extracted from the RT instead of the intermediate-pressure turbine. The superheat degree of related extracted steam would decrease and the exergy destruction of the heat transfer process in RHs would be reduced. Thermodynamic and economic analyses of the RT system in a typical 1000 MW power plant were carried out. The results showed that the heat consumption rate would be reduced as compared to the conventional configuration. When the RT isentropic efficiency is set to 85%, the heat consumption rate of the RT system could decrease by 46.9 kJ/kWh as compared to that of the reference system without RT under design condition. The heat consumption rate decrement will be greater with higher RT isentropic efficiency. Meanwhile, thermodynamic analysis of the RT system under off-design condition showed that the energy saving effect of the RT system would be decreased when the design conditions are no longer met. The exergy analysis showed that the total exergy destruction decrement of related RHs in the RT system decreases as the load drops down. Moreover, the economic analysis revealed that the net economic benefit of the RT system could reach 0.57 M$ per year, and would perform economically within a wide range fluctuations of power market conditions.

Nomenclature

Symbols

E_{total}	total energy input per unit time
P_{gen}	generated electric power
ΔE_{ex}	exergy destruction
$E_{ex,in}$	exergy input
W_{in}	power input
$E_{ex,out}$	exergy output
W_{out}	power output
$E_{ex, loss}$	exergy loss caused by energy loss
ΔI_{net}	net economic benefit
ΔI_{es}	income of energy saving
$\Delta C_{O\&M}$	annual operating and management costs
CRF	capital recovery factor
Δq	reduction in the heat consumption
P_F	fuel price
h_{eq}	equivalent operation hours per year
P_{gen}	power output of the power plant
$\eta_{i,rt}$	isentropic efficiency of regenerative turbine
k	the discounted rate
n	the equipment life span
$C_{O\&M}$	operation and maintenance cost
γ	proportionality coefficient

Abbreviations

RT	regenerative turbine
USC	ultra-supercritical

RH   regenerative heater
HPT   high-pressure turbine
IPT   intermediate-pressure turbine
LPT   low-pressure turbine
CON   condenser
DEA   deaerator
GEN   generator
THA   turbine heat acceptance
LHV   lower heating value
ICC   installed capital costs

Introduction

Coal-fired power generation provides over 40% of electricity supply worldwide and is expected to continue to play a predominant role in the global energy mix in the near future [1, 2]. In China, it provides nearly 80% of the total electricity requirement and accounts for more than half of total coal consumption. Consequently, these power plants contribute nearly 50%, 37%, 33% and 50% to the total SOx, NOx, dust, and CO_2 emission volumes, respectively, for the entire country [3, 4]. Encountering these emissions necessitates the need to implement strategies and plans to mitigate the impacts of coal-fired power plants on air quality and global environment [5]. In view of recent regulations concerning emission reduction, electric power industries have been required to improve the energy efficiency of coal-fired power plants, which could both lessen the impacts on the environment and increase the economic viability, particularly for developing countries where electric power consumption is rapidly rising [6, 7]. Improvements in efficiency would be satisfied, using larger units and higher steam parameters. As a result, a lot of large ultra-supercritical (USC) units are widely applied in recent years.

On a global scale, USC power generation technology is entering a fast-development stage. The parameters of the live steam in USC power plants has rapidly increased worldwide, resulting in improved parameters of the extracted steam with increased superheat degree. Moreover, the increased superheat degree of the extracted steam leads to large temperature difference and exergy destruction during the heat transfer process of the regenerative heater (RH) [8]. Appropriate superheat utilization of extracted steam by reducing the superheat degree is an effective approach to further improve the thermal efficiency of USC power plants. Generally, an outer steam cooler is mostly used to effectively utilize the superheat of extracted steam, which involves setting a surface-type heat exchanger before the regenerative heater [9]. The superheated extracted steam enters the outer steam cooler, and some of its superheat is used to heat the feed water, which reduces the superheat degree and improves the feed water

temperature. Liu et al. [9] investigated the thermal performance of a steam cycle employing an outer steam cooler. Li et al. [10] conducted the thermodynamic and techno-economic analysis of a double reheat power plant with an outer steam cooler. A heat circulation calculation model for a power plant employing an outer steam cooler was established in Ref. [11]. These studies have demonstrated that using the outer steam cooler is an effective superheat utilization measure in power plants. Using the outer steam cooler can reduce the superheat degree of extracted steam by more than 100°C. However, the outer steam cooler could only utilize the superheat of special extracted steam.

Another effective method for superheat utilization of extracted steam is to employ a regenerative turbine (RT). In this system, part of the exhaust steam from the high-pressure turbine (HPT) flows directly into a RT without entering the reheater [12, 13]. Several extracted steam points are set in the RT to replace those in the intermediate-pressure turbine (IPT). The extracted steam from the RT is not reheated, and thus, the superheat degree of multi-stage extracted steam is effectively reduced in this system. This could achieve a reduced exergy destruction of the regenerative heaters (RHs) and improve the thermal performance of the whole power plant. Ploumen et al. [14] compared the thermal performance between a common steam cycle and the RT system. Cai et al. [15] conducted thermodynamic analysis of a RT system in a double reheat USC power plant. Xu et al. [16] also investigated the energy effect of double reheat power plants. The results of these studies indicated that the thermal efficiency could increase by employing a RT to utilize superheat of extracted steam.

Moreover, large USC power plants often operate under partial load for peak regulation. Han et al. [17] conducted a simulation study of a lignite-fired power system integrated with flue gas drying and waste heat recovery to present performances under variable power loads. Thermodynamic analysis of a solar aided coal-fired power plant under design/off-design conditions was studied in Ref. [18]. The superheat degree of extracted steam increases when the load decreases under the sliding pressure operation mode. Thus, it is necessary to conduct thermodynamic analysis of superheat utilization under off-design conditions. The RT system needs new equipment and facilities, resulting in an increased power plant investment. Rovira et al. [19] investigated thermodynamic and techno-economic analyses of combined cycle gas turbine power plants. However, few studies have focused on thermodynamic and techno-economic analysis of the RT system to utilize the superheat of the extracted steam in 1000 MW USC power plants, especially under off-design operation conditions. In addition, the isentropic efficiency of the

RT could influence the energy saving effect of superheat utilization, few studies have investigated the RT system with different isentropic efficiency.

Against this backdrop, in the present work, thermodynamic and techno-economic analyses of superheat utilization using the RT in a 1000 MW USC power plants were conducted under design/off-design conditions, to achieve the following objectives: (1) to conduct thermodynamic evaluation of superheat utilization system with different RT isentropic efficiency under design/off-design conditions; (2) to reveal the energy saving mechanism of the RT system based on exergy analysis of superheat utilization of the extracted steam; and (3) to assess the economic performance of the superheat utilization system using the RT.

Steam Superheat Utilization Using a Regenerative Steam Turbine in a Typical 1000 MW USC Power Plant

Reference system description

A typical USC power plant is selected as the reference system in this study. The simplified process flow diagram of the USC power plant is shown in Figure 1. The plant consists of one single-flow high-pressure turbine (HPT), one double-flow intermediate-pressure turbine (IPT), and two double-flow low-pressure turbine (LPT). All the HPT, IPT, and LPT sections in the turbine are connected to the generator by a common shaft. The steam from the exhaust of the HPT is returned to the boiler for reheating and sent to the double flow IPT. The exhaust steam from the IPT flows through the LPT system and goes into the

condenser (CON). The regenerative system of the steam cycle has eight-stage regenerative heaters (RHs), including four low-pressure RHs, three high-pressure RHs, and one deaerator (DEA).

The power plant is designed to generate live steam at the nominal conditions of 26.25 MPa and 600°C. The power output of the reference system under design condition which can be also termed as turbine heat acceptance (THA) condition is 1000 MW. The reheat steam is heated to 600 °C, and the exhaust steam pressure of LPT is set to 5.75 kPa. Table 1 lists major parameters of the extracted steam in the reference system.

Configuration of the RT system

As shown in Table 1, the superheat degree of the extracted steam in RH3, DEA and RH5 is higher than other extracted steam. Obviously, the extracted steam of RH3, DEA and RH5 is extracted from the IPT. Because the steam has been reheated during the reheat process and then enter the IPT, the temperature of the extracted steam from the IPT is high. If the superheat degree of the extracted steam in RH3, DEA and RH5 can be effectively utilized, the thermal performance of the reference system will be improved effectively. Using the RT is an appropriate approach to achieve the superheat utilization of the extracted steam in the reference system.

Figure 2 depicts a simplified process flow of steam cycle in the RT system. A portion of the exhaust steam of the HPT is sent to a RT instead of the reheater, and therefore, the extracted steam in RH3, DEA and RH5 will be extracted from the RT instead of being extracted from the IPT.

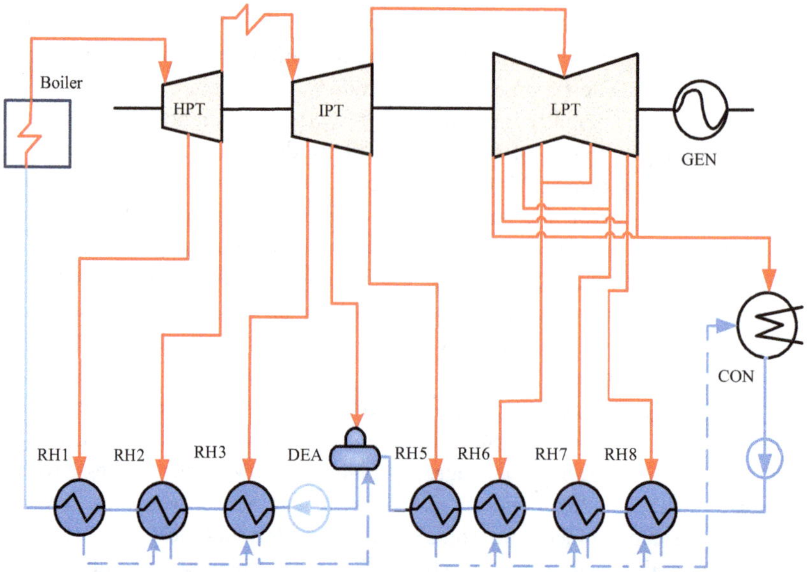

Figure 1. Process flow diagram of the steam cycle in reference system.

Table 1. Main parameters of extracted steam in reference system under design condition.

RHs	Steam pressure (MPa)	Steam temperature (°C)	Saturated temperature (°C)	Superheat degree (°C)
1	7.49	402.4	288.4	106.4
2	5.55	352.7	268.7	84.0
3	2.36	483.0	219.4	263.6
4	1.17	381.0	184.4	196.6
5	0.59	289.0	156.0	133.0
6	0.24	192.4	124.7	67.7
7	0.06	86.1	84.1	2.0

In this configuration, the superheat degree of the extracted steam in RH3, DEA and RH5 is significantly reduced because this portion of extracted steam will not be reheated. As a consequence, the temperature difference of the heat transfer process will be also dramatically reduced. Likewise, the steam entering the reheater of the boiler can be obviously reduced, and, as a result, the heat absorption of the boiler decreases with the reduction in the heat capacity of the reheater.

Thermodynamic Analysis of Steam Superheat Utilization Using RT Under Design/Off-Design Conditions

Main models considerations and thermodynamic evaluation criteria

The thermodynamic cycles of the power plant are simulated using EBSILON Professional in this study [20].

The selection of an accurate method is essential to ensure the precision and reliability of simulation results. EBSILON Professional is a widely used power plant simulation tool whose main purpose is to calculate thermodynamic quantities including enthalpies, pressures and mass flows in the steam cycle, and is restricted to thermodynamic equilibrium states to describe plant components [21–23]. Thermodynamic models of power plants, which can create a set of heat balance data that complies with mass and energy balances, are utilized in the simulation process.

In the present study, the model descriptions of the main components are as follows. The boiler with double reheat is modeled as a black box. The inlet pressure and the isentropic efficiencies are defined in the steam turbine. In most cases, the outlet pressure is determined by the inlet pressure of the following turbine stage. In the last turbine stage, the outlet pressure is determined by the inlet pressure of the condenser. For the RHs, the terminal temperature difference of the primary heater (i.e., the temperature difference between the saturated steam and the heated primary water) and the terminal temperature difference of the after-cooler (i.e., the temperature difference between the drain and the heated primary water) are to be specified. The inlet temperature and pressure of the cooling medium for the condenser are also specific.

In addition, the operation of the power plant is considered to be in a steady state and the mean isentropic efficiencies of the HPT, IPT, and LPT are equal to 0.90, 0.93, and 0.89, respectively.

The heat consumption rate and thermal efficiency are commonly used in the electric power industry to evaluate

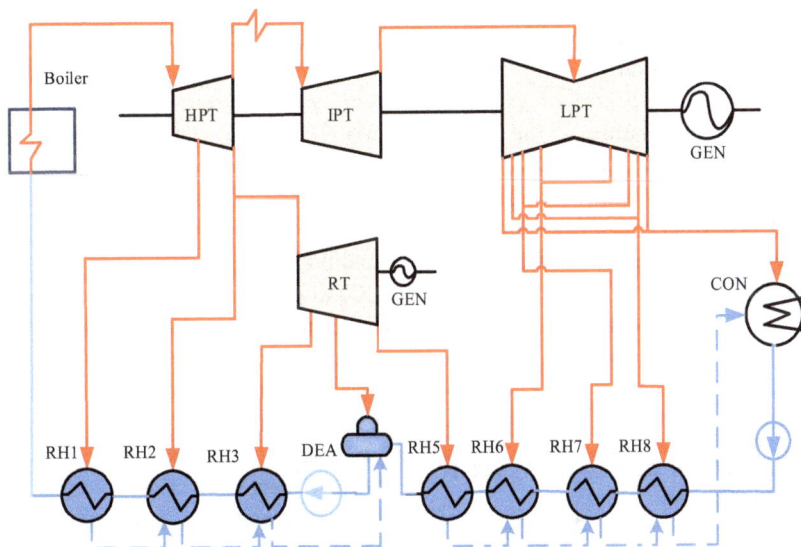

Figure 2. Process flow diagram of the steam cycle in RT system.

the thermal performance of power generation units, which can be defined as follows:

$$q = \frac{E_{total} \times 3600}{P_{gen}} = 3600 / \frac{P_{gen}}{E_{total}}, \quad (1)$$

$$\eta = \frac{P_{gen}}{E_{total}}, \quad (2)$$

where E_{total} refers to the total energy input per unit time. E_{total} is the total energy input including the chemical energy of coal, the energy of air and the energy of makeup water. To simplify the calculation, the quantitative value of E_{total} is considered as the chemical energy of coal, which is equivalent to the lower heating value (LHV) of coal input per unit time.

P_{gen} refers to the generated electric power. The number 3600 refers to 3600 kJ/kWh. The unit of heat consumption rate q is kJ/kWh.

Thermodynamic evaluation of the RT system under design condition

Thermodynamic analysis for the RT system and the reference system is conducted through the simulation. The isentropic efficiency of the RT is set to 85%. The extracted steam pressure of the RT system is equivalent to that of the reference system in simulation. Table 2 gives the major parameters of the steam cycle and thermal performance of the reference system and the RT system under design condition.

Table 2 shows that the live steam flow rate increases in the RT system because the mass flow of the extracted steam increases as the superheat degree decreases. The heat consumption rate of the RT system is significantly reduced by 46.9 kJ/kWh. The thermal efficiency of the RT system is increased by 0.27%-points.

Figure 3 compares the superheat degree of extracted steam in the two systems. The temperatures of the related extracted steam of RH3, DEA and RH5 are reduced after employing the regenerative steam turbine. Consequently, the superheat degree of the related extracted steam is also reduced. The superheat degree of the related extracted steam in RH3, DEA and RH5 extracted from the IPT is high in the reference system, which is effectively reduced by extracting from the regenerative steam turbine without being reheated in the RT system.

Thermodynamic evaluation of the RT system under off-design conditions

Considering that large USC power plants need to frequently operate under off-design operation condition, it

Table 2. Comparisons of major parameters and thermal performances of reference system and RT system under design condition.

Items	Reference system	RT system
Live steam flow rate (kg/sec)	750.0	770.3
Live steam pressure (MPa)	26.25	26.25
Live steam temperature (°C)	600.0	600.0
Reheated steam pressure (MPa)	5.0	5.0
Reheated steam temperature(°C)	600.0	600.0
Feed water pressure (MPa)	32.7	32.7
Feed water temperature (°C)	295.4	295.4
Heat consumption rate (kJ/kWh)	7969.8	7922.9
Decrease of heat consumption rate (kJ/kWh)	–	46.9
Thermal efficiency (%)	45.17	45.44
Increase in thermal efficiency (%-points)	–	0.27

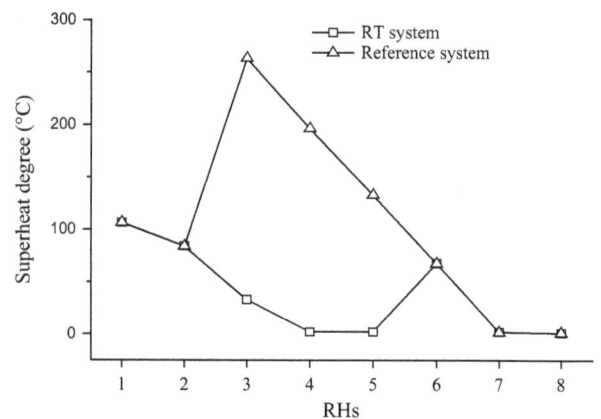

Figure 3. Superheat degree of the extracted steam in RT system and reference system under design condition.

is necessary to study the thermal performance of a 1000 MW USC power plants under off-design conditions. Four typical operation conditions, including THA load, 75% THA load, 50% THA load, and 40% THA load conditions are selected for analysis in this study. Table 3 presents the superheat degree of extracted steam in the reference system under design and off-design operation conditions.

Obviously, the superheat degree increases at off-design conditions because the large power generation units always adopt the sliding pressure operation mode [24]. That is, the live steam pressure and live steam mass flow decrease correspondingly as the power output decreases. However, live steam temperature normally remains constant when the load decreases. Therefore, the pressure of extracted steam decreases gradually as the load decreases while the temperature remains almost unchanged. This indicates that the super heat degree will even greater and the

Table 3. Superheat degree of extracted steam in reference system under different loads.

RHs	THA	75%THA	50%THA	40%THA
1#	119.2	127.4	158.1	171.4
2#	76.8	84.7	112.6	124.3
3#	284.0	299.9	319.9	329.6
4#	212.1	226.8	245.7	254.9
5#	162.2	175.7	193.2	202.9
6#	114.7	126.9	142.9	152.7
7#	59.5	70.4	84.2	94.9
8#	11.2	21.1	33.7	45.3

Table 4. Heat consumption rate of the RT system and reference system.

Load condition	Reference system	RT system	Decrement
100%THA	7969.8	7922.9	46.9
75%THA	8098.5	8060.5	38.0
50% THA	8385.4	8359.0	26.4
40%THA	8654.6	8634.1	20.5

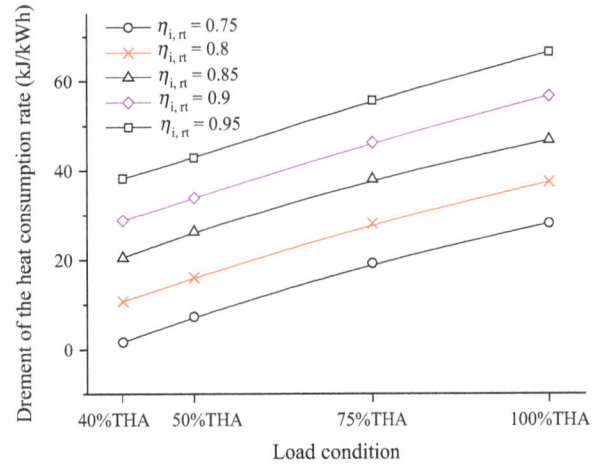

Figure 4. Decrement of heat consumption rate between RT system and reference system.

superheat utilization of extracted steam is more important under off-design condition.

Table 4 compares the heat consumption rate comparison of the reference system and the RT system under design and off-design conditions. The heat consumption rate of each system increases when the load decreases. Through adopting the RT, the heat consumption rate decreases more significantly than that of the reference system. In addition, the heat consumption rate of the regenerative system will further decrease when the isentropic efficiency of the RT increases.

Considering that the isentropic efficiency of the RT has important influence on the thermal performance of the RT system, Figure 4 illustrates the decrement of the heat consumption rate between the reference system and the RT system with different RT isentropic efficiency under design and off-design conditions. It can be observed that the decrement of heat consumption rate of each RT system increases when the load increases. Therefore, the energy saving effect of RT system is more obvious as the load increases. Moreover, the energy saving effect of the RT system becomes greater with higher RT isentropic efficiency.

Exergy analysis of RT system and reference system

As mentioned above, the thermodynamic evaluation shows that the RT system can effectively utilize the superheat of extracted steam, that is, the superheat degree of the related extracted steam is reduced obviously by adopting the RT. As a result, the energy saving effect of the RT system is considerable to achieve the superheat utilization. When the load increases, the energy saving effect of the RT system is more obvious. To further reveal the mechanism of superheat utilization, the comparison of exergy analysis of the related regenerative heaters is performed for the reference system and the RT system.

Clearly, the heat transfer temperature difference and the superheat degree of the extracted steam in RH3, DEA and RH5 will be reduced, leading to a decrease of exergy destruction of the heat transfer process in the regenerative heater.

Figure 5 illustrates the T-S diagram of the heat transfer process between the extracted steam and the feed water or condensate water in the regenerative heater. The entropy generation and the exergy destruction are the measurement indicators of the thermodynamic irreversibility. Entropy on the abscissa represents the entropy of the working medium. T on the ordinate denotes the temperature of the working medium. T_0 represents the environment temperature. The process 5-4-3 represents the exothermic process of the extracted steam, and the process 1-2 represents the endothermic process of the feed water or condensate water. The entropy increase during the heat transfer process in the regenerative heater is ΔS. When the superheat degree of the extracted steam decreases, the exothermic process of the extracted steam could be exhibited as 6-7-3 and the endothermic process of the feed water or condensate water is still shown as 1-2. Obviously, the entropy increase will be reduced by δS caused by the reduced heat transfer temperature difference. As a result, the exergy destruction will be reduced, which can be represented by the shaded part in Figure 5.

The general exergy balance of the system components can be expressed as:

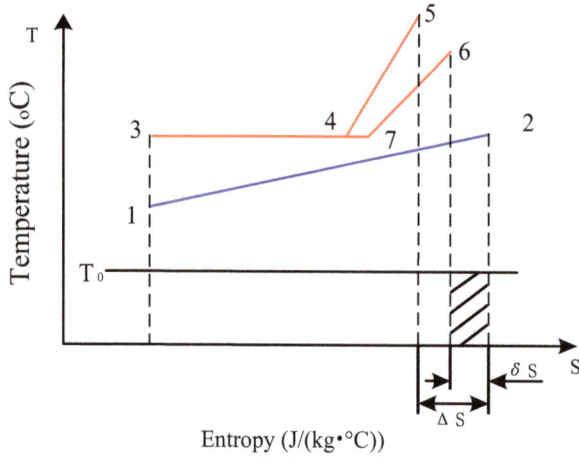

Figure 5. T-S diagram of the heat transfer process of RHs.

$$\sum E_{ex,in} + \sum W_{in} = \sum E_{ex,out} + \sum W_{out} + E_{ex,loss} + \Delta E_{ex}. \quad (3)$$

In equation (3), ΔE_{ex} refers to the exergy destruction; $E_{ex,in}$ and W_{in} refer to the exergy input and the power input, respectively; $E_{ex,out}$ and W_{out} denote the exergy output and the power output, respectively. $E_{ex,loss}$ represents the exergy loss caused by energy loss. However, in this study, there is very little heat dissipation during the energy conversion process, therefore, the simulation models of the systems are considered as adiabatic process and the heat dissipation is reasonably ignored, and $E_{ex, loss}$ in the energy conversion process is zero [25, 26].

The exergy destruction within a special component of the energy conversion system can be derived from the exergy balance equation. For the regenerative heater, there is no W_{in} and W_{out} during the heat transfer process. As a result, the general exergy balance of the regenerative heater can be expressed in the following rate form:

$$\sum E_{ex,in} = \sum E_{ex,out} + \Delta E_{ex}. \quad (4)$$

Figure 6 shows the total exergy destruction of RH3, DEA and RH5 in the RT system and the reference system under design and off-design conditions. The two solid line denotes the total exergy destruction of related RHs (RH3-RH5) in the reference system and the RT system, respectively. The total exergy destruction of RH3, DEA and RH5 in the RT system is effectively reduced as compared to that of the reference system. In addition, the total exergy destruction decrement of related RHs between the reference system and the RT system decreases as the load drops down, as shown by the dashed line in Figure 6.

To further study the reason of different energy saving effect between the RT system and the reference system

under off-design conditions, the exergy destruction decreases the coefficient of RHs, which represents the ratio of the RHs' exergy destruction difference between the RT system and the reference system to the RHs' exergy destruction of the reference system, is proposed in this study. Figure 7 gives the decrease coefficient of the exergy destruction of the RT system. It can be observed that the exergy destruction decrease coefficient of the RHs increases as the load increases in the RT system. Thus, the decrement of heat consumption rate of RT system increases when the load increases and the energy saving effect becomes more obvious as the load increases.

Economic Analysis of the RT System

Basic evaluation

As discussed above, the RT system can effectively utilize the superheat of extracted steam in related RHs. As such, it is expected that new equipment and facilities, such as RT and steam pipes, will be added to the reference system. This will lead to the increasing power plant investment with rising equipment cost and the operating and management cost.

In the RT system, the net economic benefit can be defined as follows:

$$\Delta I_{net} = \Delta I_{es} - \Delta ICC \times CRF - \Delta C_{O\&M}, \quad (5)$$

where ΔI_{net}, ΔI_{es}, ΔICC and $\Delta C_{O\&M}$ denote the net economic benefit, the income of energy saving, total additional installed capital costs and the annual operating and management costs, respectively, and CRF is the capital recovery factor.

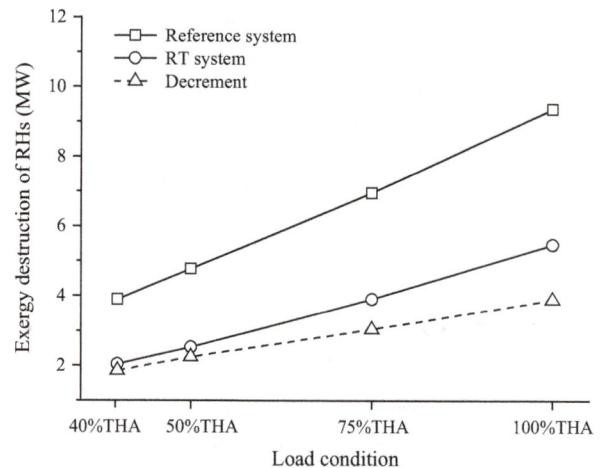

Figure 6. Exergy destruction comparisons of RHs in reference system and RT system under different load conditions.

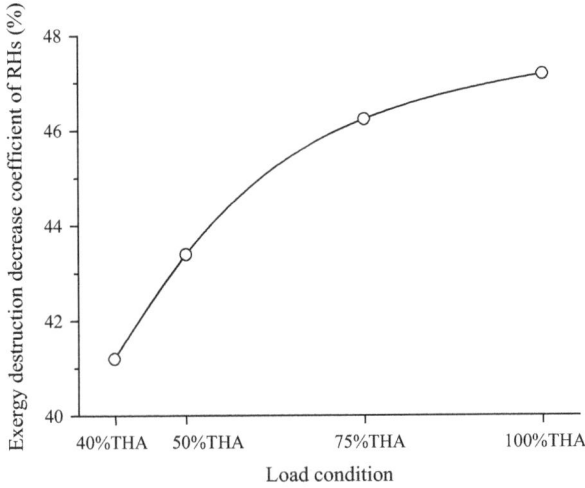

Figure 7. Exergy destruction decrease coefficient of RHs in RT system.

The ΔI_{es} is calculated as follows:

$$\Delta I_{es} = \Delta q \, P_F \, h_{eq} \, P_{gen}, \qquad (6)$$

where Δq refers to the reduction in the heat consumption (kJ/kWh), P_F refers to the price of fuel ($/J LHV), h_{eq} denotes the equivalent operation hours per year (h), and P_{gen} is the power output of the power plant (MW).

The capital recovery factor (CRF) is related to the discounted rate (k) and the equipment life span (n). Because k refers to the fraction interest rate per year, and n refers to the number of years that the capital has been borrowed over a fixed rate of interest [27]. CRF can be calculated with the following as follows:

$$CRF = [k \cdot (1+k)^n] / [(1+k)^n - 1]. \qquad (7)$$

The operating and management cost ($C_{O\&M}$) is assumed to be proportional to ICC, with the proportionality coefficient of γ set to 4% [28].

The basic assumptions for net economic benefit calculation are listed in Table 5.

The values for total additional installed capital costs (ICC) associated with added equipment and the construction of the RT system to achieve steam superheat utilization are illustrated in Table 6 [29, 30]. ICC includes the investment cost of the added equipment and related auxiliary fees (e.g., construction and installation cost) in the project. As illustrated in Table 6, when the RT is implemented, the additional ICC of the power plant increased by 3.02 M$.

Table 7 presents the results of techno-economic analysis for the RT system. The heat consumption decrement of the regenerative system is 46.9 kJ/kWh, and the ΔI_{es} of the RT system reach 1.06 M$. Eventually, the values of net economic benefit of the RT system are 0.57 M$. The RT

Table 5. Basic assumptions for net economic benefit calculation.

Investment index	Value
Fuel price (P_F)[1]	4.09 $/GJ
Annual discount rate[2] (k)	0.12
Plant economic life[2] (n)	30 years
Annual operation hours (h_{eq})	5500 h/year

[1]The coal price is based on Ref [10].
[2]Plant economic lifetime and discount rate are according to Ref. [28].

Table 6. Additional installed capital costs of the RT system.

Investment index	Values (M$)
Regenerative turbine	2.16
Pipes	0.72
Construction and installation costs	0.14
Total additional installed capital costs	3.02

Table 7. Annual economic performance of the RT system.

Performance index	Values
Decrement of heat consumption (kJ/kWh)	46.9
Extra economic benefit of energy saving (M$)	1.06
Annualized additional installed capital costs (M$)	0.37
Annualized operation and maintenance costs (M$)	0.12
Net economic benefit (M$)	0.57

system can bring considerable economic benefits by superheat utilization of extracted steam of related regenerative heaters.

Sensitivity analysis

According to the economic evaluation (cf. Section 5.1), the fluctuations of key economic parameters will influence the economic performance of the power plant with RT. Figure 8 shows the net economic benefit of the RT system as functions of percent changes in additional ICC (ΔICC), fuel price, discounted rate and annual operation hours, respectively.

Obviously, the net economic benefit of the RT system increased as fuel price and annual operation hours increased, but showed a decrease with the increase of additional ICC and discounted rate. Besides, the net economic benefit of the RT system was highly depended on the fuel price and the annual operation hours. This is due to the fact that fuel is the sole input of the power plant and thus these two factors directly influence the income of energy saving. Quantitatively, a 20% increase in fuel price would cause the net economic benefit to increase from $0.57M to $0.77M in the RT system. By contrast, when the additional ICC of the RT system increased by 20% compared to the baseline value, the net economic

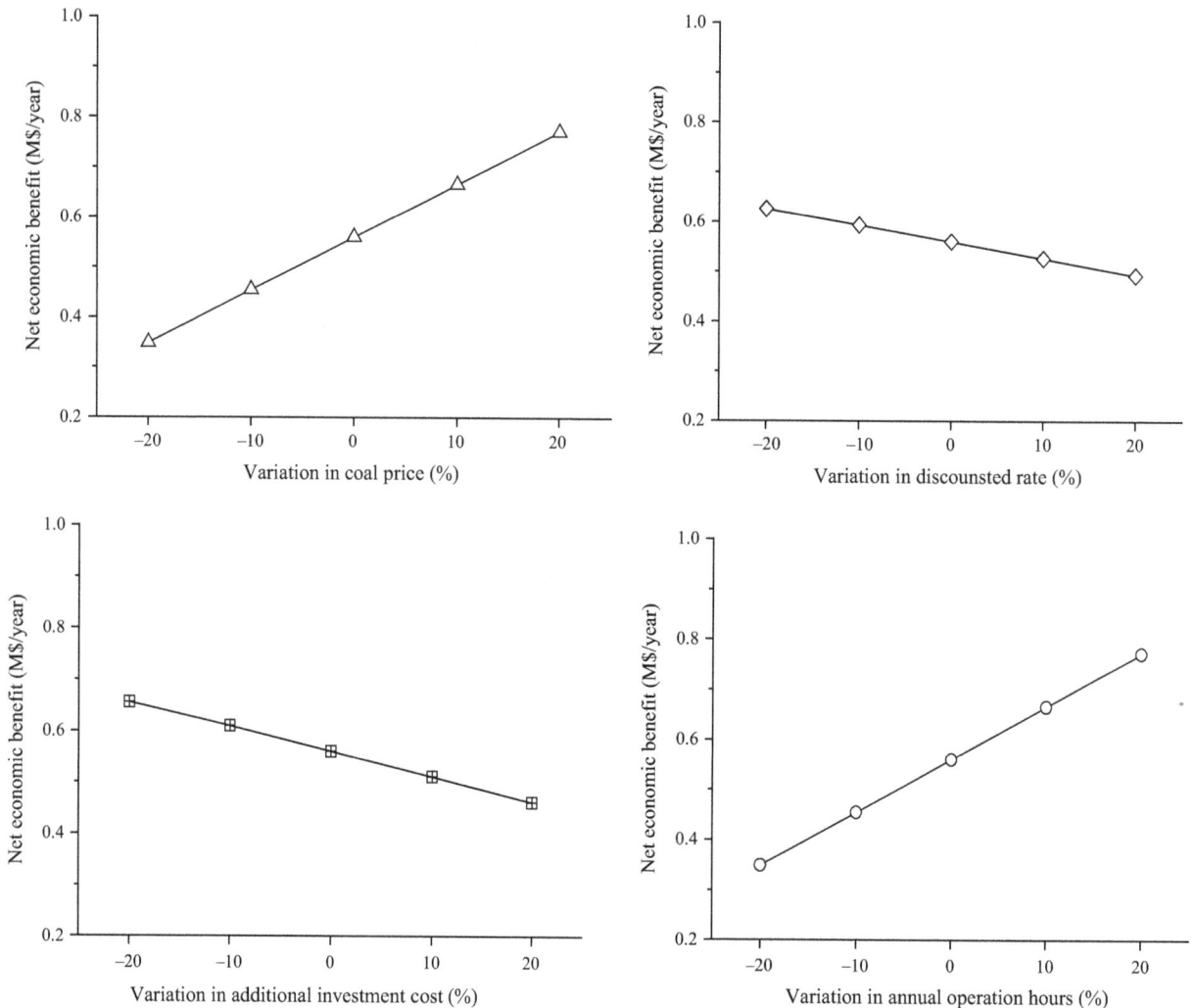

Figure 8. Influence of the in fuel price, discounted rate, additional ICC and annual operation hours on net economic benefit of RT system.

benefit just slightly decreased from $0.57M to $0.46M. This is because that the influence of the additional ICC of the RT system on the economic benefit is weakened by the factor of CRF.

In general, the sensitivity analysis revealed that the net economic benefit of the RT system were more sensitive to the fuel price and annual operation hours and relatively less sensitive to the additional ICC and discounted rate. More importantly, within a wide range fluctuations of power market conditions, the RT system will still performed economically as compared to the reference system without RT.

Conclusion

The thermodynamic and economic analyses of the regenerative steam turbine system in a 1000 MW USC power plant were conducted in this study. The RT system exhibits

better performance than the reference system under design/off design condition. The results reveal the followings.

1. The heat consumption rate of the RT system is significantly reduced by 46.9 kJ/kWh than that of the reference system under design condition, which reveals the significant energy saving effect of steam superheat utilization.
2. The RT system exhibits different energy saving effects under off-design conditions. The heat consumption rate decrement of the RT system decreases from 46.9 kJ/kWh to 20.5 kJ/kWh as the load of the power plant decreases from THA condition to 40% THA condition. However, a more obvious energy saving effect with higher isentropic efficiency of the RT can be expected no matter under design or off-design conditions.
3. Economic analysis results showed that the remarkable energy saving effect overweighed the increase in ICC, and consequently, the RT system featured considerable

economic benefit. Besides, the RT system would still perform economically within a wide range of fluctuations of the power market conditions.

Acknowledgments

This study is supported by National Basic Research Program of China (Grant No. 2015CB251504), the National Nature Science Fund of China (No. 51476053). This paper is also supported by Fundamental Research Funds for the Central Universities (No. 2017MS013).

Conflict of Interest

None declared.

References

1. Zhang, D. K. 2013. Ultra-supercritical coal power plants: materials, technologies and optimisation. Woodhead Publishing, Philadelphia, PA.

2. Tzolakisa, G., P. Papanikolaoua, D. Kolokotronisa et al. 2012. Simulation of a coal-fired power plant using mathematical programming algorithms in order to optimize its efficiency. Appl. Therm. Eng. 48:256–267.

3. China Electricity Council. 2014. The current status of air pollution control for coal-fired power plants in China 2014. China Electric Power Press, Beijing.

4. Yang, Y. P., C. Xu, G. Xu et al. 2015. A new conceptual cold-end design of boilers for coal-fired power plants with waste heat recovery. Energy Convers. Manage. 89:137–146.

5. Chaaban, F. B., T. Mezher, and M. Ouwayjan. 2004. Options for emissions reduction from power plants: an economic evaluation. Elect. Power Energy Syst. 26:57–63.

6. Nicol, K.. 2015. Application and development prospects of double-reheat coal-fired power units. IEA Clean Coal Center, London, UK.

7. Benidris, M., S. Elsaiah, and J. Mitra. 2016. An emission-constrained approach to power system expansion planning. Elect. Power Energy Syst. 81:78–86.

8. Song, Z. P., and J. X. Wang. 1985. Energy-saving theory. China Water Power Press, Beijing.

9. Liu, Z. Z., Y. Li, and X. H. Cheng. 2004. Analysis of thermal economic benefits of first stage extraction location after reheat of reheat steam turbine set with outside steam cooler. Turbine Technol. 05:382–384 [in Chinese].

10. Li, Y. Y., L. Y. Zhou, G. Xu et al. 2014. Thermodynamic analysis and optimization of a double reheat system in an ultra-supercritical power plant. Energy 74:202–214.

11. Li, J. G., X. K. Yang, L. P. Li et al. 2004. A study of circulation heat calculating model for thermal system with outer steam cooler. Turbine Technol. 46:344–346 [in Chinese].

12. Kjaer, S., and F. Drinhaus. 2010. A modified double reheat cycle. Proceeding of the ASME 2010 Power Conference, Pp. 45–47.

13. Blum, R., S. Kjaer, and J. Bugge. 2007. Development of a PF Fired High Efficiency Power Plant (AD700). Energy Solutions for Sustainable Development Proceedings.

14. Ploumen, P., G. Stienstra, and H. Kamphuis. 2011. Reduction of CO_2 emissions of coal fired power plants by optimizing steam water cycle. Energy Proc. 4:2074–2081.

15. Cai, X. X., Y. P. Zhang, Y. Li et al. 2012. Design and exergy analysis on thermodynamic system of a 700°C ultra-supercritical coal-fired power generating set. J. Chin. Soc. Power Eng. 32:971–978.

16. Xu, G., L. Zhou, S. Zhao et al. 2015. Optimum superheat utilization of extraction steam in double reheat ultra-supercritical power plants. Appl. Energy 160:863–872.

17. Han, X. Q., M. Liu, J. S. Wang et al. 2014. Simulation study on lignite-fired power system integrated with flue gas drying and waste heat recovery-Performances under variable power loads coupled with off-design parameters. Energy 76:406–418.

18. Peng, S., H. Hong, Y. J. Wang et al. 2014. Off-design thermodynamic performances on typical days of a 330 MW solar aided coal-fired power plant in China. Appl. Energy 130:500–509.

19. Rovira, A., C. Sánchez, M. Munoz, M. Valdés et al. 2011. Thermoeconomic optimisation of heat recovery steam generators of combined cycle gas turbine power plants considering off-design operation. Energy Convers. Manage. 52:1840–1849.

20. STEAG Energy Services GmbH, EBSILON Professional; 2010.

21. Cifre, P. G., K. Brechtel, S. Hoch, H. Garcia, N. Asprion, H. Hasse et al. 2009. Integration of a chemical process model in a power plant modelling tool for the simulation of an amine based CO_2 scrubber. Fuel 88:2481–2488.

22. Erlach, B., B. Harder, and G. Tsatsaronis. 2012. Combined hydrothermal carbonization and gasification of biomass with carbon capture. Energy 45:329–338.

23. Bruhn, M. 2002. Hybrid geothermal–fossil electricity generation from low enthalpy geothermal resources: geothermal feed-water preheating in conventional power plants. Energy 27:329–346.

24. The Development Way and the Strategic Key Energy of Energy during 12th Five Years Period. Energy Observation Nets, 2010 [in Chinese].

25. Aljundi, I. H. 2009. Energy and exergy analysis of a steam power plant in Jordan. Appl. Therm. Eng. 29:324–328.

26. Restrepo, Á., R. Miyak, F. Kleveston, and E. Bazzo. 2012. Exergetic and environmental analysis of a pulverized coal power plant. Energy 45:329–338.

27. Xu, G., H. G. Jin, Y. P. Yng et al. 2010. A comprehensive economic analysis method for power generation system with CO_2 capture. Int. J. Energy Res. 34:321–332.

28. Thomas, B., W. Robert, C. Stefano, and C. Paolo. 2005. Co-production of hydrogen, electricity and CO_2 from coal with commercially ready technology. Part B: economic analysis. Int. J. Hydrogen Energy 30:769–784.

29. China Power Engineering Consulting Group Corporation. Reference price index of thermal power engineering design. China Electric Power Press, Beijing, China. [in Chinese].

30. Xu, G., Y. Wu, Y. P. Yang et al. 2013. A novel integrated system with power generation, CO_2 capture, and heat supply. Appl. Therm. Eng. 61:110–120.

14

Review of passive heating/cooling systems of buildings

Neha Gupta[1] & Gopal N. Tiwari[2]

[1]Centre for Energy Studies, Indian Institute of Technology Delhi Hauz Khas, New Delhi 110016, India
[2]Bag Energy Research Society (BERS), 11B, Gyan Khand IV, Indirapuram, Ghaziabad 201010, Uttar Pradesh, India

Keywords
Passive cooling, passive heating, passive solar techniques, thermal comfort

Correspondence
Gopal N. Tiwari, Bag Energy Research Society (BERS), 11B, Gyan Khand IV, Indirapuram, Ghaziabad 201010, UP, India.
E-mail: gntiwari@ces.iitd.ernet.in

Funding Information
Global Technology Watch Group, Department of Science and Technology, Government of India (Grant/Award Number: "100/IFD/3118/2014-2015").

Abstract

In this review, an attempt has been made to analyze passive solar heating and cooling concepts along with their effects on performance of a building's thermal management. The concepts of Trombe wall, solarium, evaporative cooling, ventilation, radiative cooling, wind tower, earth air heat exchanger, roof pond, solar shading for buildings, and building-integrated photovoltaic thermal (BiPVT) systems are extensively covered in this review. Comparison of results by various heating and cooling concepts has been made. It has been observed that direct heating through double-glazed window saves maximum conventional fuel for thermal heating during winter months. Further, an evaporative cooling is one of the best cooling concepts which is economical too in summer period.

Highlights

1. Double-glazed window for thermal heating.
2. Combination of evaporative cooling and wind tower for passive cooling.
3. Trombe wall for both passive heating and cooling.
4. BiPVT system for thermal heating and electricity production.

Introduction

An estimated 6.7% of the global energy requirement is used for thermal management of buildings for which precise Arab data are not available [1]. As per an estimate, at least 35% of the total building's energy requirements may be satisfied by using alternative resources [1]. The world's total energy requirement may be reduced by 2.35% at an approximate incremental cost of 15% over the total construction and planning cost [1, 2]. As a proof of concept, the solar H.P. Co-operative building in India was able to reduce heat loss by 35% by using double-glazing solar passive design for thermal heating [3]. Different physical processes for providing thermal comfort for passive buildings include solar radiation, long-wave radiation exchange, radiative cooling, and evaporative cooling. Solar radiation and radiative cooling are the processes used for both thermal heating and cooling purposes [1].

Passive solar design is used as a cost and resource efficient method for achieving natural harmony between climate, architecture, and people. The building structure should be self-sustainable that is generating the energy for its own consumption.

Classification of various heating, cooling, and heating/cooling passive concepts has been done in Table 1 including building-integrated photovoltaic thermal (BiPVT) systems.

Table 1. Classification of heating and cooling concepts.

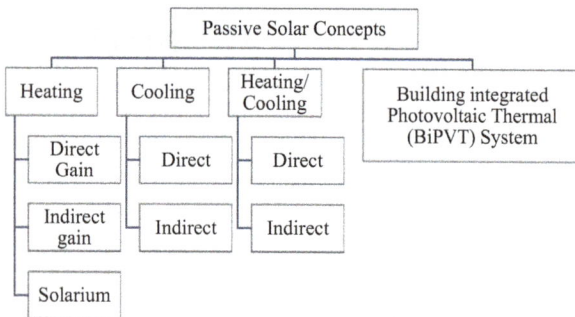

Passive solar design strategies for passive house

Passive design strategies are related to various aspects of a building like shape, size, orientation, form, site, layout etc. Their impact can be easily seen in the performance of the building's energy use. This is because the massing and layout of the buildings can generate self-shading effects and can enhance the ventilation and natural lighting. At hardly any additional cost in adapting design strategies like building orientation, we can significantly achieve an effect on the building's thermal performance. Depecker et al. [4] investigated the relationship between shape coefficient and heating loads. The study revealed that the heating load in cold climatic conditions was directly proportional to the shape coefficient, the reason being that the solar gain through the glazing was low. The study also suggested that opaque walls do not have any association with either heating load or shape coefficient, thus bearing less importance in mild and sunny climatic conditions. Stevanović [5] reviewed the heat flow across a building envelope and revealed that an optimized building form can reduce the heating load up to 12% per total volume of the building. Aldawoud [6] discussed different geometries of the atrium for the energy saving performance and concluded that square-shaped atrium has the best performance. The results also suggested that more energy savings can be achieved in low rise structure with an atrium which has a larger glazing to roof ratio for temperate and cold climatic condition. However, for hot and dry climatic conditions, high structure with a low glazing to roof ratio is preferred. An algorithm based on self-shading building envelopes for reduction in cooling and lighting loads integrated with different façade types, glazing, orientations, shapes, and life cycle costs has been developed [7, 8]. Square shape was found to give better results for all climate types. Stevanović [5] discussed that the detached units placed in curved road configurations requires large heating and cooling when compared to the attached units placed in straight layout. The latter saving

up to 30% cooling and 50% heating requirements. Energy savings of 1–5% may be achieved just by changing the orientation, aspect ratio, and shape factors [9, 10]. The study concluded that the aspect ratio has very less influence on the building's energy use. The study also concluded that with increase in the size of south-facing window, there is corresponding decrease in the total annual load in cold climatic conditions and the same increases in warm climatic conditions.

Passive Heating Concepts

The various heating concepts in brief have been summarized in Table 2 with remarks. Few concepts have been discussed in detail as follows:

Direct gain

In this case, solar radiation is directly transmitted through glazed window into the living room for thermal heating. During the day, the whole building structure collects, absorbs, and stores the heat and releases the heat at night for thermal heating as shown in Figure 1.

The rate of heat transfer into the room can be expressed as follows,

$$\dot{Q} = \left(\frac{1}{U_t} + \frac{L}{K} + \frac{1}{h_i} \right)^{-1} (T_{sa} - T_r) = U_L (T_{sa} - T_r) \quad (1)$$

where,

$$T_{sa} = \frac{\tau I(t)}{U_t} + T_a \text{ and } U_t = \left(\frac{1}{h_0} + \frac{1}{h_i} \right)^{-1}$$

Balcomb et al. [11] suggested that single-glazed south-oriented system without storage mass is comparatively incompetent, whereas double-glazed system along with night insulation and soundly storage mass heat capacity proves to be efficient in meeting the heating requirements. Percentage of solar annual heating with storage heat capacity of 400 kJ/m^{2}gC for single-glazed system, double-glazed system, single-glazed system with night insulations and

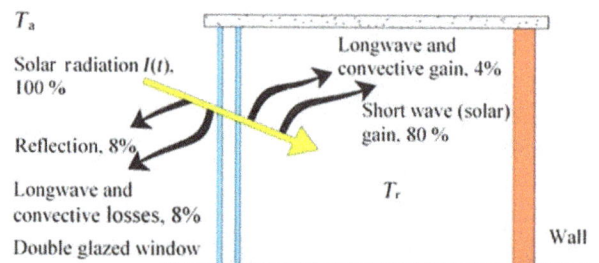

Figure 1. Direct gain during sun shine hour.

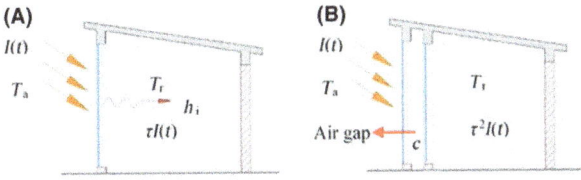

Figure 2. (A) Single glazing. (B) Double glazing.

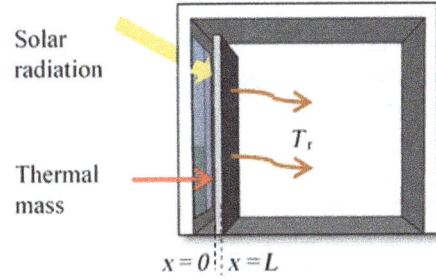

Figure 3. Indirect gain during sunshine hours.

double-glazed system with night insulation was observed to be about 10, 65, 80, and 90%, respectively. For 45° tilt, percentage of annual storage heating turns out to be nearly 87, 80, 58, and 18% for active system, double-glazed system with night insulation, double-glazed system and single-glazed system, respectively, with ratio of glass area to house area of 0.4 [1]. This means that double-glazed with night insulation is as good as active system tilted near the optimum angle. According to the authors, double glazing should be used in order to reduce heat loss from room to outside air, and windows should be covered by insulation at night to solve the same problem. Single-glazed system and double-glazed system are shown in Fig. 2A and B, respectively.

Rate of useful energy for single-glazed system,

$$\dot{q}_u = \tau I(t) - U_t(T_r - T_a) \qquad (2a)$$

where,

$$U_t = \left[\frac{1}{h_0} + \frac{L_g}{K_g} + \frac{1}{h_i}\right]^{-1} = \left[\frac{1}{5.8} + \frac{1}{2.8}\right]^{-1}$$

$$= 1.88 \, \text{W/m}^2, \left(\text{neglecting} \frac{L_g}{K_g}\right)$$

Rate of useful energy for double-glazed system,

$$\dot{q}_u = \tau^2 I(t) - U(T_r - T_a) \qquad (2b)$$

where,

$$U = \left[\frac{1}{h_0} + \frac{1}{c} + \frac{1}{h_i}\right]^{-1} = \left[\frac{1}{5.8} + \frac{1}{4.8} + \frac{1}{2.8}\right]^{-1} = 1.35 \, \text{W/m}^2$$

Reduction of losses can be calculated as $\left[\dfrac{1.88 - 1.35}{1.88}\right] = 28\%$

$$(2c)$$

The air gap reduces the heat transfer by conduction since air is a poor conductor. This proves that a reduction of 9% in heat gain and a reduction of 28% in losses can be achieved by using double-glazed system when compared with single-glazed system.

Indirect gain

In indirect gain, the heat is allowed to enter through glazing and is stored in the thermal mass. The heat is then transferred to the room via conduction and convection (Fig. 3).

Indirect gain concepts include Trombe wall, water wall and trans wall.

The rate of heat transfer into the room can be expressed as follows,

$$\dot{Q} = \left(\frac{1}{U_t} + \frac{L}{K} + \frac{1}{h_i}\right)^{-1}(T_{sa} - T_r) = U_L(T_{sa} - T_r) \qquad (3)$$

where,

$$T_{sa} = \frac{\alpha \tau I(t)}{U_t} + T_a$$

The thickness, surface area, material, and thermal properties of the thermal mass controls the heat flow inside the room. The thermal mass is dark colored for maximum efficiency and promote absorption of solar radiation. Reduction in building's heating demands with use of passive heating concepts has been reported to be about 25% annually in various studies [12].

Solarium

Solarium is a unification of direct gain and thermal storage concepts. Solarium consists of three sections namely

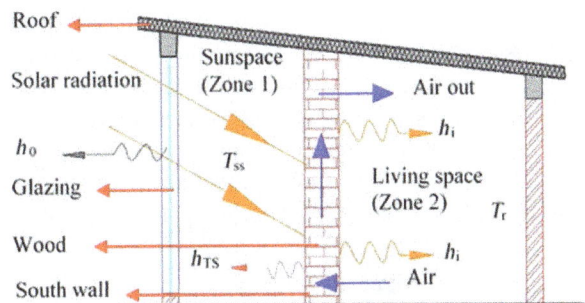

Figure 4. Cross-section view of solarium-cum-passive solar house.

sunspace with thick mass wall on the south side (for Northern hemisphere), linking space, and living space as shown in Figure 4. The thermal wall between the living space and sunspace helps in heat retention and distribution, thus improving the efficiency. The sunspace collects the energy through the glazing, absorbs it, and prewarm air for the living space. The sunspace works on the direct gain principle, in which the heat is used to maintain the temperature suitable for its transfer to the living space.

The rate of heat transfer into the living zone can be expressed as follows,

$$\dot{Q}=\left(\frac{1}{h_m}+\frac{L}{K}+\frac{1}{h_i}\right)^{-1}(T_{sa}-T_r)=U(T_{sa}-T_r)\quad(4)$$

where,

$$T_{sa}=\frac{\alpha\tau I(t)}{U_t}+T_a \text{ and } h_m=\left(\frac{1}{h_0}+\frac{1}{h_{TS}}\right)^{-1}$$

Tiwari and Kumar [13] have presented a thermal analysis of a solarium-cum-passive house as represented in Figure 4. The findings of the study were based on the energy balance of different components, link walls (like air collector, trans wall, water wall, metallic sheet). The findings have been based on best Thermal Load Levelling (TLL).

Variation in room air temperature and ambient temperature were noticed due to the solar intensity necessitating an estimate of the room temperature variations. TLL can be calculated using the following expression:

$$\text{TLL}=\frac{(T_{r,\max}-T_{r,\min})}{(T_{r,\max}+T_{r,\min})}\quad(5)$$

Decrement factor (f) can be defined as the reduction ratio of inside surface temperature of a room to the outside surface temperature of the room and can be expressed as follows:

$$f=\frac{(T|_{x=L})_{\max}-(T|_{x=L})_{\min}}{(T|_{x=0})_{\max}-(T|_{x=0})_{\max}}\quad(6)$$

Temperature fluctuations in the living room were noticed in case of air collector as link wall and resolved using an additional thermal mass. With use of air collector as link wall, along with effect of isothermal mass in living space and presence of movable insulation, drop in maximum temperatures of both the zones were found to be 3–4°C.

Based on Table 2, one can observe that direct gain is more convenient for sunshine hours heating (office) and rest of the concepts are used for residential buildings. Solarium will be useful for both the applications.

Passive Cooling Concepts

To maintain comfortable indoor environment, there should be reduction in rate of solar heat gains (by using solar devices, insulation, appropriate building materials and colors, decrease in thermal heat gains by lighting controls etc.) and removal of excess heat from the building via convection, evaporative cooling, radiative cooling, air movement, cool breezes, earth coupling, reflection of radiation etc. Passive cooling concepts channel the airflow, thus removing the heat. The various cooling concepts in brief have been summarized in Table 3 with remarks. Few concepts have been discussed as follows:

Wind tower

Hot ambient air is allowed to enter the wind tower during the day. Transferring heat to the walls upon contact creating a cool downward draft. The heat stored in the walls warms the cold night air. Low pressure at the top leads to an upward draft.

Saffari and Hosseinnia [18] suggested that drop of 12°C in indoor temperature and relative humidity increased by 22% can be achieved with use of wet columns 10 m in height. Hughes et al. [19] underlined various cooling techniques integrated with wind towers. The key parameters considered include the ventilation rate and temperature to determine the feasibility of implementing the devices for their respective use. The temperature decrease was found to be in the range of 12–15°C by incorporating evaporative cooling with wind towers. The addition of the cooling devices reduces the air flow rates and the overall efficiency of the wind tower. This temperature drop was found to be greater than a solitary wind tower arrangement. Montazeri et al. [20] examined two- sided wind tower. The maximum performance was achieved during an experiment at 90° angle. Chaudhry et al. [21] investigated a novel closed-loop thermal cycle embedded inside a circular wind tower with internal cross-sectional area of 1 m^2 with 1 m height installed at the roof top to achieve internal thermal comfort. Louvers present at the wind tower openings were angled at 45°. The study concluded that the exit temperature using traditional cooling was increased up to 4°C without any impact of the height in case of the proposed heat pipe design with water and ethanol as working fluids.

Evaporative cooling

The interior spaces are cooled by passing the hot ambient air over damped surface to cool an air stream (by evaporating the water) before its introduction to the interior spaces. Amer [22] reported that evaporative cooling is one of the most oldest, effective, and efficient technique with the potential to reduce indoor temperature by about 9.6°C. Heat transfers for evaporative cooling have been shown in Figure 5.

Table 2. Various passive heating concepts (for thermal heating).

S. No.	Concepts	Results	Ref.	Climatic conditions	Remarks	Applications
2.1	Direct gain (Figs. 1, 2A and B)	i 80% solar gain for double-glazed window. ii Use of double-glazed system leads to reduction of losses by 28% when compared with single glazed system (Equations 2a–2c).	[14]	Very cold.	i No phase change. ii Day lighting	Office building
2.2	Indirect gain (Fig. 3)	25% reduction in heating load. (Equation 3)	[12]	South China.	i Phase change. ii No day lighting	Household
2.3	Solarium (Fig. 4) (Equation 4)					
a.	Link walls iAir collector iiTrans wall iiiWater wall ivMetallic sheet	Maximum room temperature achieved: i Air Collector: 23–24°C (Zone 2) ii Trans wall: 21–22°C (Zone 2): better performance iii Water wall: 17–18°C (Zone 2) iv Metallic sheet: 30–31°C (Zone 1)	[13]	Srinagar, India.	i During the day time, the temperature of zone 1 is greater than living room but reduces at night	Sun space: Glass construction (except for the roof) Living space: Wooden construction
b.	Water wall as link wall	i The variation in living space was observed to lie within the range of 18–20°C. ii For a 5 cm thick water wall, the average temperature for Zone 1 was found to be 35°C and 30°C for Zone 2 with water temperature of about 40°C. iii Best TLL in the living space achieved	[15]	Srinagar, India	The thickness of the water wall was found to be inversely proportional to the temperatures of the sunspace and the living space	i Glass construction for wall and roof with movable insulation east and west face of sunspace. ii Only valid for wooden construction
c.	Combination of water wall and Phase Change Component Material as link wall	i Best TLL along with comfortable temperature (approx. 20°C) for the zone 2 was achieved. ii With decrease in thickness of PCCM from 10 to 5 cm, the heat flux transmitted from the link wall to the living space increased by about 35%	[16]	North America (Boulder)	Zero ventilation rate	Household
d.	An attached solarium with motorized shading devices	i Reduction of about 76% in the heating demands. ii 134 kWh/m² of excess heat can be collected and stored by solarium. iii 70% reduction in case plants are grown inside the solarium	[17]	Montreal	Combined interior and exterior motorized shading	i Greenhouses. ii Solar houses. iii High-efficiency buildings where solar gains are priority

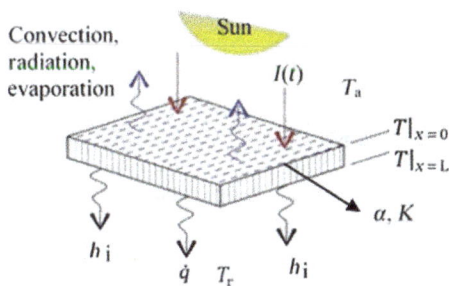

Figure 5. Evaporative cooling: Wetted surface exposed to solar radiation.

The rate of thermal energy per unit area can be expressed as follows,

$$\dot{q} = U_L(T_{sa} - T_r) \tag{7}$$

where,

$$T_{sa} = \frac{\alpha I(t)}{h_1} + T_a - \frac{\varepsilon \Delta R}{h_1}, h_1 = h_{rw} + h_{ew} + h_{cw} \text{ and}$$

$$U_L = \left(\frac{1}{h_1} + \frac{L}{K} + \frac{1}{h_i} \right)^{-1}$$

In the Middle East, evaporative cooling is combined with wind towers to channel the cool wind passed over water cisterns into the interiors to produce cooling and refreshing effect [23]. The evaporative cooling is extensively used in the form of dessert coolers in arid areas in sun light hours (ambient temperature between 37 and 42°C), despite leading to increased indoor air humidity. The method is limited to the regions with low outdoor humidity with 0.3–0.5 m³ water per dwelling availability [1]. Kamal [24] discussed a passive downward evaporative cooling system consisting of downward tower with wetted cellulose pads installed at the top of the tower with inside temperatures ranging between 29 and 30°C while the outside temperature ranged from 43 to 44°C with 6–9 number of air changes. Roof surface evaporative cooling was also discussed where the water was sprayed over suitable water-retaining materials like gunny bags laid over the roof. The results showed that the wetted roof temperature was 40°C with ambient temperature of 55°C. Qingyuan and Yu [25] concluded that the potential of evaporative cooling is subjective to the difference between humidity ratio of outdoor air and wet bulb temperature at saturation.

Solar shading techniques

Solar shading devices reduce the heat gains and thus provide comfortable indoor temperature, reducing the cooling costs. They can also function as an esthetic element aside from day lighting if properly designed. Drop of almost 6 °C in the room temperature has been observed [26]. Kumar et al. [27] evaluated various passive cooling techniques and found that solar shading alone is responsible in reducing the inside temperature by about 2.5–4.5°C. Further drop of 4.4–6.8°C was observed with insulation and controlled air exchange rate.

Shading devices

They can be classified as vertical (vertical louvers, projecting fins), horizontal devices (canopy, awnings, horizontal louvers, and overhangs), egg crate devices (concrete grille blocks, metal grills), and screenings (venetian blinds, double glass

windows, window quilt shade, movable insulation curtains, natural vegetation etc.). Various roof shading techniques have been shown in Figure 6. Horizontal shading devices are best suited for south-oriented openings, whereas vertical shading devices for east- and west-facing facades.

Kima et al. [28] studied the impact of four shading types – overhangs, blind system, light shelf, and experimental shading device. Maximum cooling energy saving was determined to be 11% for experimental shading with 76° of solar altitude. Grynning et al. [29] conducted a study in which it was found that solar shading systems are necessary to reduce the need of artificial cooling of an office block. Simulations for north- and south-oriented cubicles with different floor areas, openings, and shading schemes were run. The simulations revealed that the shading systems contribute toward lowered transmittance value of the window. It was found that the cooling load increases with increase in the window size from 41% to 61%, therefore, there was a decrease in the heating demand. Energy demands were found to be larger in north-facing offices as compared to south facing. Shading devices, if improperly used, can increase the energy demands by 5%, hence they should not be installed on north-oriented workspaces. The results also show that by using a proper and correct shading technique, energy demands can be reduced for south-facing facades by 9%, whereas an improper technique may lead to an increase of 10%. It was also found that four pane glazing is beneficial as compared to two or three pane glazing and can reduce the energy demands as high as 20% and 7% if two pane and three pane glazing is replaced by four pane glazing, respectively.

Shading devices have been in use since ancient times. The Mughal architecture (India) used these in the form of inclined and deep shades to cover more surface area with deep carvings on building exteriors. These carvings help create mutual shading and the extended surface helps to increase the convective heat transfer [30] (Fig. 11).

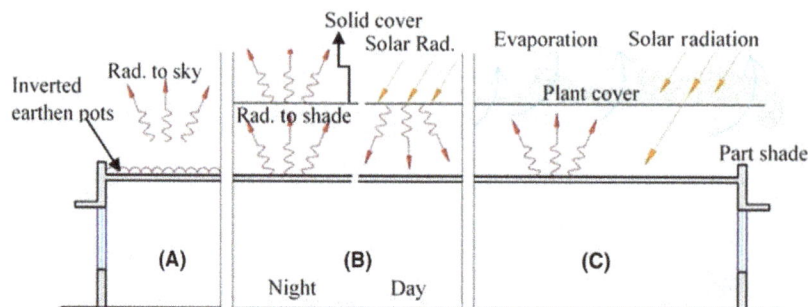

Figure 6. Roof shading by (A) earthen pots, (B) solid cover and (C) plant cover.

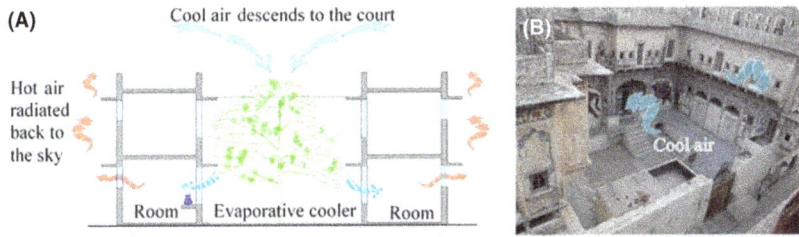

Figure 7. Radiative cooling (A) Schematic section and (B) Haveli in Shekhawati, Rajasthan.

Radiative cooling

Effective radiation from the exposed horizontal surface to ambient air via convective and radiative heat transfer is termed as radiative cooling. Roof transfers heat to the night sky via long wave radiation exchange and convection (Fig. 7A).

The radiant heat exchange between sky and a body/surface can be expressed as:

$$\dot{q}_r = \varepsilon\sigma(T_{sky}^4 - T^4) \qquad (8)$$

Rate of long-wavelength radiation exchange between ambient air and sky can be expressed as:

$$\Delta R = \sigma[(T_a + 273)^4 - (T_{sky} + 273)^4] = 60\,\text{W/m}^2 \qquad (9)$$

Courtyard planning

Localized heating within buildings may be reduced to large extent by exploiting the thermal interaction due to the difference in temperature of courtyard and the building core depending upon the aspect ratio of the court, wind speed, and direction [31] (Fig. 7A and B). In vernacular architecture, the courtyards were integrated with vegetation and water bodies to enhance the humidity, evaporative cooling, and provision of shade.

Infiltration/Ventilation

Energy conservation and natural ventilation should be the prime concern of any building design as the buildings are often planned as sealed and well insulated, with low heat gain or loss, the extreme use of HVAC systems to improve air quality and to dilute the VOCs emitted by the building materials and furniture [32]. In buildings, there are two types of airflows induced due to pressure difference leading to natural air flow. Infiltration occurs from adventitious openings that are present in every building in the form of interfaces, cracks, or any gaps etc. Infiltration can be minimized with draught sealing, air locks, airtight and quality construction of doors and windows. Wang et al. [33] suggested that airtightness alone can have a significant impact on the heating and cooling performance of the building. In the study, infiltration of hot and humid air led to an increase of 9.4% and 56% in total cooling load and latent load, respectively. Reduction of 1.4% in heating load was observed due to higher outside temperature. The other type of airflow is ventilation. Windows play an important role for air circulation within the building premises with recommended value of air movement being 0.2 and 0.4 m/s during winters and summers, respectively [14].

The ventilation losses can be given by:

$$Q_v = 0.33\,NV(T_r - T_a) \qquad (10)$$

Rate of heat transfer from roof bottom to room air is,

$$M_a C_a \frac{dT_r}{dt} = h_i\left(T_{|x=L} - T_r\right)A_r + A_{win}U_t\left(T_{sa,win} - T_r\right) \\ - 0.33\,NV(T_r - T_a) \qquad (11)$$

Based on Table 3, one can observe that the combination of evaporative cooling and wind towers is able to reduce the temperature by up to 12–17°C. Evaporative cooling is the most economical concept for cooling of a building.

Passive Heating and Cooling Concepts

These are the concepts that can be used for both seasons for heating and cooling purposes. The various heating/cooling concepts in brief have been summarized in Table 4 with remarks. Few concepts have been discussed as follows:

Trombe wall

Trombe wall is a large (at least 400 mm thick) mass (material – stone, brick, reinforced cement concrete, mud, etc.), exposed to sunlight through south-facing exterior glazing in northern hemisphere. The thick walls transfer

Table 3. Various passive cooling concepts.

S. No.	Concepts	Results	Ref.	Climatic Conditions	Remarks	Applications
3.1	Wind Towers					
a.	Proposed improvements: i Integrated with evaporative cooling concept. ii Tower heads capable of trapping wind from any direction. iii One-way damper with large screen openings	i 306 kg of mass storage material is used per cubic meter of tower, capable of storing 36 m² heat. ii The above proposed design increases the efficiency of heat transfer by 5–10 times without a penalty in the total mass of the thermal material used	[34]	Hot arid areas (Areas with variable wind directions)	i Wind towers are capable of releasing air at higher flow rates to the buildings. ii An energy storing system (long conduits made of baked non-glazed clay) was also introduced to increase the heat transfer area	i Residential. ii Commercial. iii Analysis may be used as guideline for natural ventilation and passive cooling systems
b.	Two novel designs: i Wetted columns with cloth curtains. ii Wetted surface with evaporative cooling pads	i High wind conditions: Wetted column tower design. ii Low wind conditions: Wetted surface tower design. iii The design released air at much lower temperature to the interior space as compared to the conventional design. iv Wet columns with 10 m height can reduce the indoor temperature by 12°C and RH by 22% [35]	[35–37]	Hot arid regions (Middle East)	Integration of cooling devices with the conventional wind towers is beneficial with air exiting the towers at significantly lower temperature than the outside temperature	–
e.	Proposed design includes: i Clay conduits. ii Water pool	i Air flow induced by the tower has direct impact on the reduction of internal temperature. ii Increasing the number of conduits results in better efficiency than a wetted column. iii Wetted surface deign is able to reduce the temperature by up to 17.6°C	[38]	Hot dry region of Ouargla (maximum temperature 47–52°C)	i Clay conduits mounted to improve mass and heat transfer. ii Water pool introduced to increase the humidification	i Residential. ii Bio climatic housing
f.	Wind towers installed at the roof top.	i The proposed tower can rotate and align itself in the direction of the predominant wind to compensate for low wind speeds. ii Natural day light inside the premise can be improved by using transparent construction material for the wind catchers	[39]	Windy regions	The proposed wind towers could be used in various configurations such as wind tower with windows, a wind tower and a solar chimney/heater or two wind towers in different directions	i Residential. ii Closed arenas. iii Commercial buildings. iv Administrative buildings
3.2	Evaporative cooling (Fig. 5; equation 7)					
a.	Passive evaporative cooling.	Effective for metal ceiling and thus reduces the room temperature.	[40]	Hot humid climate (Thailand)	Higher solar radiation and ambient temperature gave better results	All building types

Table 3. Continued

S. No.	Concepts	Results	Ref.	Climatic Conditions	Remarks	Applications
b.	Solar air heater design along with the concept of evaporative cooling	Responded well during the winters but was not effective for summer cooling	[41]	Composite climate	Changes (like addition of wetted south wall collector and roof duct at the top) were made to the model in order improve performance in both seasons	–
c.	Performance of various roof materials on evaporative cooling effect	Siliceous shale itself is capable of reducing the roof surface temperature up to 8.63°C due to its high evaporation rate as compared to mortar concrete (0°C). This material releases more latent heat while silica sand, volcanic ash and pebbles yield more sensible heat of about 0.62, 0.18 and 0.18, respectively	[42]	–	Siliceous shale with adsorption rate of 0.07 kg/m²/h is found to have the greatest evaporation performance (0.3 kg/m²/h)	All building types
d.	Passive evaporative cooling wall based on various porous layers	Porous pipe surface temperature and the air flow of about 4–6°C and 3–5°C, respectively below the ambient can be attained, thus meeting the cooling demand	[43]	Shanghai, China (October).	i Porous ceramic pipes with high water sucking ability were used as construction material for the evaporative wall. ii Not suitable for extreme humid climate and locations with shortage of water for evaporation.	Hot and dry climates or locations while the design of outdoor or semi enclosed spaces in parks, pedestrian areas and residential courtyards aims at controlling increased surface temperature.
e.	Indirect evaporative cooling	i There was drop of 1°C in mean daily temperature [44], ii There was an average drop in indoor temperature of up to 2.5°C compared to the outside temperature [44], iii This system is capable of reducing the thermal discomfort due to excess of heat in about 95–100% of the year [44], iv Capable of reduction of the energy demand of HVAC by 20% in next 20 years [45]	[44, 45]	Maracaibo, Venezuela (average daily temperature ranged between 26.5–27.6°C) [44], Dry and hot climate [45]	i Metallic water tank with thickness of 3–4 cm was laid over concrete slabs. Insulated high reflectance sheets of 1 cm thickness were used as roof element and fans were used in order to enhance the evaporation [44]. ii Strong potential for improving indoor comfort conditions	The system can be employed in various Brazilian territories, irrespective of the arid climate [44]
f.	Evaporative cooling through a fountain	i The resultant inside temperature was observed to fall within the comfortable zone of 20°C. ii There is a potential for reduction of the cooling load by 9% and the annual energy consumption by 23.6%	[46]	Hot and arid regions (Dubai)	Outside temperature was about 45°C	All building types

Table 3. Continued

S. No.	Concepts	Results	Ref.	Climatic Conditions	Remarks	Applications
3.3	Solar Shading Techniques					
a.	Shading of roof	High indoor temperature can be controlled by providing a roof cover made from locally available material like hay, terracotta tiles, inverted earthen pots, solid cover (sheets), plants, thermal insulation. (Fig. 6)	[24]	–	Masonry and RCC constructions tend to make the indoor temperature as high as 41°C when the roof top temperature is about 65°C	All building types
b.	Shading by trees and vegetation	i The ambient temperature near the outer wall may be substantially reduced by 2–2.5°C without excess use of supplementary energy [26]. ii Deciduous trees should be used on the south and southwest of buildings [47]. iii Evergreen trees should be used on the south and west side of the façade [47]. iv Shading and evapotranspiration from trees can reduce the surrounding temperature as much as 5°C [47]. v The air temperature can be reduced by 2°C in the surrounding area with presence of a nearby park [48].	[26, 47, 48]	–	Deciduous trees provide summer shading and daylight in winters by shedding the leaves	i Energy savings. ii Overall benefits for environment (reduced emissions). iii All building types
c.	Tree buffering as solar control in a south-east oriented building	i The solar irradiation peak in the non-shaded area and the shaded area at the same time was observed to be 600 W/m² and 100 W/m², respectively. ii Solar radiation exceeded on the shaded wall on the mid-day and reached 180 W/m² as the sun was at high horizon	[49]	Agricultural University of Athens	Parameters like air and wall surface temperatures, wind speeds, humidity, heat exchange between the wall surface and surrounding environment were measured in the shaded (by deciduous trees) and unshaded areas for a hot summer period	i Energy savings. ii Overall benefits for environment (reduced emissions). iii All building types
d.	Roof solar collector	Roof thermal comfort cannot be effectively achieved by only roof solar collector system and it should be installed along with Trombe wall	[50]	Hot and humid climate	The roof solar collector design used: CPAC Monier concrete tiles and gypsum board	Housing
e.	Ventilated roof	As the ventilated roof provides insulation along with the protection against the solar gains, it is more advantageous during the summer season. No clear improvement in winters	[51, 52]	–	The proposed setup consisted of air gap sandwiched between the reinforced concrete prefabricated slab and the insulation for both the seasons	–

Table 3. Continued

S. No.	Concepts	Results	Ref.	Climatic Conditions	Remarks	Applications
f.	Cool roof (37 roof designs)	i About 10–40% reduction in air conditioning energy. ii Flat roof: Heat gain and average indoor temperature were 414 kWh and 32.5°C, respectively. iii Domed roof: Heat gain and average indoor temperature were 310 kWh and 32.2°C, respectively. iv Vaulted roof: Drop of 1.5°C in the average indoor temperature. Fall of 53% and 826 kWh savings during summers in discomfort hours with rim angle of 70° with high albedo coating as compared to the reference case of the conventional noninsulated roof	[53]	Cairo, hot dry climate	The typical roof temperature can reach up to 37°C above the ambient for hot dry climate and can exceed by about 20°C when compared to the surroundings covered by vegetation	Low and medium rise residential buildings
3.4	Radiative Cooling (Fig. 7A and B; equations 8 and 9)					
a.	Correlation between the radiative cooling power and the temperature difference between the ambient and the sky	Radiative cooling has the potential to save up to 25% of the power consumption independent of all location	[54]	Tropical climate (Malaysia)	The cooling power decreases with decrease in the difference between the ambient and sky temperature	–
b.	Radiative cooling applications	The specific cooling power measurements ranges from 20 to 80 W/m²	[55]		Movable insulations, air-based systems and open or closed water-based systems	–
c.	Water-based radiative cooling plate	Net cooling power of 81 W/m² was achieved by removing the cover for 7 h of operation	[56, 57]	Israel	Flat plate collector attached to a storage tank was used	Conventional space cooling
d.	Open water-based system	The plant showed a specific cooling power of 120 W/m² as achievable	[58]	Würzburg, Germany	Cooling outputs are higher for the closed system because of the absence of the thermal resistance between the water and the ambient	–
e.	Regional characteristics	With increase in elevation, there is a decrease in the cooling load but the potential of radiative cooling is large	[59]	China	The long wave terrestrial radiation shows no correlation with the elevation while the short wave incoming radiation shows a proportionate decrease at the normal lapse rate. Thus, leading to an increase in the value of radiative cooling	Residential buildings

Table 3. Continued

S. No.	Concepts	Results	Ref.	Climatic Conditions	Remarks	Applications
3.5	Infiltration/Ventilation (Equations 10 and 11)					
a.	Building design	i Window to ground ratio for residential buildings lies in the range of 0.33–0.58 (average 0.44). ii The mean window to wall ratio for like dining/living, bed room and are 0.34 and 0.27, respectively	[60]	Hong Kong	–	High- rise residential buildings
b.	Hybrid ventilation and night ventilation	i Natural ventilation should not be disturbed by the additional mechanical airflow in hybrid ventilation. ii Minimize HVAC system, solar loads and artificial lighting	[61]	Germany	Architectural solutions were used which solutions included thermal insulation, moderate window dimensions, central atrium, shading systems and cross ventilation	Office buildings

the absorbed solar energy to the interiors by conduction and convection. Trombe wall reduces the heat flux from the exposed outer surface wall to the interiors of the building. Thermal mass reduces the temperature fluctuation peaks (decrement factor) and shifts the peak to a later time than the air temperature peaks (time lag) due to a lower overall heat transfer coefficient.

Energy balance for Trombe wall are as follows:

Solair temperature,

$$T_{sa} = \frac{\propto I(t)}{h_0} + T_a - \frac{\in \Delta R}{h_1} \qquad (12)$$

Heat flux (W/m^2),

$$\dot{q} = U_L(T_{sa} - T_r) \qquad (13)$$

where,

$$U_L = \left(\frac{1}{h_0} + \frac{L}{K} + \frac{1}{h_i} \right)^{-1}$$

Energy balance for bare surface (Trombe wall) with insulation of thickness L_i, having thermal conductivity of K_i during the night time can be expressed as follows:

$$\dot{q} = U_L(T_{sa} - T_r) \qquad (14)$$

where,

$$U_L = \left(\frac{1}{h_0} + \frac{L}{K} + \frac{L_i}{K_i} + \frac{1}{h_i} \right)^{-1}$$

Trombe wall may have two different types of exposed surfaces for passive buildings:

a). Bare exposed surface (unglazed Trombe wall) with absorptivity, $\alpha \leq 0.4$. This surface is used for thermal cooling (Figs. 8 and 11; equation 16).

b). Blackened and exposed surface (glazed Trombe wall) with absorptivity ≥ 0.9. This surface is used for thermal heating (Fig. 9; equation 15).

Trombe wall – Passive heating

Blackened and glazed surface (Fig. 9) of the Trombe wall should be considered for passive heating.

Energy balance for blackened and glazed surface are as follows:

Heat flux (W/m^2),

$$\dot{q} = U_L(T_{sa} - T_r) \qquad (15)$$

where,

$$T_{sa} = \frac{\propto \tau I(t)}{U_t} + T_a, \text{and} U_L = \left(\frac{1}{U_t} + \frac{L}{K} + \frac{1}{h_i} \right)^{-1}$$

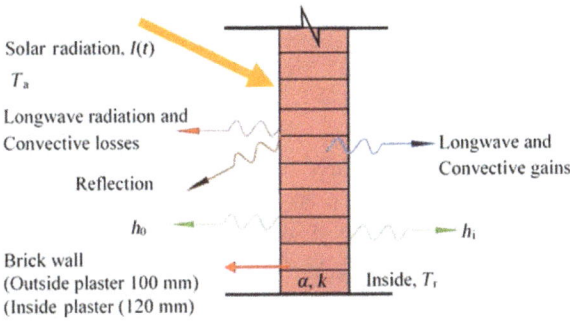

Figure 8. Energy flow diagram for Trombe wall (unglazed).

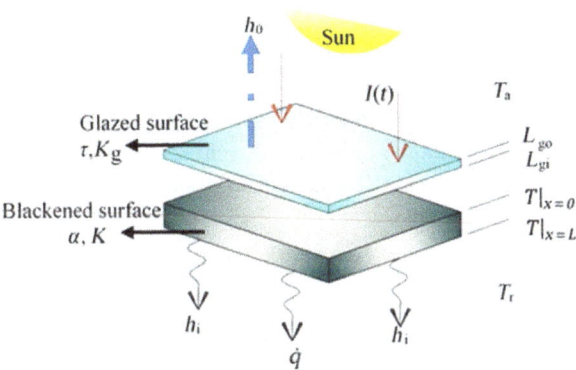

Figure 9. Blackened and glazed surface.

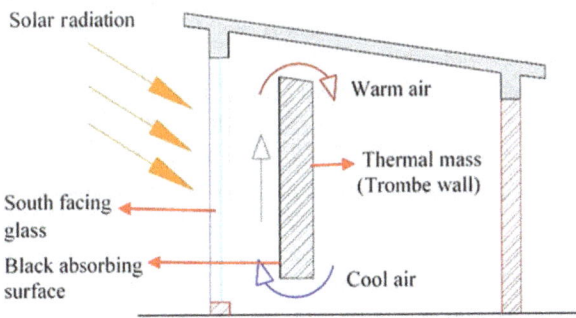

Figure 10. Trombe wall with vent openings.

Figure 10 shows simple passive heating technique using Trombe wall, with vent openings provided near the floor and the ceiling to allow convective heat transfer. These concepts have been applied in Pyrenees Orientales district of France and in the U.S. southwest.

For heating purposes, thermal resistance of glazing is proven to be more beneficial because glazing minimizes the heat loss through itself. A solar house based on this concept has also been constructed in Jordan (1983–1984) with two sections, namely the heated section and the non-heated space. Ta'ani et al. [62] reported that it is possible to meet 54% of the building's heating demands by solar energy with collector array efficiency with proper retrofitting. Tyagi and Buddhi [63] proposed a filling of phase change materials (PCM) to enhance the storage of latent heat in the masonry wall. These units are lighter in weight and require less space as compared to the massive thermal mass. Nwachukwu et al. [64] found that heat storage capacity along with the heat transfer through a Trombe wall can be improved by coating the exteriors of the thermal mass with superior absorption vigor. Balcomb and McFarland [65] found that Trombe walls with vented openings proved to be 10–20% more efficient, especially in extreme climatic conditions like Boston. Saadatian et al. [66] discussed different types of Trombe walls, viz. classical and modified, zigzag, composite, fluidized, solar water wall, solar trans wall, solar hybrid wall, Trombe wall with PCM and PV-Trombe wall. The efficiency of Trombe wall can be improved by 8% by using a fan. The optimal size (i.e., ratio of Trombe walls area to area of rooms with other walls) should be 37% with 300–400 mm thickness. The insulation increases the efficiency by 56% and also decreases the size of the thermal wall. The energy and environmental performance has been compared with and without Trombe wall in the study by Bojić et al. [67]. The research is based on two cases: the first case, without any Trombe wall and the second case, with installations of two Trombe walls at the south side of Mozart house located in Lyon, France. A 450 mm layer of clay brick 1220 has been used as the Trombe core material. The second case uses the solar energy captured to save around 14% of all electricity and 20% of annual energy as compared to the first case. The energy ratio is around 6 with the energy payback time of around 8 years for the electrical heating with optimum core thickness. The above values are around 3 and 18 years, respectively, for natural gas heating. The savings can further increase with increase in the thickness of Trombe wall layer. The optimum thickness of clay brick is around 0.35 m and 0.25 m for electrical heating and natural gas heating, respectively. The study also reveals that lower density material like clay brick save more operating energy as compared to higher density material like concrete.

Trombe wall – Passive cooling

A thick thermal mass used as the exterior facade reduces the decrement factor and leads to a time lag. A heavy structure is preferred for thermal cooling as the mass acts as an insulator and a heat storage medium. The unglazed Trombe wall may be constructed with different materials like stone, brick, reinforced cement concrete, mud, etc. This is a very old technique and is clearly

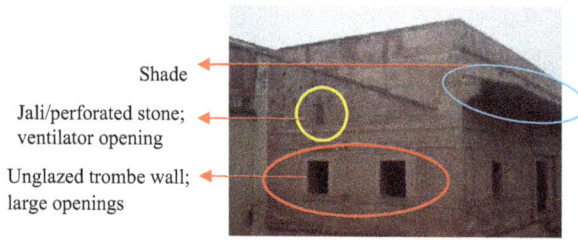

Figure 11. Trombe wall, openings, shades (passive cooling) Fatehpur Sikri, Agra, India.

visible in historical buildings like Humayuns Tomb, Qutub complex, Fatehpur Sikri (Fig. 11) etc. The ventilation system, as marked, has been arranged above the lintel level of the windows and the inside temperature is found to be comfortable all year round.

A bare surface (Fig. 8) should be considered for passive cooling purpose. Expression of solair temperature for bare surface Trombe wall can be written as:

$$T_{sa} = \frac{\alpha}{h} I(t) + T_a - \frac{\varepsilon \Delta R}{h} \qquad (16)$$

An innovative approach of combining unglazed Trombe wall (Bare surface) with storage facility was evolved by Tiwari et al. [68] as shown in Figure 12A and B.

Trombe wall – Passive heating and cooling

Shen et al. [69] discussed the installation of adjustable dampers at the glazing and adjustable vents at the wall. This system is favorable for both heating and cooling purposes. During winters, the upper damper closes while the lower damper and both vents are open, although in summers, lower damper along with upper vent are closed. It was also discussed that the composite wall has better energetic performances than the classical wall in cold and/ or cloudy climatic conditions. Krüger et al. [70] analyzed the heating and cooling potential of Trombe wall by

installing two test cells of 5.4 m^3 internal volume and 2.6 m^2 floor area for a subtropical location. One of them was a naturally ventilated Trombe wall and the other one was without it (reference test cell). Higher performance was seen in the reference test cell under cold conditions. For heating purpose, three operation modes were tested – air vents 1 and 3 closed (case 1); air vent 3 closed (case 2); and air vents 1 and 3 closed with dampers installed in the storage wall openings (case 3). For weekly averages and clear sky conditions, the smallest average difference was recorded in case 3 and the highest in case 2, from the control test cell. Case 2 was considered the best configuration for the weekly period data, but case 1 proved to be better in clear day conditions with slightly higher temperature difference to the exterior (5.2°C against 4.5°C). Case 2 was considered to be a better option for winters. For summers, four operation modes were used – all air vents shut (operation mode 1); air vents 1 and 4 shut (operation mode 2); air vent 2 shut (operation mode 3); air vents 2 and 4 shut (operation mode 4). Mode 3 is considered to be the best option in summers.

Roof pond

In earlier studies, it has been seen that about 50% of the heat gains for single-story buildings is received via roof [71]. The conventional approaches to reduce the heat flux via roof includes false ceilings, insulations, increasing the roof thickness, roof shading, or using roof coatings. Roof pond is another technique found in 1920s and was first investigated at the University at Texas [72]. Sutton [73] observed that the roof surface temperature might reach 65.6°C without any treatment. By installing an open roof pond, the above temperature can be dropped down to 42.2 and 39.4°C with 0.05 and 0.15 m depths of the pond, respectively [74]. A drop of 20°C in the room temperature can be achieved by using the roof pond in arid regions [75]. Kharrufa and

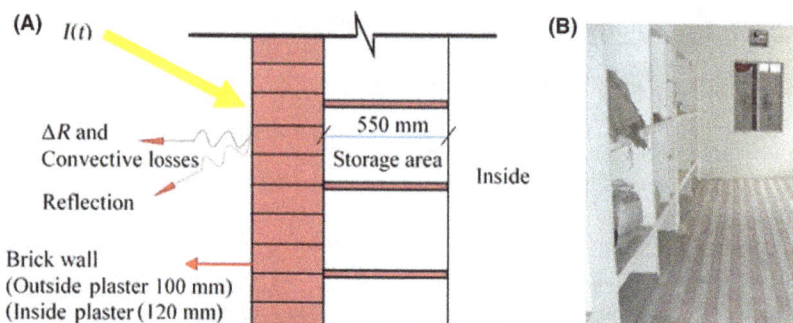

Figure 12. Unglazed Trombe wall (Bare surface) with storage (A) Section and (B) Sodha BERS Complex, Varanasi.

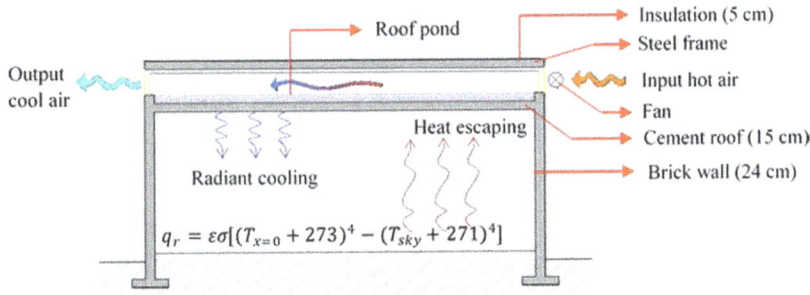

Figure 13. Schematic section of roof pond with forced electric ventilation.

Adil [76] tested one room (4 × 7 × 2.75 m) to check the effectiveness of the roof pond which was mechanically ventilated (Fig. 13) for cooling in hot dry summer of Baghdad.

Earth air heat exchanger

The air is allowed to pass through a tunnel or a buried pipe at least 1–3 m below the surface for both heating and cooling purposes. A depth of 2–3 m is generally considered since at that depth, it is cooler than the outside in summers and warmer in winters [35]. Figure 14 shows the scheme of Earth air heat exchanger (EAHE).

Thermal energy gained by the flowing air can be expressed as:

$$\dot{Q}_u = \dot{m}_a C_a (T_{fo} - T_{fi}) \tag{17}$$

$$\dot{Q}_u = \dot{m}_a C_a (T_0 - T_{fi}) \left[1 - e^{\frac{-2\pi r h_c}{\dot{m}_a C_a} L} \right] \tag{18}$$

where,

T_0 is the ground temperature outside pipe (°C).

For the given design parameters by Tiwari et al. [77], a drop of 5–6°C in the outlet air temperature in summers

was noticed, using five air changes with 100 mm diameter and 210 mm length of the pipe.

Combination of various passive heating and cooling concepts

Various passive solar strategies have been discussed and it is evident that the energy use can be reduced to some extent with the above discussed strategies individually. To achieve a high level of energy performance, a combination of different passive solar techniques is necessary. The various heating/cooling concepts in brief have been summarized in Table 4 with remarks.

Based on Table 4, one can observe that unglazed Trombe wall can be used for both heating and cooling, whereas glazed Trombe wall can be used for heating purpose only. Combination of Trombe wall, cool roof, and thermal insulation proves to be very effective and can achieve 46% and 80% of savings in winters and summers, respectively.

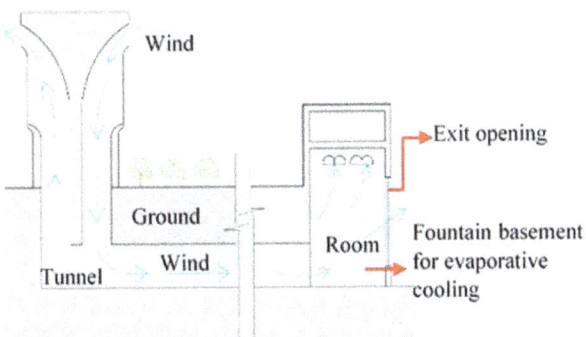

Figure 14. Scheme of Earth air heat exchanger.

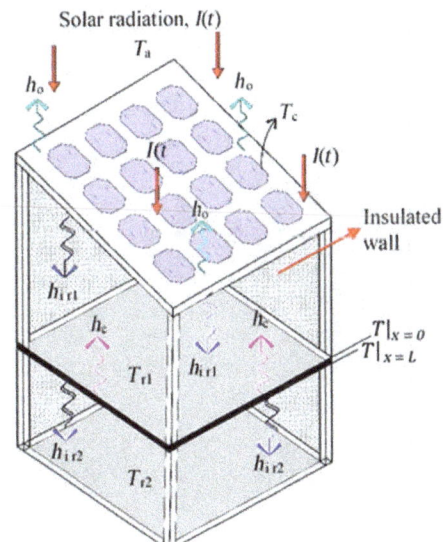

Figure 15. Building-integrated semitransparent photovoltaic thermal (BiSPVT) system.

Table 4. Passive heating and cooling concepts.

S. No.	Concepts	Results	Ref.	Climatic Conditions	Remarks	Applications
4.1	Trombe wall (Fig. 8; equations 12–14)					
4.1.1	Trombe wall – Passive heating (Fig. 9; equation 15)					
a.	PV- Trombe wall Case 1: with openings and heat storage. Case 2: With DC fan. Case3: Performance evaluation by varying air flow for 3 types of glazing. Case 4: Installation over the glazing. Case 5: Thermal insulation	Case 1 [78]: i Indoor temperature can be increased by 7.7°C. ii The Daily average electrical efficiency can reach up to 10.4%. Case 2 [2]: i Indoor temperature can be increased by 14.42°C. ii The daily average electrical efficiency can reach up to 10–11%. Case 3 [79]: i The effect of airflow on the performance stagnates when the air velocity reaches the value of 1 m/s. ii The maximum PV efficiency (of about 0.185) was achieved in the double glass with argon. iii Highest reduction in cooling load at air velocity of 1.75 m/s was achieved in double glass with argon (about 160 W/m²). iv The study concluded that double glass filled with argon and double glass PV-Trombe wall is preferable over the single glass at air velocity of 1.5 m/s. Case 4 [80]: i The thermal performance of the Trombe wall can be reduced by up to 17%. Case 5 [81]: i The room temperature can be increased by 2.36°C with use of thermal insulation in cold climatic conditions and decreased by 2.47°C in hot climatic conditions. For the same electrical efficiency decreases by <2%. ii The room temperature can be decreased by 2°C with use of curtain shading in hot climatic conditions and electrical efficiency increases by 1%	[2, 78–81]	Case 1 and 2: China. Case 3: Universiti Teknologi Petronas. Case 4: South China. Case 5: Composite climate	Case 2: Reduces the PV temperature simultaneously. Case3: i PV- Trombe wall glazing types: single glass, double glass and double glass with Argon. ii Double glass shows 18% less reduction in cooling load as compared to double glass with argon. iii Ventilated double glass offers more insulation Case 5: It is recommended that thermal insulation in both winter and summer and appending a shading curtain in summer are adopted, especially for the diurnally used PV-Trombe wall.	PV Trombe wall can provide feasible solutions for high energy consumption and environmental degradation especially for tropic region
b.	Opening and closing modes in the management of air vents	i The best time to open the air vent is 2–3 h after sunrise and the best time to close it is 1 h before sunset. ii The heat storage capacity of the Trombe wall is fully released at 7:30 a.m. and it reaches its maximum value of 10.6 MJ/m² at 4:00 p.m. At 7:30 a.m., the heat release reaches its maximum value of 10.4 MJ/m²	[82]	GangCha country, China	Conclusion i is appropriate for almost all type of Trombe wall heating system. Conclusion ii is case specific	–

Table 4. Continued

S. No.	Concepts	Ref.	Climatic Conditions	Remarks	Results	Applications
4.1.2	Trombe wall- Passive cooling (Figs. 8 and 11) (Equation 16)					
a.	Thermal behavior of Trombe wall in summers	[83]	Mediterranean climate	A comparative analysis of difference between the thermal behaviors of two Trombe walls was carried by varying the screening, ventilation, occupancy and internal heat gains	i A decrease of 1.4°C in the internal surface temperature of the wall was observed and 0.5 MJ/m² in daily heat gains was observed with the use of rolling shutters as screening. ii The heat gains of the screened Trombe wall were about 18 times lower than those of the unscreened. iii A reduction of 72.9% and 65% in cooling energy needs was observed with the combination of overhangs, rolling shutters and cross ventilation, respectively, in comparison with unvented Trombe wall without solar shadings	Low or highly insulated building
4.2	Roof pond					
4.2.1	Roof pond – Passive cooling					
a.	Experimental roof pond building	[84]	Baghdad (July and August)	i 3 × 2.5 × 2 m room with pool depth of 60 mm. ii There is a need to design the vapor leakages from the pond to prevent them from winter heating and rains	i The vapors were discharged from the pond in order to improve the building's summer performance whereas the same will reduce the efficiency of the system in winters. ii Cooling load (500 KJ/m²) is at its maximum at noon because the most significant component of the cooling load is the direct heat gained through the south glazing. iii There is a significant sensible heat contribution at 11.00 am, which is responsible for the cooling. At noon, cooling is due to the latent heat of evaporation	Increased effectiveness of roof ponds in high- humidity regions
b.	Site experiments for: Case 1: roof with moist soil Case 2: Walkable roof pond along with night water circulation	[85]	Hot dry climate of Saudi Arabia	Case 1: Roof was shaded by 10 cm of pebbles. Case 2: Roof pond was filled with pebbles with an insulation layer and thin tiles over the layer	Case 1: Reduction of 5°C in indoor temperature. Case 2: Reduction of 6°C was seen in the indoor temperature compared to the outdoor temperature	Not suitable for low cost housing since dead load of water needs to be considered in the roofing structure
c.	Roof pond with gunny bags (RPWGB) floating on water surface	[72]		The technique is better than RPWWGB because of thermal satisfaction inside the pond during daytime irrespective of the building type	i This technique is better than the roof pond with wetted gunny bags (RFWWGB) in terms of cooling performance of room temperature and heat flux through the roof into the pond. ii This technique has better performance when compared to movable insulations. iii The optimum water depth for the RPWGB should be 200 mm and 50 mm for concrete and metal-decked roofs, respectively	Further field examina-tions are required before putting this system into practical use

Table 4. Continued

S. No.	Concepts	Results	Ref.	Climatic Conditions	Remarks	Applications
d.	Roof pond with forced electric ventilation (Fig. 14)	i The indoor temperature was better stabilized (with fan) as compared to the pond without fan and cover. The variations became limited to 3.5°C without a fan and 3°C with one. ii The average temperature reduction with the pool alone lowered the room temperature by 3.36°C compared to the roof without a pond. iii A substantial reduction of 6.0°C in the peak external temperature at 15:00 between the room without a pond and the room with a ventilated pond was achieved. iv There was also a substantial reduction of 6.5°C in the peak internal temperature at 6:00. v With a larger percentage of roof to wall, the roof pond cooling will be more effective and the drop in the cooling load will be around 29%		Hot dry climate (Baghdad, Iraq)	–	–
e.	Ventilated pond protected with a reflecting layer	i There was a reduction of 30% in the maximum indoor air temperature compared to the corresponding temperature of a building without any roof cooling technique. ii The proposed system has a 24 h cooling effect because the ceiling temperature is higher than that of the water. At night, the water better prevents the heat loses compared to a bare concrete roof	[86]	Crete, Greece	In the setup, a pond with 1.10–1.12 m depth was filled with water and covered with an aluminum layer at 1.15 m height from the free water surface	Small buildings
4.2.2	Roof pond – Passive heating and cooling.					
a.	Skytherm	Room temperature achieved was 27°C and 22°C when the outside temperature was 37°C and 5°C for summers and winters, respectively	[1]	Different weather types with various means of modulating ambient conditions	Use of metallic plate at bottom of the pond with blowers in summers and pond covered with transparent plastic in winters	–
4.3	Earth Air Heat Exchanger (EAHE) (Fig. 14; equations 17 and 18)					
4.3.1	Earth Air Heat Exchanger (EAHE) – Passive heating					
a.	EAHE coupled with a solar air heating duct	i The proposed design was connected at the exit end with a solar air-heating duct to increase the heating capacity by 1217.625–1280.753 kWh. ii There was a substantial rise in the inside temperature and the coefficient of performance (COP) of 1.1–3.5°C and 4.57, respectively. iii With 34 m of tunnel length, more than about 82–85% iv of total rise in air temperature can be achieved which means that by reducing the length of the tunnel to 34 m, the cost of the installation was optimized. v The COP increased up to 4.57 when assisted with solar air heating duct	[87]	Northwestern India, arid climate of Ajmer during winters	Thermal performance of EAHE with a 60 m long, 100 mm diameter horizontal polyvinyl chloride pipe buried 3.7 m deep configurations	–

Table 4. Continued

S. No.	Concepts	Results	Ref.	Climatic Conditions	Remarks	Applications
4.3.2	Earth Air Heat Exchanger (EAHE) – Passive cooling					
a.	Analytical model to calculate the cooling potential	i The results validated that EAHE can provide 30% of the cooling energy demand. ii There was a reduction o 1700 W in the peak cooling load and 2.8°C in the indoor temperature during summers	[88]	Dessert climate (hot and arid)	Thermal resistance of the material of the pipe was neglected	Domestic buildings
b.	Explore the effect of geometrical and dynamical parameters on thermal performance of EAHE	i With increase in the length of the pipe: • The air outlet temperature decreases. • The daily mean efficiency increases. • The coefficient of performance falls. • When pipe length changes from 10 to 30 m: • For an inlet temperature of 29°C, there was a reduction of 2°C in the outlet temperature. The reduction rate was not found to be constant. • The COP falls by 10.5% and the daily mean efficiency rises by 142%. iii With increase in cross section and air velocity: • The air outlet temperature increases. • The daily mean efficiency decreases. • A rise of 1.2°C was noticed in the ambient temperature and a drop of 31.6% in the daily mean efficiency was observed when the air velocity changed from 1 to 3 m/s	[89]	Algerian Sahara conditions	i A pipe with 5 mm thickness was buried in soil with 22.27°C temperature. ii The ambient temperature was the same as the air inlet temperature i.e., 29°C	–
4.3.3	Earth Air Heat Exchanger (EAHE) – Passive heating and cooling					
a.	EAHE assisted by a wind tower (Fig. 14)	i Unlike the pipe dimensions (length and diameter), the height and cross section of the tower had no influence on the performance. ii A tower with 5.1 m height and 0.57 m² of cross-sectional area generates 592.61 m³/h of airflow. iii The air velocity increases and the maximal gradient of temperature decreases with the increase in the pipe diameter. iv With a pipe length of 70 m, the daily cooling potential reached a maximum of 30.7 kWh and that the daily cooling potential was proportional to the pipe diameter. v This scheme is more efficient than the conventional	[35]	Hot and arid regions of Algeria	Ambient temperature more than 45°C	–

Table 4. Continued

S. No.	Concepts	Results	Ref.	Climatic Conditions	Remarks	Applications
b.	Energy conservation potential of EAHE	i A 19 kW cooling potential was recorded to maintain an average room temperature 27.65°C for the proposed design with 80 m of pipe length, 0.53 m³ of cross-sectional area and 4.9 m/s air flow velocity ii An auxiliary energy load of 1.5 kW for winter season is required in achieving comfort conditions affecting an average room temperature of 24.48°C	[90]	Mathura, India	–	i Non- air conditioned building. ii Can be coupled to greenhouse and building simulation codes
c.	Photovoltaic- thermal collector accompanied with an EAHE	i Capable of increasing the indoor temperature by 7–8°C at night during winters. ii The hourly thermal energy generated, during day and night is 33 MJ and 24.5 MJ, respectively. iii The yearly thermal energy generated has been calculated to be 24728.8 kWh, while the net annual electrical energy savings is 805.9 kWh and the annual thermal exergy energy is 1006.2 kWh	[91]	New Delhi, India	–	Green house
d.	Improvement of the EAHE thermal potential	The thermal performance of an EAHE can be improved by 73% and 11% for cooling and heating purpose, by increasing the number of ducts, while keeping the area occupied by the ducts and the mass flow rate of air fixed	[92]		In all proposed installations the thermal potential for cooling was higher than the thermal potential for heating	–
e.	Evaluation of EAHE	80 m long tunnel with 0.528 m² of cross-sectional area has 512 kWh and 269 kWh cooling and heating capacity, respectively	[93]	India	The heating capacity was found to be inadequate for providing the necessary comfort conditions	Hospital complex
4.4	Combination of various passive heating and cooling concepts.					
a.	Trombe wall, ventilated walls, glazed walls, fenestration, green roof, PV roofs, evaporative roof cooling along with thermal insulation and PCM	i Thermal mass is more effective when the outside ambient temperature difference between days and nights are high. ii Size of the mechanical devices can be reduced by incorporating a holistic energy efficient design which compensates for the additional capital investment for the energy efficiency features	[94]	–	Various building envelope components were studied to review the potential of passive energy savings	–
b.	Window size on each façade, shadings on south façade and thermal insulation on both roof and walls	About 25.31% of the energy consumption and 11.67% of the lifecycle costs can be reduced using these variables	[95]	Mediterranean region	Jaber and Ajib performed a study to minimize the energy consumption and lifecycle cost	Residential building

Table 4. Continued

S. No.	Concepts	Results	Ref.	Climatic Conditions	Remarks	Applications
c.	Building orientation, thickness of external wall, air infiltration, wall and roof insulation, lighting, windows to wall ratio, glazing type	About 50% of the annual energy use can be minimized when compared to the existing design practices. This can be achieved mostly by integrating the roof insulation, reducing the air infiltration and using energy efficient lighting and other electrical equipment	[96]	Tunisia	Ihm and Krarti performed a research in order to reduce the energy consumption and life cycle costs	Single family houses
d.	Wall and roof insulation, shading style, thermal mass, night ventilation, air change rate	Use of an optimal solution can reduce the building energy requirement by 94% when compared to the actual design practices adopted in Sydney	[97]	Sydney	Design optimization for lifecycle heating and cooling costs	Low energy home (detached, single story house)
e.	Thermal insulation, Trombe wall and cool roof	i 46% and 80% of savings can be achieved in winters and summers, respectively, in comparison to an ordinary building. A reduction of 5.41 kW was also achieved in the peak cooling load. ii About 20.7% was contributed alone by the storage capacity of the Trombe wall. iii 60.3% and 47.7% savings in heating and cooling, respectively, can be achieved by integrating the movable overhangs, internal curtains and low emissive argon coating loads	[98]	Tunisia	Analyzed the impact of architectural characteristics and passive techniques on its energy requirements	–
f.	Passive cooling techniques: Ground cooling, solar control, night ventilation, evaporative cooling, night sky infrared radiation, infiltration control and thermal insulation	i The results confirmed the effectiveness of the earth pipes. ii With the application of a conductive thermal control in the west wall and the roof slab, a reduction of temperature fluctuation from 24.07 of the outside conditions to an average DBT of about 21°C was noticed. iii The DBT inside the setup for evaporative cooling remained in a comfortable zone i.e., 22.86°C. iv The shading devices blocked any direct sun gain and avoided overheating the interior improving the indoor comfort conditions and thus reducing the energy consumption	[99]	Hot and dry regions (Mexico)	To achieve hygrothermal conditions for occupants and reduce the energy consumption of air-conditioning	New and existing buildings

Table 4. Continued

S. No.	Concepts	Results	Ref.	Climatic Conditions	Remarks	Applications
g.	Louver shading devices, double-glazed, natural ventilation: wind catcher, cross ventilation, green roofing, evaporative cooling via fountain, insulation, indirect radiant cooling, light color coatings with reflection	i The solar heat gain coefficient was about 17% and a reduction of 55% in energy was noticed with the use of solar control film. ii A reduction of 23.6% in annual energy consumption can be achieved with use of solar passive cooling strategies. iii A 220 m² green roof with sprinklers was used and it was estimated that the green roof reduces the energy demand by 6% and drops the roof temperature by 30°C	[100]	Hot and arid regions (Dubai)	To study the effectiveness of various passive cooling techniques to enhance the thermal performance and to decrease the energy consumption	Residential buildings
h.	Night ventilation, roof insulation, shading devices, courtyard planning, micro climate modifications	i The indoor temperatures of the Chinese houses were higher by 1°C than the outdoor temperatures during the day under open window conditions and by 2°C at night under closed window conditions. ii The outdoor temperature of Malay house was recorded to be 1.7°C higher than the terraced house. iii The indoor temperature was 5°C lower than the outdoor temperature in case of courtyard planning at the peak period whereas the values for both indoor and outdoor temperatures were same during the night. iv The periods of indoor operative temperatures exceeding the 80% comfortable upper limit in Malay houses, Chinese shop houses, daytime ventilated and night ventilated terraced houses were 47%, 7–8%, 91% and 42%, respectively	[101]	Hot and humid (Malaysia)	Field studies were conducted in two traditional timber Malay houses and two traditional masonry Chinese shop houses	Modern terraced houses
i.	Orientation, cross ventilation, day lighting, unglazed Trombe wall, earth sheltering, wind towers, solar water heater system, roof top PV system and PV thermal greenhouse dryer	i During the extreme weather conditions in both summers and winters, all the floors were in comfortable range of temperature. ii The temperature recorded for the basement (earth sheltering) was 28°C, for ground and first floor it was recorded at approximately 18–20°C. iii The basement temperature was found to be 7–8.65°C lower than the ambient temperature during harsh summers. iv A rise of 2–3°C in the indoor temperature was found during harsh winters on the first floor due to the presence of a clerestory window. v The total energy saving due to the thermal heat gain and day lighting to be 34,445.568 kWh, out of which 5852.93 kWh accounts for day lighting only	[68]	Composite climate (India)	–	All building types

Table 5. Building integrated Photovoltaic Thermal (BiPVT) system.

S. No.	Concepts	Results	Ref.	Climatic conditions	Remarks	Applications
5.1	Building integrated Opaque Photovoltaic Thermal System (BiOPVT) System					
a.	Integrated with façade with air duct	Minimum 2 3°C rise in room temperature	[107]	Srinagar, India with (although valid for different climatic conditions)	i Rise in room temperature is comparatively lesser than SPVT because of low conductivity of tedlar. ii Ambient temperature 4.4°C	i Electricity generation. ii Thermal heating
b.	Installed on the roof top	45% energy produced	[111]	Brazil	Installation on the roof top yields more energy than on any vertical façade	i Multi- family dwellings. ii Thermal and electricity generation
c.	Installed on roof	3.19 kW annual electrical energy can be saved	[112]	Las Vegas	South orientation	Residential buildings
5.2	Building-integrated Semitransparent Photovoltaic Thermal (BiSPVT) System (Fig. 15; equations 19–23)					
a.	Installed on the roof top without air duct	i Maximum 18°C rise in room temperature [107]. ii Air mass flow rate (0.85–10 kg/s) through duct increases the room air temperature from 9.4 to 15.2°C [107]. iii 1203 MWh of electricity can be saved annually [113]	[107, 113]	Srinagar, India (although valid for different climatic conditions)	i Non- packing area (i.e., glass area) increases the heat gain. ii Double glazing reduces the heat loss. iii Better performance when compared to BiOPVT system	i Day lighting. ii Thermal energy. iii Electricity production. iv Office buildings
b.	Integrated to a roof	i 47°C achieved at first floor. ii The optimum roof thickness for the above study was found to be 300–400 mm to minimize the decrement factor	[102]	Varanasi, India	The effect of number of air changes per hour through the room on the inside temperature, decrement factor and TLL of the building for thermal comfort were considered	Crop drying

Building-Integrated Photovoltaic Thermal (BiPVT) System

Solar energy is converted to electrical energy by photovoltaic (PV) modules with efficiency of 10–15%. The rest of the energy is radiated back to the atmosphere or absorbed as heat. Photovoltaic thermal system (PVT) refers to the extraction of this absorbed energy and bringing it into use. Integration of PVT with a building (façade, roof, windows etc.) is referred to as building integrated photovoltaic thermal system (BiPVT). The efficiency of BiPVT system is much larger, since it produces electricity and also provides the building with thermal energy. In case, the PV modules are opaque type, the system is termed as building-integrated opaque photovoltaic thermal system and if the PV modules used are semitransparent, the system is referred as BiSPVT (Fig. 15).

Following [102] (Fig. 15), energy balance for solar cell is,

$$\alpha_c \tau_g I(t)\,\beta A_m = \left[U_{tca}\left(T_c - T_a\right) + U_{bcr1}\left(T_c - T_{r1}\right)\right] A_m$$
$$+ \tau_g I(t)\beta A_m \eta_c \tag{19}$$

Energy balance for roof of room 2 at $x = 0$ is

$$\alpha_R \tau_g^2\left(1-\beta\right) I(t) A_m = h_c A_R \left(T|_{x=0} - T_{r1}\right) - k A_r \left.\frac{dT}{dx}\right|_{x=0} \tag{20}$$

Energy balance for room 1 air temperature:

$$M_{a,1}C_{a,1}\frac{dTr_1}{dt} = h_c \left(T|_{x=0} - T_{r1}\right) A_R + U_{br1}\left(T_c - T_{r1}\right) A_m \tag{21}$$
$$- 0.33 N_1 V_1 (T_{r1} - T_a)$$

Energy balance for roof to room 2 at $x = L$ is

$$-K \left.\frac{dT}{dx}\right|_{x=L} = h_{ir2}(T|_{x=L} - T_{r2}) \tag{22}$$

Energy balance for room 2 air temperature is

$$M_{a,2}C_{a,2}\frac{dTr_2}{dt} = h_{ir2}\left(T|_{x=L} - T_{r2}\right) A_{r2} - 0.33 N_2 V_2 (T_{r2} - T_a) \tag{23}$$

Joshi et al. [103] have found that exergy efficiency of PVT system (11.6–16%) is higher than that of the PV system (8–14%) by using Petela's formula [104]. Joshi and Tiwari [105] have found that the monthly energy and exergy of a 1.2 m PVT module lies between 35–60 kWh and 7–16 kWh, respectively, for different months and cities of India. Chen et al. [106] have found that BiPVT system with 64 m² surface area can produce 8.5–10 kWh of thermal output.

Vats and Tiwari [107] have derived analytical expressions for room air temperature of BiSPVT and BiOPVT system integrated to the roof of a room and facade with

and without air duct. It was found that the room air temperature for façade and roof in SPVT and OPVT systems with air duct is lower than without an air duct. The reason behind the finding was that there was indirect heating due to the presence of insulated façade/roof between the PV module and the room air. The difference between the room air temperature in SPVT and OPVT façade with and without air duct was found to be 1.46°C and 9.80°C, respectively. The same for SPVT and OPVT roof with and without air duct was found to be 1.13°C and 9.55°C, respectively. Further, the results have been tabulated in Table 4. Vats and Tiwari [108] evaluated the energy and exergy performance of BiSPVT system with roof and found that HIT PV module (heterojunction comprised of a thin amorphous silicon PV cell on top of a crystalline silicon cell) has maximum overall thermal energy of 2497 kWh, maximum annual electricity of 810 kWh, and maximum exergy of 834 kWh. The authors also found that the efficiency of HIT and a-Si is 16% and 6%, respectively, varying inversely with the solar cell temperature, while maximum annual thermal energy of 464 kWh was observed in case of thin amorphous silicon cell (a-Si). An a-Si has maximum thermal energy of 79 kWh with packing factor of 0.62 [109].

Vats et al. [110] evaluated the energy and exergy performance of BiSPVT system to roof with and without ducting for cold climatic conditions. It was observed that HIT accounts for maximum overall thermal energy and a-Si for minimum in both cases with and without an air duct. Annual overall exergy in HIT with duct was 643 kWh and without duct was 610 kWh. The study concluded that with duct, approximately 15% overall thermal energy is greater than without duct. Vats et al. [109] studied the influence of packing factor of SPV module integrated to the roof on the room temperature, module, and electrical efficiency of the module. The study concluded that the temperature of the module decreases with decrease in the packing factor. With decrease in the packing factor, there is an increase in its electrical efficiency and rise in the room air temperature by 3°C due to increase in the nonpacking factor. The authors found that HIT PV module has maximum annual electrical energy of 813 kWh and a-Si has maximum thermal energy of 79 kWh with packing factor of 0.62. Efficiency increases by 0.2–0.6% with a corresponding decrease in the module temperature by 10°C if packing factor is reduced from 0.83 to 0.62. The results of few studies have been summarized in Table 5.

Based on Table 5, one can observe that BiPVT system can be used for thermal heating and electricity generation. In addition, semitransparent PV module is also an effective heating technique to sustain design, which not only produces electricity and provides thermal energy but also allows day lighting, reducing the energy demands.

Conclusions and Recommendations

- For passive heating, direct gain is more convenient for sunshine hours heating (office) and rest of the concepts are used for residential buildings. Solarium will be useful for both the applications. Use of double-glazed system leads to reduction of 9% of heat gain and reduction of losses by 28% compared to single-glazed system. Exposed walls should be double glazed to trap maximum solar radiation inside the room with minimum U-value.
- For passive cooling, the combination of evaporative cooling and wind tower proves to be very effective and can reduce the temperature by up to 12–17°C. Evaporative cooling is the most economical concept for cooling of a building.
- For passive heating/cooling, combination of Trombe wall, cool roof, and thermal insulation can achieve 46% and 80% of savings in winters and summers, respectively.
- BiSPVT system gives better result in terms of efficiency, thermal environment, space heating, day lighting, and electricity use. Photovoltaic systems are among the most promising alternative energy source. Building-integrated photovoltaic systems can provide savings in electricity costs, reduce pollution, and also add to the architectural appeal of the building.

Nomenclature

A	Area	m^2
C_a	Specific heat of air	J/kg K
c	Air conductance	W/m K
h_0	Outside heat transfer coefficient	W/m^2 K
h_1	Total heat transfer coefficient	W/m^2 K
h_c	Convective heat transfer coefficient	W/m^2 K
h_{cw}	Convective heat transfer coefficient (wetted surface)	W/m^2 K
h_{ew}	Evaporative heat transfer coefficient (wetted surface)	W/m^2°C
h_i	Inside heat transfer coefficient	W/m^2 K
h_m	Top losses of solarium	W/m^2 K
h_r	Radiative heat transfer coefficient	W/m^2 K
h_{rw}	Radiative heat transfer coefficient (wetted surface)	W/m^2 K
h_{TS}	Convective and radiative heat transfer coefficient from wall's outer surface to sunspace	W/m^2 K
$I(t)$	Solar intensity	W/m^2
k	Thermal conductivity	W/m K
L	Thickness	m
M_a	Mass of air	kg
$\dot{m}_a$	Mass flow rate of air in pipe (EAHE)	kg/s
N	Number of air changes	–
T	Temperature	°C
T_{fo}	Air temperature at the outlet of EAHE	°C

Continued.

T_{fi}	Air temperature at the inlet of EAHE	°C
T_p	Temperature of metallic surface of water containers	°C
T_r	Room air temperature	°C
T_{sa}	Solair temperature	°C
T_{ss}	Temperature of sunspace (solarium)	°C
T_w	Water temperature	°C
U_{bcr1}	Overall heat transfer coefficient from solar cell to room 1 through glass cover	W/m^2 K
U_L	Overall heat transfer coefficient	W/m^2°C
U_t	Total heat transfer coefficient	W/m^2 K
U_{tca}	Overall heat transfer coefficient from solar cell to ambient through glass cover	W/m^2 K
V	Volume of room	m^3
$\dot{Q}$	Rate of heat transfer	W
$\dot{q}$	Rate of useful energy gain	W/m^2
$\dot{q}_r$	Radiant heat exchange between sky and a surface	W/m^2
Q_v	Ventilation losses	W

Greek symbols

α	Absorptivity	–
β	Packing factor	–
ΔR	Rate of long wavelength radiation exchange between ambient air and sky	W/m^2
ε	Emittance	–
σ	Stefan-Boltzmann constant	$W/m^2/K^4$
τ	Transmissivity	–

Subscript

1	Room 1
2	Room 2
a	Ambient air
c	Solar cell
g	Glass
m	PV module
R	Roof
r	Room
win	Window

Acknowledgements

The authors are thankful to Bag Energy Research Society, Indirapuram, Ghaziabad for their cooperation and help in the research work.

Conflict of Interest

None declared.

References

1. Agrawal, P. C. 1989. A review of passive systems for natural heating and cooling of buildings. Solar Wind Technol. 6:557–567.

2. Jie, J., Y. Hua, P. Gang, J. Bin, and H. Wei. 2007. Study of PV- Trombe wall assisted with DC fan. Build. Environ. 42:3529–3539.

3. Chandel, S. S., and R. K. Aggarwal. 2008. Performance evaluation of a passive solar building in Western Himalayas. Renewable Energy 33:2166–73.

4. Depecker, P., C. Menezo, J. Virgone, and S. Lepers. 2001. Design of buildings shape and energetic consumption. Build. Environ. 36:627–635.

5. Stevanović, S. 2013. Optimization of passive solar design strategies: a review. Renew. Sustain. Energy Rev. 25:177–196.

6. Aldawoud, A. 2013. The influence of the atrium geometry on the building energy performance. Energy Build. 57:1–5.

7. Capeluto, I. G. 2003. Energy performance of the self- shading building envelope. Energy Build. 35:327–336.

8. Tuhus-Dubrow, D., and M. Krarti. 2010. Genetic-algorithm based approach to optimize building envelope design for residential buildings. Build. Environ. 45:1574–1581.

9. Mingfang, T. 2002. Solar control for buildings. Build. Environ. 37:659–664.

10. Inanici, M. N., and F. N. Demirbilek. 2000. Thermal performance optimization of building aspect ratio and south window size in five cities having different climatic characteristics of Turkey. Build. Environ. 35:41–52.

11. Balcomb, J. D., J. C. Hedstrom, and R. D. McFarland. 1977. Simulation analysis of passive solar -heated buildings- Preliminary results. Sol. Energy 19:277–282.

12. Liu, Y. W., and W. Feng. 2011. Integrating passive cooling and solar techniques into the existing building in South China. Adv. Mater. Res. 368–373:3717–3720.

13. Tiwari, G. N., and S. Kumar. 1991. Thermal evaluation of solarium-cum-passive solar house. Energy Convers. Manage. 32:303–310.

14. Tiwari, G. N. 2012. Solar energy- fundamentals, design, modelling and applications. Narosa Publishing House Pvt. Ltd, Delhi.

15. Tiwari, G. N., and Y. P. Yadav. 1988. Analytical model of a solarium for cold climate- A new approach. Energy Convers. Manage. 28:15–20.

16. Tiwari, G. N., Y. P. Yadav, and S. A. Lawrence. 1988. Performance of a solarium: an analytical study. Build. Environ. 23:145–151.

17. Bastien, D., and A. K. Athienitis. 2012. A control algorithm for optimal energy performance of a solarium/greenhouse with combined interior and exterior motorized shading. Energy Procedia 30:995–1005.

18. Saffari, H., and S. Hosseinnia. 2009. Two-phase Euler-Lagrange CFD simulation of evaporative cooling in a Wind Tower. Energy Build. 41:991–1000.

19. Hughes, B. R., J. K. Calautit, and S. A. Ghani. 2012. The development of commercial wind towers for natural ventilation: a review. Appl. Energy 92:606–627.

20. Montazeri, H., F. Montazeri, R. Azizian, and S. Mostafavi. 2010. Two-sided wind catcher performance evaluation using experimental, numerical and analytical modeling. Renewable Energy 35:1424–1435.

21. Chaudhry, H. N., J. K. Calautit, and B. R. Hughes. 2015. Computational analysis of a wind tower assisted passive cooling technology for the built environment. J. Build. Eng. 1:63–71.

22. Amer, E. H. 2006. Passive options for solar cooling of buildings in arid areas. Energy 31:1322–1344.

23. Miyazaki, T., A. Akisawa, and T. Kashiwagi. 2006. The effects of solar chimneys on thermal load mitigation of office buildings under the Japanese climate. Renewable Energy 31:987–1010.

24. Kamal, M. A. 2012. An overview of passive cooling techniques in buildings: design concepts and architectural interventions. Acta Tech. Napocen. Civil Eng. Architect. 55:84–97.

25. Qingyuan, Z., and L. Yu. 2014. Potentials of passive cooling for passive design of residential buildings in China. Energy Procedia 57:1726–1732.

26. Bansal, N. K., G. Hauser, and G. Minke. 1994. Passive building design- A handbook of natural climate control. Elsevier Science, Amsterdam.

27. Kumar, R., S. N. Garg, and S. C. Kaushik. 2005. Performance evaluation of multi-passive solar applications of a non air-conditioned building. Int. J. Environ. Technol. Manage. 5:60–75.

28. Kima, G., H. S. Lim, T. S. Lim, L. Schaefer, and J. T. Kim. 2012. Comparative advantage of an exterior shading device in thermal performance for residential buildings. Energy Build. 46:105–111.

29. Grynning, S., B. Time, and B. Matusiak. 2014. Solar shading control strategies in cold climates- Heating, cooling demand and daylight availability in office spaces. Sol. Energy 107:182–194.

30. Ali, A. 2013. Passive cooling and vernacularism in Mughal buildings in North India: a source of inspiration for sustainable development. Int. Trans. J. Eng. Manag. Appl. Sci. Technol. 4:15–27.

31. Littlefair, P. J., M. Santamouris, S. Alvarez, A. Dupagne, D. Hall, J. Teller et al. 2000. Environmental site layout planning: solar access, microclimate and passive cooling in urban areas. Construction Research Communications Ltd., BRE Publications, London.

32. Gallo, C., M. Sala, and A. M. M. Sayigh. 1988. Ventilation. Pp. 158–165 in C. Gallo, M. Sala and A.

M. M. Sayigh, eds. Architecture- comfort and energy. Elsevier Science Ltd, Oxford.

33. Wang, Z., Y. Ding, G. Geng, and N. Zhu. 2014. Analysis of energy efficiency retrofit schemes for heating, ventilating and air-conditioning systems in existing office buildings based on the modified bin method. Energy Convers. Manage. 577:233–242.

34. Bahadori, M. N. 1985. An improved design of wind towers for natural ventilation and passive cooling. Sol. Energy 35:119–129.

35. Benhammou, M., B. Draoui, M. Zerrouki, and Y. Marif. 2015. Performance analysis of an earth-to-air heat exchanger assisted by a wind tower for passive cooling of buildings in arid and hot climate. Energy Convers. Manage. 91:1–11.

36. Bahadori, M. N. 1994. Viability of wind towers in achieving summer comfort in the hot arid regions of the middle east. Renewable Energy 5:879–892.

37. Bahadori, M., M. Mazidi, and A. Dehghani. 2008. Experimental investigation of new designs of wind towers. Renewable Energy 33:2273–2281.

38. Bouchahm, Y., F. Bourbia, and A. Belhamri. 2011. Performance analysis and improvement of the use of wind tower in hot dry climate. Renewable Energy 36:898–906.

39. Dehghani-sanij, A., M. Soltani, and K. Raahemifar. 2015. A new design of wind tower for passive ventilation in buildings to reduce energy consumption in windy regions. Renew. Sustain. Energy Rev. 42:182–195.

40. Chungloo, S., and B. Limmeechokchai. 2007. Application of passive cooling systems in the hot and humid climate: the case study of solar chimney and wetted roof in Thailand. Build. Environ. 42:3341–3351.

41. Raman, P., S. Mande, and V. Kishore. 2001. A passive solar system for thermal comfort conditioning of buildings in composite climates. Sol. Energy 70:319–329.

42. Wanphen, S., and K. Nagano. 2009. Experimental study of the performance of porous materials to moderate the roof surface temperature by its evaporative cooling effect. Build. Environ. 44:338–351.

43. Chen, W., S. Liu, and J. Lin. 2015. Analysis on the passive evaporative cooling wall constructed of porous ceramic pipes with water sucking ability. Energy Build. 86:541–549.

44. Cruz, E. G., and E. Krüger. 2015. Evaluating the potential of an indirect evaporative passive cooling system for Brazilian dwellings. Build. Environ. 87:265–273.

45. Zhiyin, D., Z. Changhong, Z. Xingxing, M. Mahmud, Z. Xudong, A. Behrang et al. 2012. Indirect evaporative cooling: past, present and future potentials. Renew. Sustain. Energy Rev. 16:6823–6850.

46. Qiu, G., and S. Riffat. 2006. Novel design and modelling of an evaporative cooling system for buildings. Int. J. Energy Res. 30:985–999.

47. Kamal, M. A. 2003. Energy conservation with passive solar landscaping. in Proceedings on National Convetion on Planning for Sustainable Built Environment. M.A.N.I.T, Bhopal.

48. Ca, V. T., T. Asaeda, and E. M. A. Abu. 1998. Reductions in air conditioning energy caused by a near by park. Energy Build. 29:83–92.

49. Papadakis, G., P. Tsamis, and S. Kyritsis. 2001. An experimental investigation of the effect of shading with plants for solar control of buildings. Energy Build. 33:831–836.

50. Khedari, J., W. Mansirisub, S. Chaima, N. Pratinthong, and J. Hirunlabh. 2000. Field measurements of performance of roof solar collector. Energy Build. 31:171–178.

51. Dimoudi, A., A. Androutsopoulos, and S. Lykoudis. 2006. Summer performance of a ventilated roof component. Energy Build. 38:610–617.

52. Dimoudi, A., S. Lykoudis, and A. Androutsopoulos. 2006. Thermal performance of an innovative roof component. Renewable Energy 31: 2257–2271.

53. Dabaieh, M., O. Wanas, M. A. Hegazy, and E. Johansson. 2015. Reducing cooling demands in a hot dry climate: a simulation study for non- insulated passive cool roof thermal performance in residential buildings. Energy Build. 89:142–152.

54. Hanif, M., T. Mahlia, A. Zare, T. Saksahdan, and H. Metselaar. 2014. Potential energy savings by radiative cooling system for a building in tropical climate. Renew. Sustain. Energy Rev. 32:642–650.

55. Cavelius, R., C. Isaksson, E. Perednis, and G. E. F. Read. 2005. Passive cooling technologies. Austrian Energy Agency, p. 125.

56. Juchau, B. 1981. Nocturnal and conventional space cooling via radiant floors. in International Passive and Hybrid Cooling Conference, Miami beach.

57. Erell, E., and Y. Etzion. 1992. A radiative cooling system using water as a heat exchange medium. Architect. Sci. Rev., 35:39–49.

58. Beck, A., and D. Büttner. 2006. Radiative cooling for low energy cold production. in Proceedings of the Annual Building Physics Symposium.

59. Zhang, Q., K. Asano, and T. Hayashi. 2002. Regional characteristics of heating loads for apartment houses in China. Trans. AIJ 555:67–69.

60. Wan, K., and F. Yik. 2004. Building design and energy end-use characteristics of high-rise residential buildings in Hong Kong. Appl. Energy 78:19–36.

61. Pfafferott, J., S. Herkel, and M. Wambsganß. 2004. Design, monitoring and evaluation of a low energy

office building with passive cooling by night ventilation. Energy Build. 36:455–465.

62. Ta'ani, R., H. El-Mulki, and S. Batarseh. 1986. Jordan solar house-second testing year. Solar Wind Technol., 3:315–318.

63. Tyagi, V. V., and D. Buddhi. 2007. PCM thermal storage in buildings: a state of art. Renew. Sustain. Energy Rev. 11:1146–1166.

64. Nwachukwu, N. P., and W. I. Okonkwo. 2008. Effect of an absoptive coating on solar energy storage in a Thrombe wall system. Energy Build. 40:371–374.

65. Balcomb, J., and R. McFarland. 1978. Simple empirical method for estimating the performance of a passive solar heated building of the thermal storage wall type. 2nd National Solar Conference, ISES, USA.

66. Saadatian, O., K. Sopian, C. H. Lim, N. Asim, and M. Y. Sulaiman. 2012. Trombe walls: a review of oppurtunities and challenges in research and development. Renew. Sustain. Energy Rev. 16:6340–6351.

67. Bojić, M., K. Johannes, and F. Kuznik. 2014. Optimizing energy and environmental performance of passive Trombe wall. Energy Build. 70:279–286.

68. Tiwari, G., A. Deo, V. Singh, and A. Tiwari. 2016. Energy efficient passive building: a case study of SODHA BERS COMPLEX. Renewable Energy, 1:1–30.

69. Shen, J., S. Lassue, L. Zalewski, and D. Huang. 2007. Numerical study on thermal behaviour of classical or composite Trombe solar walls. Energy Build. 39:962–974.

70. Krüger, E., E. Suzuki, and A. Matoski. 2013. Evaluation of a Trombe wall system in a subtropical location. Energy Build. 66:364–372.

71. Nahar, N. M., P. Sharma, and M. M. Purohit. 1999. Studies on solar passive cooling techniques for arid areas. Energy Convers. Manage. 40:89–95.

72. Tang, R., and Y. Etzion. 2004. On thermal performance of an improved roof pond for cooling buildings. Build. Environ. 39:201–209.

73. Sutton, G. E. 1950. American social heating ventilation engineers guide. p. 131.

74. Tiwari, G. N., A. Kumar, and M. S. Sodha. 1982. A review- Cooling by water evaporation over roof. Energy Convers. Manage. 22:143–53.

75. Jain, D. 2006. Modeling of solar passive techniques for roof cooling in arid regions. Build. Environ. 44:277–287.

76. Kharrufa, S. N., and Y. Adil. 2008. Roof pond cooling in buildings in hot arid climates. Build. Environ. 43:82–89.

77. Tiwari, G., V. Singh, P. Joshi, Shyam, A. Deo, Prabhakant et al. 2014. Design of an Earth Air Heat Exchanger (EAHE) for climatic condition of Chennai, India. Open Environ. Sci., 8:24–34.

78. Jie, J., Y. Hua, P. Gang, and L. Jianping. 2007. Study of PV- Trombe wall installed in a fenestrated room with heat storage. Appl. Therm. Eng. 27:1507–1515.

79. Irshad, K., K. Habib, and N. Thirumalaiswamy. 2015. Performance evaluation of PV- Trombe wall for sustainable building development. Procedia CIRP 26:624–629.

80. Liu, Y., and W. Feng. 2012. Integrating passive cooling and solar techniques into the existing building in South China. Adv. Mater. Res. 368–373:3717–3720.

81. Ji, J., H. Yi, W. He, and G. Pei. 2007. PV-Trombe wall design for buildings in composite climates. J. Sol. Energy Eng. 129:431–437.

82. Liu, Y., D. Wang, C. Ma, and J. Liu. 2013. A numerical and experimental analysis of the air vent management and heat storage characteristics of a Trombe wall. Sol. Energy 91:1–10.

83. Stazi, F., A. Mastrucci, and C. D. Perna. 2012. Trombe wall management in summer conditions: an experimental study. Sol. Energy, 86:2839–2851.

84. Ahmad, I. 1985. Improving the thermal performance of a roof pond system. Energy Convers. Manage. 25:207–209.

85. Al-Hemiddi, N. A. M. 1995. Passive cooling systems applicable for buildings in hot dry climate of Saudi Arabia. Graduate School of Architecture and Urban Planning. UCLA, Los Angeles, CA.

86. Spanaki, A., D. Kolokotsa, T. Tsoutsos, and I. Zacharopoulos. 2014. Assessing the passive cooling effect of the ventilated pond protected with a reflecting layer. Appl. Energy 123:273–280.

87. Jakhar, S., R. Misra, V. Bansal, and M. S. Soni. 2015. Thermal performance investigation of earth air tunnel heat exchanger coupled with a solar air heating duct for northwestern India. Energy Build. 87:360–369.

88. Al-Ajmi, F., D. L. Loveday, and V. I. Hanby. 2006. The cooling potential of earth-air heat exchangers for domestic buildings in a desert climate. Build. Environ. 41:235–44.

89. Benhammou, M., and B. Draoui. 2015. Parametric study on thermal performance of earth-to-air heat exchanger used for cooling of buildings. Renew. Sustain. Energy Rev. 44:348–355.

90. Kumar, R., S. Ramesh, and S. C. Kaushik. 2003. Performance evaluation and energy conservation potential of earth-air-tunnel system coupled with non-air-conditioned building. Build. Environ. 38:807–813.

91. Nayak, S., and G. N. Tiwari. 2009. Theoretical performance assessment of an integrated photovoltaic and earth air heat exchanger greenhouse using energy and exergy analysis methods. Energy Build. 41:888–896.

92. Rodrigues, M. K., R. D. S. Brum, J. Vaz, L. A. O. Rocha, E. D. D. Santos, and L. A. Isoldi. 2015.

Numerical investigation about the improvement of the thermal potential of an Earth-Air Heat Exchanger (EAHE) employing the Constructal Design method. Renewable Energy, 80:538–551.

93. Sodha, M., A. Sharma, S. Singh, N. Bansal, and A. Kumar. 1985. Evaluation of an earth-air tunnel system for cooling/heating of a hospital complex. Build. Environ. 20:115–122.

94. Sadineni, S. B., S. Madala, and R. F. Boehm. 2011. Passive building energy savings: a review of building envelope components. Renew. Sustain. Energy Rev. 15:3617–3631.

95. Jaber, S., and S. Ajib. 2011. Optimum, technical and energy efficiency design of residential building in Mediterranean region. Energy Build. 43:1829–1834.

96. Ihm, P., and M. Krarti. 2012. Design optimization of energy efficient residential buildings in Tunisia. Build. Environ. 58:81–90.

97. Bambrook, S. M., A. B. Sproul, and D. Jacob. 2011. Design optimisation for a low energy home in Sydney. Energy Build. 43:1702–1711.

98. Soussi, M., M. Balghouthi, and A. Guizani. 2013. Energy performance analysis of a solar-cooled building in Tunisia: passive strategies impact and improvement techniques. Energy Build. 67:374–386.

99. Chávez, J. R. G., and F. F. Melchor. 2014. Application of combined passive cooling and passive heating techniques to achieve thermal comfort in a hot dry climate. Energy Procedia 57:1669–1676.

100. Taleb, H. M. 2014. Using passive cooling strategies to improve thermal performance and reduce energy consumption of residential buildings in U.A.E. buildings. Front. Architect. Res., 3:154–165.

101. Toe, D. H. C., and T. Kubota. 2015. Comparative assessment of vernacular passive cooling techniques for improving indoor thermal comfort of modern terraced houses in hot- humid climate of Malaysia. Sol. Energy 114:229–258.

102. Tiwari, G., H. Saini, A. Tiwari, N. Gupta, P. S. Saini, and A. Deo. 2016. Periodic theory of Building integrated Photovoltaic Thermal (BiPVT) System. Sol. Energy 125:373–380.

103. Joshi, A. S., I. Dincer, and B. V. Reddy. 2009. Performance analysis of photovoltaic systems: a review. Renew. Sustain. Energy Rev. 13:1884–1897.

104. Petela, R. 2003. Exergy of undiluted thermal radiation. Sol. Energy 74:469–488.

105. Joshi, A. S., and G. N. Tiwari. 2007. Monthly energy and exergy analysis of hybrid photovoltaic thermal (PV/T) system for the Indian climate. Int. J. Ambient Energy 28:99–112.

106. Chen, Y., A. Athienitis, and K. Galal. 2010. Modelling, design and thermal performance of BIPV/T system thermally coupled with a ventilated concrete slab in a low energy solar house: Part 1. BIPV/T System and house energy concept. Solar Energy 84:1892–1907.

107. Vats, K., and G. Tiwari. 2012. Performance evaluation of a building integrated semitransparent photovoltaic thermal system for roof and facade. Energy Build. 45:211–218.

108. Vats, K., and G. Tiwari. 2012. Energy and exergy analysis of a building integrated semitransparent photovoltaic thermal (BiSPVT) system. Appl. Energy 96:409–416.

109. Vats, K., V. Tomar, and G. N. Tiwari. 2012. Effect of packing factor on the performance of a building integrated semitransparent photovoltaic thermal (BISPVT) system with air duct. Energy Build. 53:159–165.

110. Vats, K., R. Mishra, and A. Tiwari. 2012. A comparative study for a building integrated semitransparent photovoltaic thermal (BiSPVT) system integrated to roof with and without duct. J. Fundament. Renewable Energy Appl. 2:1–4.

111. Ordenes, M., D. L. Marinoski, P. Braun, and R. Rüther. 2007. The impact of building-integrated photovoltaics on the energy demand of multi-family dwellings in Brazil. Energy Build. 39:629–642.

112. Sadineni, S., T. France, and R. Boehm. 2011. Economic feasibility of energy efficiency measures in residential buildings. Renewable Energy 36:2925–2931.

113. Li, D., T. Lam, W. Chan, and A. Mak. 2009. Energy and cost analysis of semi-transparent photovoltaic in office buildings. Appl. Energy, 86:722–729.

15

Influence of different adsorbates on the efficiency of thermochemical energy storage

Tobias Kohler & Karsten Müller

Institute of Separation Science and Technology, Friedrich-Alexander-Universität Erlangen-Nürnberg, Egerlandstr. 3, 91058 Erlangen, Germany

Keywords

Adsorption, efficiency, energy density, thermal energy storage

Correspondence

Karsten Müller, Institute of Separation Science and Technology, Friedrich-Alexander-Universität Erlangen-Nürnberg, Egerlandstr. 3, 91058 Erlangen, Germany.
E-mail: karsten.mueller@fau.de

Funding Information

No funding information provided.

Abstract

The main influencing parameter on the efficiency of adsorptive thermochemical energy storage is the efficiency of the desorption process, which is influenced by the process conditions, for example, desorption time and desorption temperature, and the working pair (adsorbent–adsorbate). Due to constrained process requirements, for example, hours of sun shine and low desorption temperatures available from a flat plate solar collector (333–373 K), the only possibility to increase the efficiency is to change the working pair. The reference working pair water–zeolite 13X needs high desorption temperatures of 500 K and high heat inputs per mass adsorbent (1080 kJ kg^{-1}) in the desorption process to reach the maximum efficiency of 79 % and maximum energy density of 844 kJ kg^{-1}. Therefore, the goal is to reach efficiencies in the same range as the maximum efficiency of water–zeolite 13X for desorption temperatures lower than 500 K with the usage of different adsorbates. Four systems of alcohol as adsorbate on activated carbon are compared with the reference working pair. The usage of alcohols on activated carbon allows for highly efficient adsorptive storage even at low desorption temperatures between 360 and 450 K. The maximum efficiency is shifted to higher desorption temperatures with increasing carbon chain length of the alcohol. At low desorption temperatures, the energy density and efficiency of methanol, ethanol, and propanol are higher than the energy density of the reference system. Hence, the alcohol systems on activated carbon are viable alternative approaches for regulating these process parameters.

Introduction

Thermal energy storage can be divided in three main categories: the storage of sensible heat, latent heat, and thermochemical heat [1]. Thermochemical heat storage is divided into storage by adsorption and storage by reaction [2]. In this work, the focus lies on the thermochemical adsorptive heat storage, which follows the mechanism shown in Scheme 1, where two components A (adsorbent) and B (adsorbate) either interact physically or react chemically [3]. The interaction or reaction of the surface A and the adsorbate B to the adsorbed species AB is exothermic and therefore releases heat which can be utilized. If energy in form of heat has to be stored, the endothermic reverse process of AB to A and B is performed.

All examined adsorption systems in this work follow the mechanism of physisorption only. The energy is stored through the enthalpy of adsorption, which results from attractive forces between the adsorbate and the adsorbent (plus contributions from interactions between molecules of the adsorbate). Under the assumption of an adiabatic adsorbent fixed bed, the only loss in the process is the sensible heat, which is needed to heat the adsorbent in the adsorption and desorption process. This loss is mainly effected by the isobaric heat capacity of the adsorbent. The operation of an adsorptive energy storage system is shown in Scheme 2. If energy, in this case heat, is needed, a gas stream which is saturated with the adsorbate flows through the adsorber (A). The adsorption of the adsorbate releases heat, which heats the gas stream. The hot and

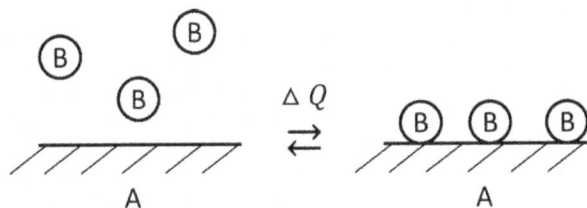

Scheme 1. Reaction/interaction system of a thermochemical adsorption energy storage process.

dry gas stream at the outlet can then be used directly or the heat can be transferred via a heat exchanger. The process of adsorption can be considered as the discharging of the energy storage. If heat is available for regeneration, a hot gas stream flows through the loaded adsorber (B). The gas stream is cooled by the desorption, which can be considered as the loading process of the storage [2].

The efficiency of the energy storage is influenced by a number of factors, for example, the isobaric heat capacity of the materials involved (i.e. primarily adsorbent and adsorbate) and the heat of adsorption. However, as shown in our previous work, the most important factor for a high efficiency is the efficiency of the desorption process. This implies an adsorption characteristic (isotherm) that allows for more complete desorption at the same desorption conditions. This can be achieved by a steeper rise of bed loading at lower partial pressure ratios. The higher desorption efficiency results in a higher amount of heat that can be released in the following adsorption and therefore high efficiencies and high energy densities of the overall process. The efficiency is defined as the ratio of the heat released in the adsorption process to the heat needed in the desorption step. The energy density is defined as the heat released in the adsorption process per mass of the adsorbent.

Desorption is mainly affected by the desorption temperature, the desorption time, and the working pair (adsorbate–adsorbent). As desorption time and temperature are process parameters that are determined by the respective application, the only parameter that can be changed to achieve higher efficiencies is the working pair. Much previous work conducted focused on using alternative adsorbents in order to achieve higher energy densities [4,5,6]. However, the efficiency of the desorption, which can only

be optimized with the change of the isotherm, is rarely considered. The variety of commonly used adsorbent classes for adsorptive energy storage systems is limited. Mostly used are activated carbon, zeolite, and silica gel [5]. New potential adsorbent materials include SAPO and ALPO [7] as well as MOFs (metal organic frameworks) like MIL-101, which net high water uptakes [4, 8]. Although these adsorbents have potential for thermochemical energy storage, most of them have only been used in laboratory scale or are implemented in systems that are just starting in market deployment [4]. A working pair that is highly studied and already used in commercial adsorption heating systems is water on zeolite 13X [5, 9]. This adsorption pair is for example implemented in a building heating system in Munich [10] Furthermore, the cycle stability of the working pair is known [11]. Therefore, this working pair is used as the reference adsorption pair in this work.

However, the progression of the isotherm is influenced by the adsorbent and the adsorbate. In order to have a higher variability to find the ideal isotherm for each desorption temperature and time (i.e., corresponding to specific heat power), the focus has to lie on the change in the adsorbate. For example, the usage of methanol instead of water on zeolite 13X leads to higher efficiencies in the desorption process at 120 °C and 6-h desorption, which is equal to a heat input of 562 kJ kg^{-1}. In comparison with other parameters that influence the efficiency like the heat of adsorption or the isobaric heat capacities of the adsorbent and the adsorbate, an isotherm, which shows a steep rise in bed loading at lower partial pressure ratio, increases the efficiency of the whole process by far the most. That result indicates that the influence of the adsorbate on the efficiency of the process is significant. There is some literature about the usage of different adsorbates in adsorption heating and cooling applications. The most commonly used adsorbates are water, ammonia, methanol, and ethanol [12–15]. However, to the best of our knowledge, a systematic way to analyze the effect of different adsorbates on the efficiency and the energy density is not presented in any of the published work investigated.

In this work, the process of adsorptive energy storage is modeled in a detailed one-dimensional simulation.

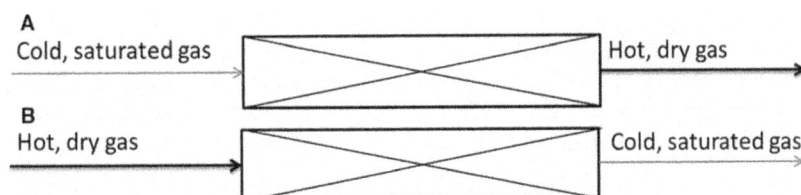

Scheme 2. Flow scheme of the adsorption (A) and the desorption (B) in the adsorptive energy storage process.

Afterwards, the two key performance indices efficiency and energy density are calculated. Methanol and ethanol are already mentioned in literature as potential adsorbates on the adsorbent activated carbon. Therefore, the first four members of the homologous series of alcohols (methanol, ethanol, propanol, butanol) on BPL activated carbon are used for a systematic evaluation of the influence of the adsorbate on efficiency and energy density. The efficiencies and energy densities of these systems are then compared to the reference system water on zeolite 13X.

Modeling Section

The process has been modeled using the software MATLAB® in the Version 8.3 (R2014a). All partial differential equations are solved by discretization of the process into several time and length steps. The time step was set to 0.01 sec and length step to 0.1 m. The mass balance of the adsorbate and of the adsorbent are calculated using equations (1) and (2). In this model, no dispersion terms are considered as they do not influence the efficiency and the energy density in the way calculated in this work. The dispersion would only change the differential temperature levels at different time steps, but does not have an influence on the integral heat balance. Furthermore, Simo et al. [16] stated that dispersion can be neglected in large-scale adsorbent fixed beds. We validated this by calculating Bodenstein number, which is 800 for the modeled process. The released and needed heats of adsorption and desorption are proportional to the temperature difference between the inlet and outlet gas stream over the whole process duration (eq. 10).

$$\varepsilon_G \rho_G \frac{\partial q_G}{\partial t} = -\dot{m}_{GS} - \dot{m}_G \frac{\partial q_G}{\partial z} \qquad (1)$$

$$\varepsilon_s \rho_s \frac{\partial q_S}{\partial t} = \dot{m}_{GS} \qquad (2)$$

ε_G and ε_S stand for the gas and solid volume fractions of the adsorbent bed, respectively. ρ_G and ρ_S stand for the density of the gas phase and of the adsorbent, respectively. $\dot{m}_G$ describes the mass flow of the gas phase and q_G, and q_S stand for the gas and bed loading, respectively. The mass exchange between the gas and the adsorbent $\dot{m}_{GS}$ was calculated with the linear driving force (LDF) approximation (eq. 3) [17].

$$\dot{m}_{GS} = \beta a_s (q_G - q_{G,theoretical}(T, q_s)) \qquad (3)$$

In equation (3), β stands for the mass transfer coefficient, which was calculated with the correlation by Kast [18] (eq. 4). In equation (4), α_{GS} stands for the heat transfer coefficient and $c_{P,G}$ for the isobaric heat capacity of the gas phase.

$$\beta = \frac{\alpha_{GS}}{c_{P,G}\rho_G} \qquad (4)$$

a_S describes the specific surface area of the adsorbent. The theoretical gas loading $q_{G,theoretical}$ at each temperature and bed loading, which is needed for the LDF, was calculated with the potential theory of adsorption by Polányi [19]. $q_{G,theoretical}$ equals the gas loading at equilibrium. In this model, the isotherms are transformed to a temperature independent form called the "characteristic curve of adsorption." The gas loading is transformed to change in free energy ΔF and the bed loading to the adsorption volume W. The progression of the curve is described with the equation by Dubinin and Astakhov (eq. 5) [20]. The parameters W_0 (saturation adsorption volume), E (adsorption energy), and n (empirical parameter) were fitted to the characteristic curves of adsorption calculated from the isotherm data taken from literature at different temperatures for the examined adsorption systems.

The fundamentals of the adsorption model and of the characteristic curve of adsorption can be taken from Polányi [19] and Dubinin [21]. The energy balances for the adsorbate and the adsorbent are given in equations (6) and (7).

$$W = W_0 \cdot e^{\left(-\frac{\Delta F}{E}\right)^n} \qquad (5)$$

$$\varepsilon_G \rho_G c_{P,mix,G} \frac{\partial T_G}{\partial t} = -\dot{m}_G c_{P,mix,G} \frac{\partial T_G}{\partial z} + \alpha_{GS} a_s (T_S - T_G) \qquad (6)$$

$$\varepsilon_s \rho_s c_{P,mix,S} \frac{\partial T_S}{\partial t} = \dot{m}_{GS} h_{ads} - \alpha_{GS} a_s (T_S - T_G) \qquad (7)$$

In equations (6) and (7), T_G and T_S are the temperatures of the gas stream and of the adsorbent, respectively. $c_{P,mix,G}$ and $c_{P,mix,S}$ are the mixed isobaric heat capacities of the gas phase and the solid phase, which were calculated from the heat capacity of the carrier gas or the adsorbent and the heat capacity of the adsorbate in the gas phase or adsorbed on the adsorbent. The heat transfer coefficient α_{GS} was estimated using a Nusselt correlation for the flown through pebble bed (eq. 8). [11]

$$\alpha_{GS} = \frac{Nu \lambda_G}{d_p} \qquad (8)$$

In equation (8), λ_G is the thermal conductivity of the gas phase and d_p the diameter of the adsorbent. The enthalpy of adsorption h_{ads} depends on the temperature and on the bed loading [22]. The calculation was performed with the equation proposed by Bering et al. [23] where α' is the expansion coefficient of the adsorbate and Δh^{LV} the enthalpy of vaporization (eq. 9).

$$h_{ads}(T_S, q_S) = \Delta h^{LV} + \Delta F - T_S \alpha' W \left. \frac{\partial \Delta F}{\partial W} \right|_T \qquad (9)$$

Examined Adsorption Systems and Process Evaluation Criteria

In this study, five adsorption systems were examined. The reference working pair zeolite 13X–water was compared to four systems of alcohols on activated carbon. The experimental data for the adsorption of water on zeolite 13X contains five adsorption isotherms measured up to the normal boiling point of water (0°C, 25°C, 50°C, 75°C, 100°C) [24]. The data sets for alcohols on BPL activated carbon contain adsorption isotherms at four temperatures (25°C, 50°C, 75°C, 100°C) [25] for methanol, ethanol, propanol, and butanol on BPL activated carbon. All material data needed for the simulation are published in the Appendix. In order to evaluate the process, two main evaluation criteria were chosen. First, the efficiency of the adsorption/desorption process is studied (eq. 10). In equation (10), $T_{gas,in}$ and $T_{gas,out}$ stand for the temperature of the gas stream entering and leaving the adsorber.

$$\eta = \frac{Q_{Ads}}{Q_{Des}} = \frac{t \dot{m} c_{P,Gas} \left(T_{Gas,Out} - T_{Gas,In} \right)}{t \dot{m} c_{P,Gas} \left(T_{Gas,In} - T_{Gas,Out} \right)} \qquad (10)$$

As described above, the efficiencies contain the heating of the adsorbent as well as a nonconstant adsorption enthalpy. Furthermore, the heat transport between the adsorbent and the gas phase is taken into account. No peripheral plant components, for example, pumps and compressors and no heat losses over the wall of the adsorber are considered. As the heat capacity of the gas stream in the investigated temperature range only changes from 1038 kJ kg^{-1} K^{-1} (298 K) to 1065 kJ kg^{-1} K^{-1} (500 K), it is considered independent from temperature. The process parameters of the simulation are shown in Table 1. These parameters are very close to an existing sorption energy storage system, which heats a school building in the city of Munich.

Another important parameter for the evaluation of thermochemical energy storages is the energy density ρ_{Qm}, which is calculated for each adsorption system (eq. 11). In equation (11), the released heat Q_{ads} is related to the mass of the adsorbent in the adsorber $m_{adsorbent}$.

Table 1. Simulation parameters.

Parameter	Value
Volume flow gas	6000 m^3 h^{-1}
Volume adsorber	3.5 m^3
Length adsorber	2 m
Carrier gas	Nitrogen
Dew point adsorption	298.15 K
Particle diameter adsorbent	2 mm

$$\rho_{Q_m} = \frac{Q_{Ads}}{m_{adsorbent}} = \frac{t \dot{m} c_{P,Gas} \left(T_{Gas,Out} - T_{Gas,In} \right)}{m_{adsorbent}} \qquad (11)$$

Results and Discussion

For reasons of comparison, the efficiencies and energy densities of the reference system water on zeolite 13X as a function of desorption temperature and time are presented first. Normally other adsorbates than water cannot be used in open adsorption energy storage systems. However, most literature data on energy densities and efficiencies for adsorption energy storage systems which are available on the market are published for open storage systems. In order to be able to compare the presented results with experimental values, the calculations are performed for an open storage system. All results shown in this work are valid for the process parameters and adsorber dimensions shown in Table 1, but the overall trend of the calculated efficiencies and energy densities can be transformed to other process parameters via the given specific heat inputs per mass adsorbent. Figure 1 shows the efficiency plotted against desorption time (for the range between 3 and 6 h) and desorption temperature (for the range between 340 K and 550 K). The efficiency of the process rises with rising desorption time and desorption temperature until the maximum efficiency is reached, when complete desorption is achieved. With rising desorption time, which equals the charging time of the energy storage, more adsorbate desorbs from the adsorbent as the thermal energy is introduced longer into the adsorbent. With rising desorption temperature (e.g., charging temperature), the driving force of the desorption rises. Therefore, more adsorbate is desorbed in the same desorption time. A first look at the efficiencies in Figure 1 shows that for desorption times below 4 h, even desorption temperatures of 550 K are not sufficient for total desorption. The maximum possible efficiency for the system water–zeolite 13X for a 3-h desorption (specific heat input of 1014 kJ kg^{-1}) is only 59%. In comparison with the desorption time of 3 h, the bed loading of zero can be reached within the shown temperature range for desorption times of 4, 5, and 6 h, which equal specific heat inputs of more than 1080 kJ kg^{-1}.

The adsorbate is fully removed at temperatures of 540 K after 4-h desorption, at 520 K after 5-h desorption, and at 500 K after 6-h desorption. These desorption temperatures result in the maximum possible efficiencies of 77%, 78%, and 79% for the respective desorption time. If the desorption temperature is further increased (for fixed desorption times), efficiency goes down, because no lower bed loading can be achieved in the desorption process and therefore, the amount of energy stored stays constant.

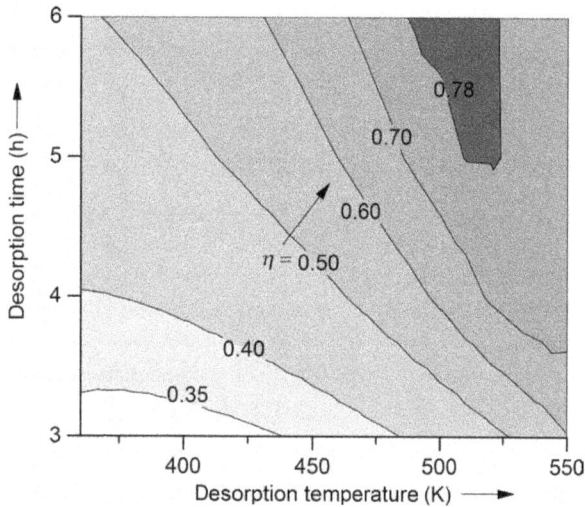

Figure 1. Efficiency of the adsorption system water–zeolite 13X for desorption temperatures between 340 and 550 K and desorption times between 3 and 6 h.

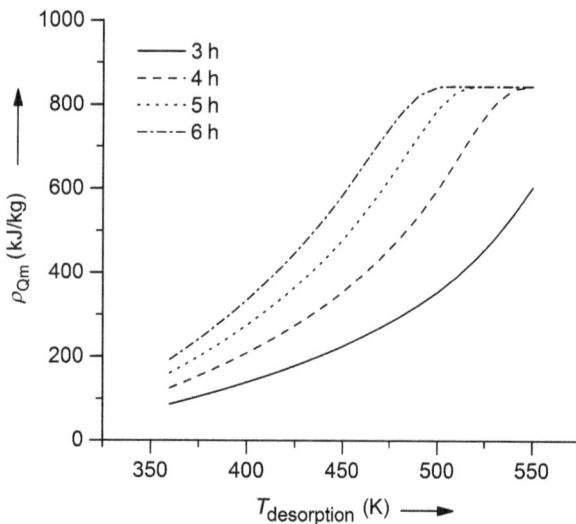

Figure 2. Energy densities of the adsorption pair water–zeolite 13X for different desorption times (3–6 h) and temperatures (350–550 K).

In order to analyze the net heat released per mass adsorbent during the adsorption process, Figure 2 shows the energy densities of the system zeolite 13X as a function of desorption temperature for different desorption times. If the adsorbent is fully regenerated in the desorption process, the energy density of the system water on zeolite 13X is calculated to 844 kJ kg^{-1}, which is by far the highest value for any of the tested adsorption systems. The calculated value is in good agreement with experimental values from literature (e.g., Ahlefeld et al. 900 kJ kg^{-1}). [5, 10, 26] This energy density can be reached within the tested desorption temperatures for desorption times of 4, 5, and 6 h. If the desorption lasts for only

3 h, the maximum energy density cannot be reached for desorption temperatures under 550 K.

The results of the efficiency and energy density calculations show, that the system water–zeolite 13X is extremely efficient and nets high energy densities at long desorption times of 4–6 h and high desorption temperatures of at least 500 K. However, the efficiency of the process is significantly lower if desorption time, which corresponds to the specific heat input, and temperature are not high enough. The problem of the low efficiency at low desorption temperatures can be overcome with the usage of a different adsorption system.

The efficiencies of the systems methanol, ethanol, propanol, and butanol on BPL activated carbon for different desorption times are given in Figure 3A–D. In order to classify these results, the efficiency of water on zeolite 13X is also shown. In Figure 3A–D it can be seen that the maximum efficiencies of the systems based on the adsorption of alcohols are achieved at lower desorption temperatures compared to the reference system. The shorter the chain length of the alcohol, the deeper can the temperature of maximum efficiency be decreased. The maximum efficiency of the alcohol–activated carbon systems slightly increases with increasing chain length from methanol to propanol. These efficiencies are with exception of the system butanol–BPL activated carbon in the same range as the maximum efficiency of water on zeolite. In case of butanol, the maximum efficiency drastically decreases, which mainly results from the relatively low enthalpy of adsorption of butanol on BPL activated carbon. The main reason for the lower heat of adsorption is the low heat of vaporization of butanol.

From the results shown in Figure 3A, a temperature range can be identified for each of the systems (except for butanol on activated carbon) in which this system reaches the highest efficiency of all systems under consideration. The system butanol on activated carbon has by far the lowest efficiencies of all tested systems and therefore is never the ideal system. As can be seen by comparison of Figure 3A–D, the desorption time influences the maximum efficiencies possible. The shorter the desorption lasts, the lower are the maximum efficiencies, as the desorption process cannot be completed. Furthermore, the desorption temperatures at which the maximum efficiency is observed rise to higher temperatures, as the same amount of the adsorbate is desorbed in a shorter desorption time.

For example, for the system methanol on activated carbon the maximum efficiency is reached at a specific heat input in the desorption of 506 kJ kg^{-1}, which equals a desorption temperature of 350 K for 6-h desorption, 355 K for 5-h desorption, 370 K for 4-h desorption, and 390 K for a 3-h desorption.

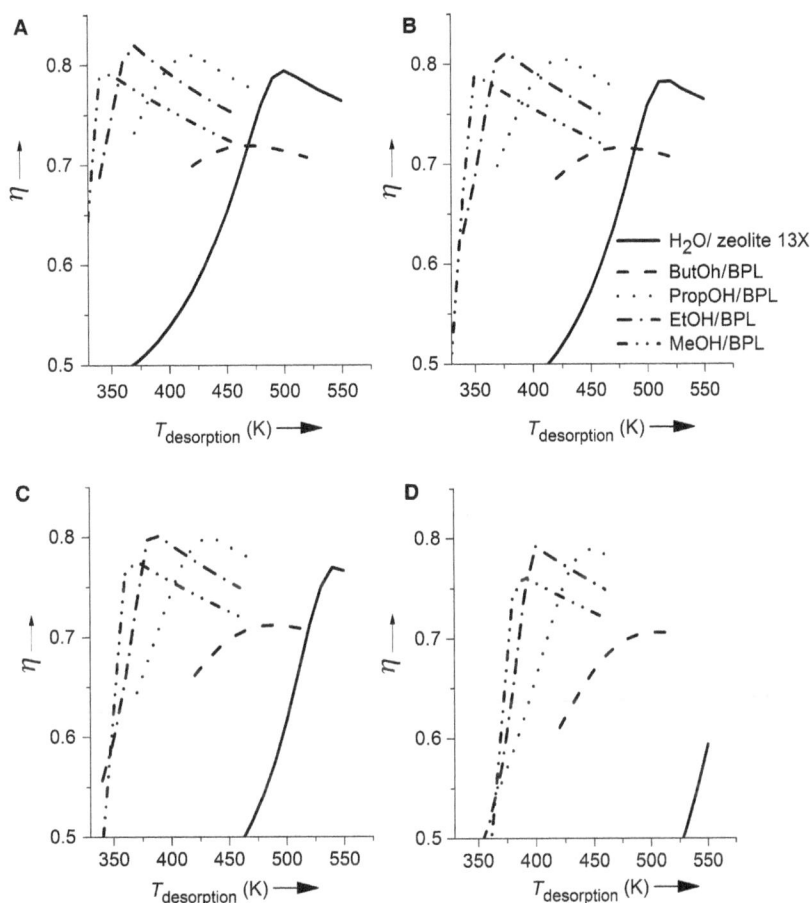

Figure 3. Efficiencies of different desorption times (A 6 h, B 5 h, C 4 h, D 3 h) and temperatures for water on zeolite (—) butanol (--), propanol (•••), ethanol (-•-), and methanol (-••-) on BPL activated carbon.

Concerning efficiency, the systems based on adsorption of short-chain alcohols on activated carbon represent a possible alternative to the reference system water on zeolite 13X. This holds especially, if the heat that has to be stored is only available for a short time or at low temperatures. However, the energy density is a further criterion of importance. In Figure 4, the energy densities of the examined working pairs for a six (4A)-, five (4B)-, four (4C)-, and three (4D)-hour desorption at desorption temperatures between 350 and 550 K are shown. The maximum energy density of the system water on zeolite 13X (840 kJ kg^{-1}) is significantly higher than the values for all alcohols on activated carbon systems. From these systems, the system based on methanol shows the highest energy density with 402 kJ kg^{-1} followed by propanol (320 kJ kg^{-1}), butanol (316 kJ kg^{-1}), and ethanol (310 kJ kg^{-1}). The big difference in energy densities between the water and the alcohol systems results mainly from the high enthalpy of adsorption of the system water on zeolite 13X.

Furthermore, the energy density is influenced by the amount of mass adsorbed and the heat capacities of the adsorbate and the adsorbent. If these components are not taken into account and the energy is only calculated out of the heat of adsorption, the energy densities would be significantly overestimated. In Table 2, the energy densities calculated in the simulation $\rho_{Qm,\textbf{model}}$ are compared to the energy densities that are calculated with constant enthalpy of adsorption and without considering the energy demand for heating the adsorbent and the adsorbate. The change in enthalpy of adsorption with bed loading has a big influence on the energy density. To demonstrate this effect, two different constant values of the differential enthalpies of adsorption were used to additionally calculate the energy density. $\rho_{Q_m},h_{\max}$ is the energy density calculated with the maximum differential enthalpy of adsorption, which occurs at zero bed loading. $\rho_{Q_m},h_{\min}$ is the energy density at a bed loading, where the differential enthalpy of adsorption reaches a saturation level.

From the results in Table 2, it can be seen that the energy densities which are calculated just with the maximum differential enthalpy of adsorption are far higher

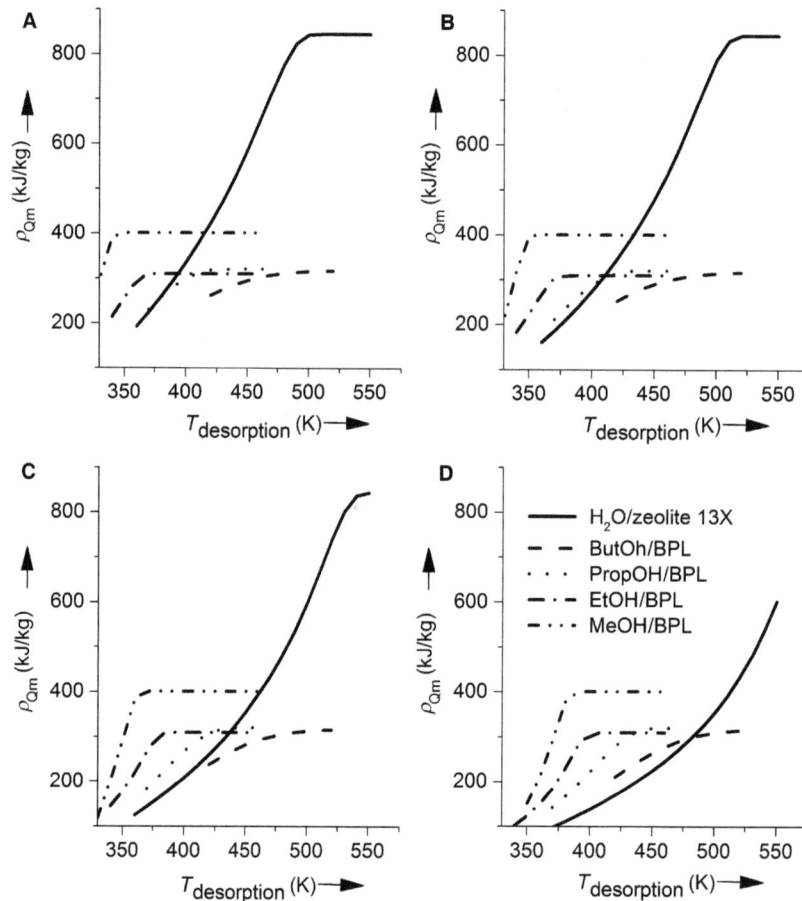

Figure 4. Energy densities for different desorption times (A 6 h, B 5 h, C 4 h, D 3 h) and temperatures for water on zeolite (—) butanol (– –), propanol (•••), ethanol (-•-), and methanol (-••-) on BPL activated carbon.

Table 2. Comparison of calculated energy densities from the detailed simulation and the calculation with constant enthalpy of adsorption and without the consideration of the isobaric heat capacities of the adsorbent and the adsorbate.

Adsorption pair	$\rho_{Qm,model}$/kJ kg^{-1}	$\rho_{Qm,h_{max}}$/kJ kg^{-1}	$\rho_{Qm,h_{min}}$/kJ kg^{-1}
Water–zeolite 13X	840	1272	597
Methanol–BPL activated carbon	402	667	475
Ethanol–BPL activated carbon	310	561	344
Propanol–BPL activated carbon	320	579	355
Butanol–BPL activated carbon	316	638	336

than the results from the detailed simulation. Even the values for the assumption of the lowest differential enthalpy of adsorption are higher for all systems, due to the lack of consideration of energy demand for heating of the materials. These results should remind of the importance of considering the heat capacities and the nonconstant enthalpy of adsorption for process simulations. A closer look at the dependency of energy density on desorption temperature for all examined desorption times makes clear that the systems with alcohols as adsorbates

reach higher energy densities than the reference system at low desorption temperatures.

However, the energy density of the system methanol on BPL activated carbon is higher than for the other alcohol systems for every desorption temperature. The discrepancy between the high efficiency and the lower energy density of the alcohol on activated carbon systems results from the definition of efficiency (eq. 9), where the heat of desorption and adsorption is calculated with the temperature difference of the gas stream at the inlet and outlet of the adsorber.

If heat is available for a given time that cannot be used otherwise, the efficiency of the process plays a less important role and therefore, the more reasonable goal is to net high amounts of energy in the adsorption process.

At low desorption temperatures or low desorption times and therefore low specific heat inputs, the energy density of the adsorbate methanol on activated carbon is higher than the energy density of the reference system. If energy has to be stored at these process conditions, the adsorbate methanol on the adsorbent activated carbon results in significantly higher efficiencies and higher energy densities than the reference system water on zeolite 13X and therefore represents a real alternative for the thermochemical energy storage at low desorption temperatures and low specific heat inputs in the desorption process.

Conclusion

The efficiency and the energy density of thermochemical adsorptive energy storage with different working pairs (adsorbate–adsorbent) was examined for different desorption temperatures (350–550 K) and specific heat inputs. The state of the art working pair water on zeolite 13X nets the highest maximum energy density of the tested systems with 844 kJ kg^{-1}, but needs desorption temperatures above 500 K to reach this value and its maximum process efficiency. The other working pairs under consideration were methanol, ethanol, propanol, and butanol on activated carbon. These systems reach their highest efficiencies and therefore maximum energy densities at lower desorption temperatures and specific heat inputs than the state of the art system. Therefore, they offer a high potential for low-grade solar energy storage systems, which only reach charging temperatures below 500 K. The maximum efficiency of the process rises with increasing chain length of the alcohol to higher desorption temperatures. The working pair methanol on activated carbon has the highest energy density (402 kJ kg^{-1}) of the alcohol systems over the whole examined range of desorption temperature and is therefore best suited to be an alternative to the reference system at low desorption temperatures and times. The efficiency and the energy density of this working pair is significantly higher for low desorption temperatures or low specific heat inputs.

Acknowledgments

The authors wish to thank Prof. Wolfgang Arlt for fruitful discussions.

Conflict of Interest

None declared.

References

1. Zhang, H., J. Baeyens, G. Cáceres, J. Degrève, and Y. Lv. 2016. Thermal energy storage: Recent developments and practical aspects. Prog. Energy Combust. Sci. 53:1–40.
2. Zhang, H., K. Huys, J. Baeyens, J. Degrève, W. Kong, and Y. Lv. 2016. Thermochemical Energy Storage for Power Generation on Demand. Energy Technol. 4:341–352.
3. N'Tsoukpoe, K. E., H. Liu, N. Le Pierrès, and L. Luo. 2009. A review on long-term sorption solar energy storage. Renew. Sustain. Energy Rev. 13:2385–2396.
4. Henninger, S., F. Jeremias, H. Kummer, P. Schossig, and H.-M. Henning. 2012. Novel Sorption Materials for Solar Heating and Cooling. Energy Procedia 30:279–288.
5. Dicaire, D., and F. H. Tezel. 2013. Use of adsorbents for thermal energy storage of solar or excess heat: improvement of energy density. Int. J. Energy Res. 37:1059–1068.
6. Jänchen, J., and H. Stach. 2012. Adsorption properties of porous materials for solar thermal energy storage and heat pump applications. Energy Procedia 30:289–293.
7. Ng, E.-P., and S. Mintova. 2008. Nanoporous materials with enhanced hydrophilicity and high water sorption capacity. Microporous Mesoporous Mater. 114:1–26.
8. Aristov, Y. I. 2015. Current progress in adsorption technologies for low-energy buildings. Future Cities Environ. 1:[e-pub ahead of print]https://futurecitiesenviro.springeropen.com/articles/10.1186/s40984-015-0011-x.
9. Hauer, A. 2007. Evaluation of adsorbent materials for heat pump and thermal energy storage applications in open systems. Adsorption 13:399–405.
10. Hauer, A. 2007. Thermal energy storage for sustainable energy consumption. Springer, Dordrecht, Netherlands.
11. Storch, G., G. Reichenauer, F. Scheffler, and A. Hauer. 2008. Hydrothermal stability of pelletized zeolite 13X for energy storage applications. Adsorption 14:275–281.
12. Mugnier, D., and V. Goetz. 2001. Energy storage comparison of sorption systems for cooling and refrigeration. Sol. Energy 71:47–55.
13. Askalany, A. A., M. Salem, I. M. Ismail, A. H. H. Ali, and M. G. Morsy. 2012. A review on adsorption cooling systems with adsorbent carbon. Renew. Sustain. Energy Rev. 16:493–500.
14. Chan, C. W., J. Ling-Chin, and A. P. Roskilly. 2013. A review of chemical heat pumps, thermodynamic cycles and thermal energy storage technologies for low grade heat utilisation. Appl. Therm. Eng. 50:1257–1273.
15. Srivastava, N. C., and I. M. Eames. 1998. A review of adsorbents and adsorbates in solid + vapour adsorption heat pump systems. Appl. Therm. Eng. 18:707–714.

16. Simo, M., S. Sivashanmugam, C. J. Brown, and V. Hlavacek. 2009. Adsorption/Desorption of Water and Ethanol on 3A Zeolite in Near-Adiabatic Fixed Bed. Ind. Eng. Chem. Res. 48:9247–9260.

17. Sircar, S. 1983. Linear-driving-force model for non-isothermal gas adsorption kinetics. J. Chem. Soc. 79:785–796.

18. Kast, W. 1988. Adsorption aus der gasphase. VCH, Weinheim.

19. Polányi, M. 1916. Adsorption von Gasen (Dämpfen) durch ein festes nichtföüchtiges Adsorbens. Verh. Dtsch. Phys. Ges. 18:55–80.

20. Dubinin, M. M., and V. A. Astakhov. 1971. Description of Adsorption Equilibria of Vapors on Zeolites over Wide Ranges of Temperature and Pressure. Am. Chem. Soc. 60:235–241.

21. Dubinin, M. M. 1967. Adsorption in Micropores. J. Colloid Interface Sci. 23:487–499.

22. Thamm, H. 1989. Calorimetric Study on the state of C1-C4 Alcohols sorbed on silicalite. J. Chem. Soc. 85:1–9.

23. Bering, B. P., M. M. Dubinin, and V. V. Serpinsky. 1966. Theory of volume filling for vapor adsorption. J. Colloid Interface Sci. 21:378–393.

24. Wang, Y., and M. D. Le Van. 2010. Adsorption Equilibrium of Carbon Dioxide and Water Vapor on Zeolites 5A and 13X and Silica Gel: Pure Components. J. Chem. Eng. Data 54:2839–2844.

25. Taqvi, S. M., W. S. Appel, and M. D. Le Van. 1999. Coadsorption of Organic Compounds and Water Vapor on BPL Activated Carbon. 4. Methanol, Ethanol, Propanol, Butanol, and Modeling. Ind. Eng. Chem. Res. 38:240–250.

26. Alefeld, G., P. Maier-Laxhuber, and M. Rothmeyer. 1981. *Proceedings of the IEA Conference on New Energy Conservation Technologies and their Commercialization*, Berlin. Pp. 796–819.

27. Bathen, D., and M. Breitenbach. 2001. Adsorptionstechnik. Springer, Heidelberg.

Appendix

Parameters of the Dubinin-Astakhov equation

Adsorption pair	W_0/cm^3 kg^{-1}	E/kJ kg^{-1}	n/-
Water–zeolite 13X	245.9	1038	2.11
Methanol–BPL activated carbon	437.3	250.5	1.92
Ethanol–BPL activated carbon	438.9	240.8	1.99
Propanol–BPL activated carbon	452.2	251.7	1.84
Butanol–BPL activated carbon	471.9	254.8	1.61

Parameter adsorbate

Parameter	Temperature dependent	Method
Density liquid	Yes	DIPPR-equation 105/116
Density gas	Yes	Redlich-Kwong EoS
c_p liquid	Yes	DIPPR-equation 107
c_p gas	Yes	DIPPR-equation 100
Vapor pressure	Yes	Wagner 25-equation
Heat of vaporization	Yes	DIPPR-equation 106
Thermal conductivity gas	Yes	DIPPR-equation 101

Parameter adsorbent (constant with temperature in the simulation) [27]

Parameter	Zeolite 13X	Activated carbon
Density/kg m^{-3}	2100	1880
Porosity/-	0.5	0.45
Particle diameter/m	0.005	0.005
Specific surface/m^2 g^{-1}	350	500
c_p solid/J kg^{-1} K^{-1}	1200	760

16

China's next renewable energy revolution: goals and mechanisms in the 13th Five Year Plan for energy

Jorrit Gosens[1] (iD), Tomas Kåberger[2] & Yufei Wang[3]

[1]Department of Technology Management and Economics, Division of Environmental Systems Analysis, Chalmers University of Technology, Vera Sandbergs Allé 8, SE-412 96 Göteborg, Sweden

[2]Department of Space, Earth and Environment, Division of Physical Resource Theory, Chalmers University of Technology, Rännvägen 6B, SE-412 96 Göteborg, Sweden
[3]Development Research Center of the State Council, Zhonghua Dizhi Dasha, Hepingjie 13 Jia 20, 100013 Chaoyang, Beijing, China

Keywords
13th Five Year Plan, China, market reform, renewable energy

Correspondence
Jorrit Gosens, Department of Technology Management and Economics, Division of Environmental Systems Analysis, Chalmers University of Technology, Vera Sandbergs Allé 8, SE-412 96, Göteborg, Sweden. E-mail: jorrit@chalmers.se

Funding Information
Svenska Forskningsrådet Formas (Grant/Award Number: '2015-294').

Abstract

Over the past few months, China has published its development plans for the 13th Five Year Plan [FYP] period [2016–2020] for energy, and separately for the electricity sector, renewable energy, hydro, wind, solar, and biomass energy. Here, we review these policies, as well as a number of key supporting policy documents that aim at increased renewable energy use in China. Presuming that China will not overshoot its growth targets for wind and PV, annual ad-ditions over the 13th FYP period will average 16 GW for wind and 13.5 GW for PV, well below the growth levels seen in recent years. The key to success in China's continued transition to renewable energy, however, does not lie in such capacity additions alone. At least as important will be the efforts at improving grid interconnectedness, flexibility of generating capacity and the grid, market mechanisms that will reduce and spread electricity demand, and better enable renewables to compete, and efforts at increasing the level of consump-tion of the renewable power generated.

Introduction

The years 2016 through 2020 make up China's 13th Five-Year-Plan [FYP] period. Here, we review the 13th FYP development plans for different energy sources, and put these goals in context by comparing with policy targets and achievements throughout the previous FYP period, and/or by explaining policy rationales by highlighting the issues that the Chinese power sector faces. We zoom in such issues for modern renewables, by including in our review a number of supporting policies that aim at increased renewable energy consumption. Together, this provides an up-to-date and comprehensive overview of the status quo and plans for the next 5 years of China's renewable power sector development.

Table 1. Development status and targets of the 12th and 13th FYP for energy.

		2010	"12th FYP" 2015 target	2015 actual	"13th FYP" 2020 target	Energy prod. target (TWh)
Total electricity consumption (TWh)	TWh	4200	6150	5693	6800–7200	
Per capita electricity consumption (kWh)	kWh/capita	3132	4529	4142	4860–5140	
Non-fossil share in primary energy cons.	%	8.6%	11.4%	12.0%	15%	
Thermal power						
Coal	GW	654	960	900.1	<1100	
Gas	GW	26.2	56	66.0	110	
Nuclear	GW	10.8	40	27.2	58	
Other thermal	GW			4.3		
West-to-East transport capacity	GW	100		140	270	
Coal cons. in coal-fired power generation	g st. coal eq.	333	323	318	<310/300	
Emissions from thermal power generation						
Sulfur dioxide	kt	9560	[6702]	5281	<2640	
Nitrogen oxides	kt	10,550	[6308]	5519	<2760	
Carbon dioxide	g/kWh	–	–	–	865/550[1]	
Carbon dioxide emission per unit of GDP, change over FYP period	%		–17%	–20%	–18%	
Renewable power						
Hydro	GW	216.1	290	319.5	380	
Conventional	GW	199	260	296.5	340	1250
Pumped storage	GW	16.93	30	23.0	40	
Wind	GW	31	100	130.8	210	420
Onshore	GW	31	95	129.7	205	
Off-shore	GW	0.15	5	1.0	5	
Solar						
Solar power total	GW	0.8	21	43.2	110	150
PV stations	GW	0.387	10	37.1	45	130
PV distributed	GW	0.413	10	6.1	60	
Solar Thermal	GW	0	1	0.001	5	20
Solar hot water heater	mln m2	168	400	440	800	325
Biopower						
All forms	GW	5.5	13	10.3	15	90
Forestry and crop residues	GW	3.6	8	5.3	7	
Waste incineration	GW	1.7	3	4.7	7.5	
Biogas power	GW	0.2	2	0.3	0.5	
Geothermal, marine, other	kt st. coal	4600	15,000	4600	–	
Biomass fuels						
Solid biomass fuels	kt	3000	10,000	8000	30,000	51[2]
Ethanol	kt	1800	4000	2100	4000	13[2]
Diesel	kt	500	1000	800	2000	10[2]
Biogas	mln m3	14,000	22,000	19,000	–	45[2]
Bio-natural gas	mln m3	–	–	–	8000	33[2]

[1]The 865 g/kWh is for coal fired power generation [20]; the 550 g/kWh target is the average for large power generation enterprises [58];
[2]Energy content of liquid fuels is recalculated to TWh for easier comparison here.
Sources: summary of [11, 20, 36, 58, 62].

Targets and Mechanisms in the 13th FYP for the Electricity Sector and Renewable Energy

Overall

The 13th FYP reiterate an earlier announced goal of a 15% share of non-fossil energy in total primary energy consumption by 2020, and 20% by 2030. Power consumption in 2020 is expected to be 6800–7200 TWh, at an average annual growth of 3.6–4.8%. Per capita consumption will be circa 5000 kWh, close to the level in "moderately developed countries". Renewable power production is targeted to grow to 1900 TWh, or 27% of total power generation. In order to prevent power shortages affecting economic development, generating capacity is to be developed "moderately ahead of demand growth", with suggested reserves of 200 TWh [1]. Targets are summarized in Table 1.

Table 2. Installations and power output for major power sources.

Coal	GW	TWh	Full-load hours[1,2]
2005	360.1	1950	–
2010	655.0	3252	5031
2011	709.3	3696	5305
2012	753.8	3785	4982
2013	795.8	3978	5021
2014	832.3	3951	4778
2015	900.1	3898	4364
2016	942.6	3906	4165

Hydro	GW	TWh	Full-load hours[1]
2005	117.4	396.4	–
2010	216.1	686.7	3404
2011	233.0	668.1	3019
2012	249.5	855.6	3591
2013	280.4	892.1	3359
2014	304.9	1060	3669
2015	319.5	1113	3590
2016	332.1	1181	3621

Nuclear	GW	TWh	Full-load hours[1]
2005	6.8	53.1	–
2010	10.8	74.7	7840
2011	12.6	87.2	7759
2012	12.6	98.3	7855
2013	14.7	111.5	7874
2014	20.1	133.2	7787
2015	27.2	171.4	7403
2016	33.64	213.2	7042

Wind	GW	TWh	Full-load hours[1]
2005	1.06	1.65	–
2010	29.58	49.40	2047
2011	46.23	74.06	1875
2012	61.42	103.1	1929
2013	76.52	138.3	2025
2014	96.57	159.8	1900
2015	130.8	185.6	1724
2016	148.6	241.0	1742

Solar	GW	TWh	Full-load hours[1]
2005	0.07	0.03[3]	–
2010	0.80	0.38	–
2011	2.22	0.68	–
2012	3.41	3.60	1423
2013	15.89	8.37	1342
2014	24.86	23.51	1235
2015	42.18	39.48	1225
2016	77.42	66.20	1107[4]

Biomass[5]	GW	TWh	Full-load hours[1]
2005	0.30	1.34	–
2010	3.41	16.10	5822[4]
2011	5.59	23.30	5178[4]
2012	7.69	31.65	4766[4]
2013	8.68	38.31	4682[4]

Table 2. (Continued)

Biomass[5]	GW	TWh	Full-load hours[1]
2014	9.80	46.14	4994[4]
2015	11.41	53.92	5085[4]
2016	13.01[6]	61.70[6]	5054[4]

[1]Full-load hours are not the same as TWh/GW in each row, as these are corrected for the period of time newly installed capacity has been on-line. Data on full-load hours, GW and TWh are all from the same statistical report [63].
[2]Full-load hours are for thermal power generation, which is predominantly but not all coal;
[3]Estimate based on similar productivity as in 2010.
[4]No productivity data is provided by the CEC; calculated as power output, divided by installed capacity last year plus half of capacity growth in current year.
[5]Refers to grid-connected plants using forestry residues, agricultural residues incl. bagasse, and waste incineration;
[6]Assuming same growth in 2016 as in 2015.
Sources: China Electricity Council for years 2010–2015 [63]. 2005 data from [64].

Coal

Different from earlier Five-Year-Plans, which had indicative growth targets, the 13th FYP for the electricity sector strives to contain coal-fired power generation capacity, to below 1100 GW [Table 1]. In large part, this is due to considerable over-capacity in coal-fired power generation built up during the 12th FYP period. Total power demand in 2015 fell 450 TWh short of what was forecasted at the beginning of the 12th FYP period [Table 1]. This slower growth is due to what is called the "new normal" of less rapid economic growth, restructuring in energy-intensive sectors, as well as considerable success in energy-saving programs [2–4]. Demand for thermal power has further been reduced by considerable growth in output from renewables [Tables 1 and 2]. The overcapacity has severely impacted operating profit of coal-fired power plants, as average productivity has sharply fallen from circa 5300 full-load hours to just over 4000 h in recent years [Table 2]. This also means the stated target of 200 TWh of power generation reserves is already well exceeded. In line with this, Yuan et al. [5] recently estimated that no additional coal-fired capacity at all is needed by 2020. Growing overcapacity has already prompted a moratorium on any additional projects starting construction in 15 provinces, until the end of 2017, with 13 of these provinces also not allowed to issue new permits [6]. The only exception are CHP plants for city-heating. The NEA has further demanded that projects under construction are reduced in size or delayed, to keep the pace with market demand growth [6, 7]. The 13th FYP for the electricity sector continues this trend, with a stated goal of canceling or slowing down the construction of at least 150 GW of coal-fired projects. The pace and scale of construction of coal-fired power bases should be adapted to power demand, and the availability of long-distance transmission capacity to remote markets [1].

For emission reduction and energy savings, 420 GW is to be retrofitted for ultra-low emissions, 340 GW to be retrofitted for energy-savings, and another 20 GW of technologically backward generation capacity taken out of operation. Total emissions of sulfur and nitrogen oxides are to be halved [Table 1].

A further key point is improved dispatching control. The three Northern regions have cold winters, and therefore a lot of CHP capacity. This is also where most wind farms are located, and more flexibility is needed. To deal with this, 133 GW of CHP units are to undergo flexibility retrofits, a further 82 GW should undergo a pure condensing retrofit, and 45 GW of peaking capacity should be added. In the rest of the country, 4.5 GW should undergo a pure condensing retrofit, and 1 GW of peaking capacity is to be added. In order to help balance renewable forms of power generation, the NEA has further demanded individual provinces to plan sufficient flexible coal-fired capacity. Such capacity should be able to throttle down to 50% or 60% of rated output, for units with capacity below or above 300 MW, respectively [8]. The measure further encourages on-site power plants in industry to contribute to flexible generation in dealing with temporary demand peaks.

Hydro

The focus of conventional hydropower development is on large hydropower bases (i.e., very large scale [collections] of dams) in Sichuan, Yunnan, and Tibet. The operational capacity of seven hydropower bases currently stands at 122 GW, to be expanded to 138 GW by 2020, and eventually to 222 GW [9]. Simultaneously, long-distance transmission capacity for the so-called "West-to-East electricity transfer" project will be expanded, to transport the hydropower to the load centers in the East [see also Table 1 and Grid construction]. In order to improve dispatch ability, there are plans for research on and implementation of joint dispatching mechanisms for cascaded hydropower, and formulation of optimal scheduling operation procedures and technical standards for cascaded hydropower [9]. Development in small and medium-sized catchment areas is to be limited, in order to preserve ecological habitats in these watersheds. In Sichuan and Yunnan provinces, where hydropower curtailment is severe, development of small and medium-sized hydropower without dispatching control will not be allowed over the 13th FYP period. Projects for poverty alleviation are exempt from this moratorium, and even

promoted. Such projects are to investigate and implement possibilities for public sharing of earnings from hydropower development in poor areas, possibly with local residents or communities receiving an equity share in such projects as partial compensation for loss of land [9, 10].

A second component is the expansion of pumped storage hydro power, with reservoirs concentrated closer to load centers in the East, Central, and North China. Over the 13th FYP period, construction should start on circa 60 GW of pumped storage hydropower, with operational capacity reaching 40 GW by 2020. The pumped storage reservoirs are not only meant as immediately dispatchable capacity for peak-shaving, but are encouraged to function as energy storage specifically for remote, large-scale wind or PV bases, and thus facilitate renewable energy consumption [9, 11].

Gas

The key policy driver for gas-fired power is flexible capacity, which currently has a limited share in generation capacity [12]. In addition to fully utilizing the flexibility of existing natural gas power plants, the plan targets 50 GW of additional flexible capacity by natural gas power plants [Table 1]. Of this, 15 GW will be tri-generation of power, heating, and cooling. The plan further asks for a number of combined cycle CHP projects, and support for projects using coal-bed methane, gasified coal, blast furnace gas, and other gasses [1].

Nuclear

Over the 13th FYP period, circa 30 GW of additional nuclear power is planned to come online, and construction shall start on a further 30 GW or more. Expansion will be in the coastal provinces, but the plan asks for feasibility studies on inland plants as well. A number of projects utilize the AP1000 reactor, a Westinghouse design, manufactured by domestic industries. Future additions will include demonstration projects using the CAP1400, a Chinese development based on the Westinghouse design, and the HPR-1000, or "Hualong one", a domestically developed Advanced Pressurised Water Reactor that China aims to compete with in global markets [1, 13, 14].

Wind

The 13th FYP shifts the development pattern of wind power. The main focus of the 11th and 12th FYP periods was the construction of large-scale wind farms, including seven wind power bases, of 10 GW or more each [12, 15]. These were built up in the Northern regions, which have plenty of sparsely populated plains, and are rich in wind resources. Currently, circa 80% of China's wind power installations is located in the North, North-West and North-East regions, or the "Three Norths". These have faced severe curtailment issues, because of relatively inflexible generation capacity in local grids, in particular during the winter heating period, when CHP district heating plants must be kept on, and because of limited long-distance transmission to larger load centers, mostly outside of the "Three Norths" region [16, 17].

The 13th FYP divides the country into two parts; the "Three Norths" and the rest of the country. Development is strongly pushed towards the latter: 60% of the circa 75 GW targeted increase over the 13th FYP period will be outside of the Norths [Table 3]. More importantly, the development target for the North is a maximum, whereas the 70 GW target for the rest of the country is considered a minimum. To stimulate the development in the rest of the country, the plan promotes improved resource surveys of the more dispersed wind resources there, more distributed wind farm development, and support for improved low speed wind turbine technology [more on this regional imbalance, and implications for policy planning in 18]. The Northern regions are allowed to develop new farms only when local markets are capable of absorbing the additional wind power production. 40 GW interprovincial transmission capacity (see also *Grid construction*) is allocated to help wind farms [and other renewables] in the North transport power to remote markets, but priority for the use of this capacity is given to the existing stock of wind farms [19].

In 2015, curtailment of wind power in the North had grown to 33.9 TWh, or 15.4% of all wind power actually produced [more in *Measures for the guaranteed minimum purchase of renewable power*]. The 13th FYP plan demands that curtailment is brought down to "reasonable levels", albeit without specifying a target. In the press conference presenting the plan, the spokesperson for the NEA indicated this would be considered as 5% or less [20]. Earlier in 2016, the NEA had already issued two separate measures to contain over-investment in regions with severe curtailment [see *Wind power investment warning mechanism*], and to guarantee that grid companies purchase all wind power generated [see *Measures for the guaranteed minimum purchase of renewable power*]. Further, a number of demonstration projects have started to use wind power for district heating. Seasonal output of wind power matches well with the heating demand, in winter periods, and heating grids are much less sensitive to short-term fluctuations than power grids [21]. Still, efficiency of pure electric boilers lags that of CHP district heating plants, and the policy

Table 3. Wind power development status and plan by province.

Region	Province	2015 status	Under construction or approved	2020 planned
East	Shanghai	610	200	500
	Jiangsu	4120	4890	6500
	Zhejiang	1040	1410	3000
	Anhui	1360	1930	3500
	Fujian	1720	2280	3000
	Eastern total	8850	10,710	16,500
Central	Jiangxi	670	2460	3000
	Henan	910	3820	6000
	Hubei	1350	2730	5000
	Hunan	1560	3940	6000
	Chongqing	230	820	500
	Sichuan	730	3210	5000
	Tibet	10	40	200
	Central total	5460	17,020	25,700
South	Guizhou	3230	3310	6000
	Yunnan	4120	5270	12,000
	Guangdong	2460	3000	6000
	Guangxi	430	3220	3500
	Hainan	310	80	300
	Southern total	10,550	14,880	27,800
	East, Central, Southern total	24,860	42,610	70,000
North	Beijing	150	100	500
	Tianjin	290	530	1000
	Hebei	10,220	5490	18,000
	Shanxi	6690	5220	9000
	Shandong	7210	5900	12,000
	West Inner Mongolia	15,270	4580	17,000
	North total	39,830	21,820	57,500
North-East	Liaoning	6390	1860	8000
	Jilin	4440	2490	5000
	Heilongjiang	5030	2130	6000
	East Inner Mongolia	8980	2690	10,000
	North-East total	24,840	9170	29,000
North-West	Shaanxi	1690	3660	5500
	Gansu	12,520	1340	14,000
	Qinghai	470	1090	2000
	Ningxia	8220	2740	9000
	Xinjiang	16,910	4640	18,000
	North-West total	39,810	13,470	48,500
	"Three Norths" region total	104,480	44,460	135,000

All values as MW. Source: [19, 65].

promoting its use suggests a limited scale, and only as a means to reduce wind power curtailment, with 1 GW of such projects suggested, spread out over Shanxi, Liaoning and Xinjiang [22]. The 210 GW target for wind is lower than the 250 GW range announced earlier in the draft version of the 13th FYP for renewable energy [23], presumably because of the challenges with effective consumption of the wind power generated.

Table 4. 2020 Off-shore wind power development plan.

Province	Grid-connected capacity (MW)	Projects under construction (MW)
Tianjin	100	200
Liaoning	–	100
Hebei	–	500
Jiangsu	3000	4500
Zhejiang	300	1000
Shanghai	300	400
Fujian	900	2000
Guangdong	300	1000
Hainan	100	350
Total	5000	10,050

Source: [19].

Offshore wind power is targeted to be at 5 GW by 2020, with another 10 GW under construction [Table 4]. The 5 GW target is the same level as what the 12th FYP plan for renewable energy originally targeted for 2015, and well below the 30 GW that plan targeted for 2020 [24]. Reasons why offshore wind power hasn't developed as quickly as policy makers had hoped include relatively immature equipment from domestic manufacturers, limited co-ordination of governmental departments involved in planning processes, and tendering procedures resulting in bids well below profitable levels [25, 26]. The 13th FYP plan for wind power speaks of improved pricing policies, but without further specification. Existing feed-in tariffs of 0.85 RMB/kWh for off-shore and 0.75 RMB/kWh for inter-tidal farms, introduced in 2014, were recently extended [19, 27].

In terms of domestic industry and technology development, the 13th FYP and the recent "Made in China 2025" Plan strive for fully domestically developed turbines of 10+ MW, including offshore turbines, by 2020, and 30 MW turbines by 2050. Three to five Chinese equipment manufacturers should be at an advanced level internationally, and have significantly improved global market shares [19, 28].

Solar power

The 110 GW target for PV, like that for wind, is well below the 150 GW target announced earlier in the draft version of the 13th FYP for renewable energy [23]. Similarly, the 13th FYP shifts the development pattern of PV, as it did for wind power. Whereas the 12th FYP period saw a rapid build-up of PV mainly in large-scale PV plants [29], the 13th FYP plan targets nearly all additional installations to be with distributed forms of PV [Table 1].

Poverty alleviation programs aim to provide PV panels to 2.8 million households, generating 3000 RMB of additional income per household [28]. For large-scale PV bases, it is stressed that these should prioritize consumption in local electricity markets, or otherwise have connection to an Ultra High Voltage [UHV] long-distance transmission cable [more in *Grid construction*]. Further construction of such bases in the North, specifically Qinghai and Inner Mongolia, can continue in a pace corresponding with local market or UHV transport capacity. In the Southwest, further construction of such bases can continue, utilizing hydropower for balancing power output, and existing UHV transmission channels for hydropower bases [28]. The principle of prioritizing local consumption is similar with that for wind, but with less restrictive language on further development.

In terms of domestic industry and technology development, the 13th FYP targets domestic development of advanced crystalline silicon PV cells with a conversion efficiency of 23% or more, as well as significantly improved and industrialized thin film PV cell production. Power generation costs should halve by 2020, becoming comparable with grid sales prices.

For solar thermal projects, the NDRC announced a Feed-in tariff of 1.15 RMB/kWh in September of 2016 [30]. By 2020, solar thermal cost should be reduced to 0.8 RMB/kWh [28]. A first batch of 20 solar thermal demonstration projects using different types of technology [tower, Fresnel, through] and total capacity of 1.35 GW was recently approved [31].

Biomass

Targets for biomass power have changed focus from crop- and forestry residue based power generation to waste incineration. The 7 GW target for the former is below the 2015 target from the 12th FYP [24], and well below the 24 GW targeted by 2020 in the 11th FYP period [32]. Problems plaguing the sector have been relatively immature technologies from domestic suppliers, with availability far below break-even levels, unexpectedly strong increases in crop residue prices, imperfect coordination of spatial planning leading to competition over fuel between neighboring plants, and a strong focus on pure electric rather than more profitable CHP projects [33–35]. The 13th FYP for biomass further stresses a need to reduce fraud with Feed-in tariff subsidies collected by biomass power plants that are mixing in coal, something which might have cooled policy makers' enthusiasm [36]. The plan further promotes CHP for new projects and retrofitting to CHP for existing projects, and an increased use of densified biomass fuels in retrofitted coal-fired boilers for heating of residential and commercial buildings.

Liquid biomass fuels, too, have fallen short of targeted growth [Table 1]. The target for bio-ethanol is well below the 2020 target of 10,000 kt from the 11th FYP period, whilst the 2020 target for biodiesel is the same [32].

Similar to biomass power generation, development of these fuels has been plagued by technological immaturity, in particular for nonfood crop biofuels, which the strongly politically preferred variety in China, as well as sensitivity to feedstock availability and prices [37, 38]. Priority areas for technological development support over the 13th FYP period are these liquid fuels, and bio-natural gas. This includes supporting bio-ethanol projects of a size of 100 kt each [36].

Grid construction

Despite a strongly centralized planning of power grid infrastructure, inter-provincial power trading has long been limited, with existing inter-provincial connections long considered backup facilities to prop up temporary shortfalls in power supply rather than these being the backbone of a unified national grid [39]. Growing regional imbalances, with energy bases for thermal power, hydro, wind, and solar mostly being built far away from the load centers in the East, have made it increasingly necessary to expand this inter-provincial and inter-regional transport capacity [29, 39, 40]. For this purpose, China has been working on the "West-to-East electricity transfer" project since the early 2000s. Over the 12th FYP, construction was started on a large UHV grid with three main North-South and three West-East corridors [40]. Current plans seek to expand this grid, including to areas further North and West [for a good map see 12].

The target of 270 GW of capacity seems set to be exceeded; total capacity of UHV projects built, under construction or approved for construction already stood at 312 GW, with a total line length of 32,000 km, halfway through 2016 [41]. Many of these lines connect with wind and solar power bases [Table 5].

Supporting Policy Mechanisms

Measures for the guaranteed minimum purchase of renewable power

The build-up of wind and PV power bases in grid regions with little flexibility in the "Three Norths" has resulted in high levels of curtailment [Tables 6 and 7]. The 38.8 TWh of wind and PV power curtailed is roughly equal to the total wind power consumption in the UK in 2015 [42]. This is one of the reasons why China, despite leading in installations, trails the US in terms of wind power consumption [43].

To combat this, the NEA has stipulated "guaranteed minimum full-load hours" for provinces where curtailment is most severe [44, 45]. Grid enterprises must sign a contract for the purchase of these amounts each year,

and award highest priority dispatch rights to these projects [Tables 6 and 7]. Renewable power projects will receive the local feed-in tariff, or the price agreed on in the tender for the project. The guarantee does not cover hours the project is not available due to maintenance, or less than forecasted output due to low winds or solar radiation. To encourage consumption beyond the guaranteed minima, owners of renewable power projects are encouraged to sign long-term power supply contracts directly with power consumers, and engage in spot-market trading [where such markets exist]. Whatever price is agreed upon in such trading mechanisms, projects will still receive subsidies equal to the locally applicable feed-in tariff minus the benchmark price for coal-fired power [44].

Distributed PV, biomass, geothermal, and marine energy projects should also receive contracts with priority rights, clear pricing, and guaranteed amounts of power purchased, but should "temporarily" not engage in such market competition [45].

The 1800–2000 full-load hours guaranteed for wind power projects, or capacity factors of 20.5–23%, are a considerable increase from current levels in provinces with high levels of curtailment [Table 6]. Still, they are well below the average potential capacity factor of 31.9% of China's existing wind farms [43].

Wind power investment warning mechanism

The imbalance of wind power production and capacity for consumption in local grids is also due to excessive wind farm development. To reduce such over-investment, the NEA has launched a monitoring and early warning mechanism for wind power investment risk [46].

The mechanism specifies very clear criteria for investment risk evaluation [Table 8]. Provinces with an orange warning level [total score of 1–1.5] are not allocated an annual development quota for further wind power construction. In provinces with a red warning level [total score below 1], local authorities should further suspend approval of new wind power projects [including for projects in outstanding annual development quota], and grid companies are to halt new grid connection procedures. For provinces where the minimum guaranteed number of full-load hours is not met [*Measures for the guaranteed minimum purchase of renewable power*], a red warning will automatically be issued. Provinces curtailment exceeding 20% or more will automatically be issued at least an orange warning.

For 2016, 7 provinces in the "Three Norths" region [Gansu, Heilongjiang, Inner Mongolia, Jilin, Ningxia, Xinjiang, and the Northern part of Hebei] were issued either a red or orange warning [46]. Because a number

Table 5. Main corridors of China's UHV grid.

Status	Stated purpose includes	From	To	Type	Length (km)	Capacity (GW/ GVA)	Operational start
Operational	UHV demonstration project	South-East Shanxi	Jingmen, Hubei	1000 kV AC	640	6	2009
Operational	UHV demonstration project	Yunnan	Guangdong	800 kV DC	1373	5	2009
Operational	Xiangjiaba Hydropower station	Sichuan	Shanghai	800 kV DC	1907	12.8	2010
Operational	Expansion of demonstration project	South-East Shanxi	Jingmen, Hubei	1000 kV AC	640	18	2011
Operational	Jinping Hydropower Station	Jinping, Sichuan	South Jiangsu	800 kV DC	2059	14.4	2012
Operational	East China coal-fired power base	Huainan, Anhui	Zhejiang - Shanghai	1000 kV AC	2 × 649	21	2013
Operational	Nuozhadu Hydropower station	Nuozhadu, Yunnan	Guangdong	800 kV DC	1413	5	2013
Operational		North Zhejiang	Fuzhou, Fujian	1000 kV AC	2 × 603	18	2014
Operational	PV bases	Hami, Xinjiang	Zhengzhou, Henan	800 kV DC	2192	16	2014
Operational	Xiluodu Hydropower Station	Xiluodu, Yunnan	West Zhejiang	800 kV DC	1653	16	2014
Operational	Ningxia wind power base; PV bases	Ningdong, Ningxia	Zhejiang	800 kV DC	1720	16	2016
Under constr.	East China coal-fired power base	Huainan, Anhui	Nanjing - Shanghai	1000 kV AC	2 × 780	12	2016
Under constr.	Xilingol South wind power base; PV bases	Xilingol, Inner Mongolia	Shandong	1000 kV AC	2 × 730	15	2016
Under constr.	Ordos East wind power base; PV bases	West Inner Mongolia	South Tianjin	1000 kV AC	2 × 608	24	2016
Under constr.	Northern Shaanxi coal-fired power base	Yuheng, Shaanxi	Weifang, Shandong	1000 kV AC	2 × 1049	15	2017
Under constr.	Jiuquan wind power base; PV bases	Jiuquan, Gansu	Hunan	800 kV DC	1119	16	2017
Under constr.	Shanxi North wind power base; PV bases	North Shanxi	Jiangsu	800 kV DC	1119	16	2017
Under constr.	Xilingol North wind power base; PV bases	Xilingol, Inner Mongolia	Taizhou, Jiangsu	800 kV DC	1620	20	2017
Under constr.	Ordos West wind power base; PV bases	Shanghai Miao, Inner Mongolia	Shandong	800 kV DC	1238	20	2017
Under constr.	Tongliao wind power base; PV bases	Jarud, Inner Mongolia	Shandong	800 kV DC	1234	20	2017
Under constr.	Zhundong wind power base; PV bases	Zhundong, Inner Mongolia	Wannan, Anhui	1100 kV DC	3400	12	2018

Capacity refers to substation capacity. Source: [19, 28, 66]. Overview includes all projects approved or under construction, of 1000+ kV AC or 800+ kV DC.

of these provinces are so far removed from attaining the guaranteed minimum hours or curtailment targets [*Measures for the guaranteed minimum purchase of renewable power*], it will likely be some time before construction here will resume. It is not clear how much of the 28.2 GW under construction or approved in the seven provinces with red or orange warnings may be stopped entirely, but it would be in line with the central government's plan to do so for a large share of those projects [see also Table 3]. The 44.5 GW of capacity under

construction or approved in the "Three Norths" region already well exceeds the 30 GW of targeted growth by 2020 [Table 3]. Simultaneously, the high amount of capacity under construction or approved in the East, South, and Central regions indicate the industry has not had much problems shifting the focus of their development operations away from the Northern regions, making it likely that annual additions to wind power capacity will remain at the level of several dozen GW a year in the foreseeable future.

Table 6. Wind power operational data and guaranteed minimum full-load hours.

Region	Area	Installed capacity (GW; 2015)	Power dispatched (GWh; 2015)	Power curtailed (GWh; 2015)	Perc. curtailed (%, 2015)	Full-load hours (2015)[2]	Guaranteed minimum full-load hours purchased[3]
North	Beijing	150	300			1703	
	Tianjin	290	600			2227	
	Hebei	10,220	16,800	1900	10.2%	1808	2000/-
	Shanxi	6690	10,000	300	2.9%	1697	1900/-
	Shandong	7210	12,100			1795	
	West Inner Mongolia[1]	15,270	25,700	5700	18.2%	1865	1900/2000
North-East	Liaoning	6390	11,200	1200	9.7%	1780	1850
	Jilin	4440	6000	2700	31.0%	1430	1800
	Heilongjiang	5030	7200	1900	20.9%	1520	1850/1900
	East Inner Mongolia[1]	8980	15,100	3400	18.4%	1865	1900/2000
North-West	Shaanxi	1690	2800			2014	
	Gansu	12,520	12,700	8200	39.2%	1184	1800
	Qinghai	470	700			1952	
	Ningxia	8220	8800	1300	12.9%	1614	1850
	Xinjiang	16,910	15,200	7100	31.8%	1571	1800/1900
East	Shanghai	610	1000			1999	
	Jiangsu	4120	6400			1753	
	Zhejiang	1040	1600			1887	
	Anhui	1360	2100			1742	
	Fujian	1720	4400			2658	
Central	Jiangxi	670	1100			2030	
	Henan	910	1200			1793	
	Hubei	1350	2100			1927	
	Hunan	1560	2200			2079	
	Chongqing	230	300			2119	
	Sichuan	730	1000			2360	
	Tibet	10	0			1760	
South	Guizhou	3230	3300			1199	
	Yunnan	4120	9400	300	3.1%	2573	
	Guangdong	2460	4100			1689	
	Guangxi	430	600			2122	
	Hainan	310	600			1914	
Total		129,340	186,300	33,900	15.4%	1728	

[1]No separate data for wind power production or curtailment was available for Inner Mongolia's Western and Eastern parts. These were calculated on the basis of reported installations, and assuming equal productivity.

[2]Full-load hours are not the same as TWh/GW here, as these are corrected for the period of time newly installed capacity had been online.

[3]Multiple values are valid for areas with different wind resource classes within the province. Some provinces have a minimum set only for some specific areas.

Source: [44, 65].

Power market reform

China's most recent round of power market reform has been ongoing since 2015, when the central government published a vision that stressed a need for more market-based electricity pricing, and direct trading mechanisms [CPC and State Council, 47, 48]. This is in contrast with the current situation, where grid companies are the sole buyer and seller of electricity, both at predetermined tariff levels. Ministries and provincial level governments have responded to this vision with a number of changes, most of these in piloting or draft phase [49].

First, these include a call to reduce cross-subsidization, reducing or ending the practice of preferential pricing for separate industries in individual provinces [50]. Reviews of such cross-subsidization levels were due by November 2016, and only three provinces have published simplified pricing structures so far, with others likely following within the next few months.

Second, measures have established priority dispatch and guaranteed purchase of power from renewable sources, see also *Measures for the guaranteed minimum purchase of renewable power* [8, 44, 45].

Third, the separation of tariffs for power consumption and transmission and distribution services [50, 51]. These have been proposed in order to ensure sufficient levels of profitability for grid companies, in a system with more

Table 7. PV power operational data and guaranteed minimum full-load hours.

Province	Installed capacity (GW)	Power dispatched[1] (MWh; 2015)	Power curtailed (GWh; 2015)	Perc. curtailed (%, 2015)	Full-load hours[2] (2015)	Guaranteed minimum full-load hours purchased[3]
Gansu	6.10	5800	2600	31%	1030	1400/1500
Xinjiang	5.66	5100	1800	26%	1110	1350/1500
Ningxia	3.09	4000	300	7%	1510	1500
Qinghai	5.64	6500	200	3%	1320	1450/1500
Inner Mongolia	4.89					1400/1500
Heilongjiang	0.02					1300
Jilin	0.07					1300
Liaoning	0.16					1300
Hebei	2.39					1400/-
Shanxi	1.13					1400/-
Shaanxi	1.17					1300/-
Total provinces without operational data	21.69	18,200			1150	
Total	42.18	39,481	4900		1180	

[1]Power output statistics for PV are not available for all provinces.
[2]Calculated as power output divided by capacity last year plus half of the capacity growth in current year. This simple method results in only a slight difference from the officially reported number of 1225 full-load hours as the national average (Table 2).
[3]Multiple values are valid for areas with different PV resource classes within the province. Some provinces have a minimum set only for some specific areas. Sources: [44, 67, 68].

Table 8. Evaluation criteria in the wind power investment risk monitoring mechanism.

Criterion	Score 2	1	0	Weight
Rate of completion of annually planned wind power development	>80%	50–80%	<50%	10%
Local policy negatively affects the development of wind power	No		Yes	10%
Proportion of relatively inflexible power generation capacity in provincial grid	<20%	20–40%	>40%	10%
Wind power curtailment				
"Three Norths" area	<10%	10–20%	>20%	30%
Other areas	<5%	5–10%	>10%	
Annual productivity (full load hours)				
Class I wind resource area	>2400	2200–2400	<2200	15%
Class II wind resource area	>2200	2000–2200	<2000	
Class III wind resource area	>2000	1800–2000	<1800	
Class IV wind resource area	>1800	1500–1800	<1500	
Trading price deficit	<5%	5–20%	>20%	15%
Share of loss-making business in sample	<10%	10–30%	>30%	10%

Source: [46].

direct and more market based electricity pricing, as well as reduced demand [more in 48]. Initially piloted in six provinces such reforms have been rolled out nation-wide by September of 2016 [48, 50].

Fourth, and related, the further promotion of demand side management mechanisms. The use of DSM is not new; at least since the 11th FYP period China has been using such mechanisms as interruptible load tariffs, voluntary load shifting, time-of-use pricing, etc., although often limited to specific industries or localities [48, 52]. Room for improvement certainly still exists, as the peak-to-valley ratio for China is circa 1:0.07, far below that in advanced economy markets [12]. The separation of

electricity and transmission and distribution charges is expected to stimulate more widespread roll-out of DSM, as it removes the previously existing perverse incentive for grid companies to keep demand high [48, 52].

Fifth, the creation of more market based electricity pricing mechanisms. Draft regulations include stipulations for direct trading between power companies and users, in the form of long-term supply contracts as well as in spot-markets [49]. Between September and December 2016, every individual province has published a market reform plan that either pilots or researches such new market mechanisms, establishes a provincial trading center as well as market regulation commissions [e.g., 53]. Two national electric power trading centers are established in Beijing and Guangzhou, focused on intra-provincial and intra-regional trading [54]. The draft regulations further indicate that experiences from such experimentation may in the future be expanded to mechanisms such as a capacity market, electricity futures, and derivatives trading [49]. The forms of power trading will exist alongside existing fixed tariffs for the foreseeable future.

The creation of spot markets in particular will be key to improving the competitive strength of intermittent renewables, which have close to zero marginal cost, and which will be receiving subsidies regardless of prices offered in spot markets [*Measures for the guaranteed minimum purchase of renewable power*]. Further, the differentiated pricing will be key to linking the generation of intermittent renewables with energy storage in pumped hydro or large-scale battery storage facilities. Maximizing both financial and environmental benefits will require that such facilities can charge with cheap, off-peak renewables and discharge at times of peak demand and pricing.

Carbon markets

China has been experimenting with carbon markets since the start of the 12th FYP period. Over 2013 and 2014, pilot markets for voluntary trading of carbon rights became operational in two provinces and five major cities [55]. The pilots cover a selection of industries and assigned carbon emission allowances, allocation mechanisms, offset mechanisms, and minimum and maximum pricing [for a good overview, see 56].

The volume traded, as a fraction of overall caps, has remained lower than in comparable schemes in the EU and California [56]. This is due in part to generous allocations for the pilot phase, but also because trading is restricted to trading within individual pilot regions, and lack of clarity on banking of pilot phase allowances for the future nationally operational scheme [56, 57]. The same issues have also suppressed pricing, with current pricing levels expected to have little influence on investment decisions in the short term [57].

The expectation was that the pilots would transition to a national scheme by 2015 [55], but this has been delayed. The State Council has recently announced a target for such a national scheme in operation by 2017 [58]. Draft regulations on the management of carbon emissions trading have been circulating since January of 2016 [59]. An earlier document specified that companies with energy consumption surpassing 10 kt of standard coal, in eight industries [electric power, petrochemical, chemical, building materials, iron and steel making, nonferrous metals, pulp and papermaking, and aviation] would be included in the national emission scheme [56, 60]. These documents thus identify the scope of the future emissions

cap-and-trade system, but do not specify targets. Concrete targets from related policy documents are 865 g/kWh for coal-fired power generation [1], and 550 g/kWh for large power generation enterprises [average of all power sources] [58]. It is not clear if these exact same targets will also be incorporated in the future cap-and-trade system.

Synopsis

Presuming that China will not overshoot its growth targets for wind and PV, annual additions over the 13th FYP period will average 16 GW for wind and 13.5 GW for PV, well below the growth levels seen in recent years. Still, even in this scenario, China's annual additions of hydro, wind, solar, and biomass combined, would average 42 GW of growth per year, or more than a third of expected global additions [61].

The key to success in China's continued transition to renewable energy, however, does not lie in such capacity additions alone. At least as important will be the efforts at improving grid interconnectedness, flexibility of generating capacity and the grid, market mechanisms that will reduce and spread demand and better enable renewables to compete, and increased levels of consumption of the renewable power generated.

This will bring challenges for energy policy analysts, as these sorts of developments may not be as easy to track as installations, which are better reported on, both in Chinese and English language sources. It will also bring new avenues for exchanges of experiences from other countries that are transitioning to high levels of renewables and are experimenting with mechanisms to integrate these into their grid.

Conflict of Interest

None declared.

References

1. NDRC & NEA. 2016. 13th FYP for the electricity sector.
2. Stern, N., and F. Green. 2015. China's "new normal": structural change, better growth, and peak emissions. Policy brief from the Grantham Research Institute on Climate Change and the Environment & Centre for Climate Change Economics and Policy.
3. Ke, J., L. Price, S. Ohshita, D. Fridley, N. Z. Khanna, N. Zhou et al. 2012. China's industrial energy consumption trends and impacts of the Top-1000 Enterprises Energy-Saving Program and the Ten Key Energy-Saving Projects. Energy Pol. 50:562–569.
4. Zhao, X., H. Li, L. Wu, and Y. Qi. 2014. Implementation of energy-saving policies in China:

how local governments assisted industrial enterprises in achieving energy-saving targets. Energy Pol. 66:170–184.

5. Yuan, J., P. Li, Y. Wang, Q. Liu, X. Shen, K. Zhang et al. 2016. Coal power overcapacity and investment bubble in China during 2015–2020. Energy Pol. 97:136–144.

6. NDRC & NEA. 2016. Circular on promoting the orderly development of coal-fired power sector.

7. NEA. 2016. Circular on further regulating the planning and construction of coal fired power plants.

8. NDRC & NEA. 2016. Provisionary measures for priority dispatch of renewable peaking power generation units.

9. NEA. 2016. 13th FYP development plan for hydro power.

10. State Council. 2016. China to use resource development for poverty relief. Available at http://english.gov.cn/policies/latest_releases/2016/10/18/content_281475469264729.htm (accessed 18 October 2016).

11. NDRC. 2016. 13th FYP development plan for renewable energy, translation Available at http://chinaenergyportal.org/en/13th-fyp-development-plan-renewable-energy/

12. Ding, N., J. Duan, S. Xue, M. Zeng, and J. Shen. 2015. Overall review of peaking power in China: status quo, barriers and solutions. Renew. Sustain. Energy Rev. 42:503–516.

13. SCMP. 2016. China promotes exports of next generation Hualong One reactor as 'competitive' with rival atomic technologies. Available at http://www.scmp.com/business/companies/article/1934163/china-promotes-exports-next-generation-hualong-one-reactor (accessed 10 April 2016).

14. Xing, J., D. Song, and Y. Wu. 2016. HPR1000: advanced pressurized water reactor with active and passive safety. Engineering 2:79–87.

15. Wang, Z., H. Qin, and J. I. Lewis. 2012. China's wind power industry: policy support, technological achievements, and emerging challenges. Energy Pol. 51:80–88.

16. Zhao, X., S. Zhang, R. Yang, and M. Wang. 2012. Constraints on the effective utilization of wind power in China: an illustration from the northeast China grid. Renew. Sustain. Energy Rev. 16:4508–4514.

17. Davidson, M. R., D. Zhang, W. Xiong, X. Zhang, and V. J. Karplus. 2016. Modelling the potential for wind energy integration on China's coal-heavy electricity grid. Nat. Energy 1:16086.

18. Xiong, W., Y. Yang, Y. Wang, and X. Zhang. 2016. Marginal abatement cost curve for wind power in China: a provincial-level analysis. Energy Sci. Eng. 4:245–255.

19. NEA. 2016. 13th FYP development plan for wind power.

20. NEA. 2016. 13th FYP for the electricity sector — Press conference transcript. Available at http://www.nea.gov.cn/xwfb/20161107zb1/index.htm (accessed 7 November 2016).

21. Zhang, N., X. Lu, M. B. McElroy, C. P. Nielsen, X. Chen, Y. Deng et al. 2016. Reducing curtailment of wind electricity in China by employing electric boilers for heat and pumped hydro for energy storage. Appl. Energy 184:987–994.

22. NEA. 2015. Tasks for using wind power for clean city-heating.

23. CEC. 2016. NEA reports 2,300 billion RMB of investment in 13th FYP development plan for renewable energy. Available at http://www.cec.org.cn/zhengcefagui/2016-02-01/148693.html (accessed 1 February 2016).

24. NDRC. 2012. 12th FYP development plan for renewable energy.

25. Zhao, X-g., and L-z. Ren. 2015. Focus on the development of offshore wind power in China: has the golden period come? Renewable Energy 81:644–657.

26. Chen, J. 2011. Development of offshore wind power in China. Renew. Sustain. Energy Rev. 15:5013–5020.

27. NDRC. 2016. Notice on adjustments to feed-in-tariffs for onshore wind and PV power.

28. NDRC MIIT & NEA. 2016. Made in China 2025 – Plan of action for energy equipment.

29. Ming, Z., X. Song, M. Mingjuan, and Z. Xiaoli. 2013. New energy bases and sustainable development in China: a review. Renew. Sustain. Energy Rev. 20:169–185.

30. NDRC. 2016. Notice on solar thermal power generation benchmark tariff policy.

31. NEA. 2016. Circular on the construction of solar thermal power generation demonstration projects.

32. NDRC. 2007. Medium and Long-Term Development Plan for Renewable Energy.

33. Xingang, Z., T. Zhongfu, and L. Pingkuo. 2013. Development goal of 30 GW for China's biomass power generation: will it be achieved? Renew. Sustain. Energy Rev. 25:310–317.

34. Liu, J., S. Wang, Q. Wei, and S. Yan. 2014. Present situation, problems and solutions of China's biomass power generation industry. Energy Pol. 70:144–151.

35. Gosens, J. 2015. Biopower from direct firing of crop and forestry residues in China: a review of developments and investment outlook. Biomass Bioenerg. 73:110–123.

36. NEA. 2016. 13th FYP development plan for biomass energy.

37. Chang, S., L. Zhao, G. R. Timilsina, and X. Zhang. 2012. Biofuels development in China: technology options and policies needed to meet the 2020 target. Energy Pol. 51:64–79.

38. Zhao, L., S. Chang, H. Wang, X. Zhang, X. Ou, B. Wang et al. 2015. Long-term projections of liquid

biofuels in China: uncertainties and potential benefits. Energy 83:37–54.

39. Li, X., K. Hubacek, and Y. L. Siu. 2012. Wind power in China–Dream or reality? Energy 37:51–60.

40. Chen, W., H. Li, and Z. Wu. 2010. Western China energy development and west to east energy transfer: application of the Western China Sustainable Energy Development Model. Energy Pol. 38:7106–7120.

41. SGCC. 2016. Jarud - Qingzhou UHV DC transmission project was approved by the State Development and Reform Commission. News item from the State Grid Corporation of China. Available at http://www.sgcc.com.cn/xwzx/gsyw/2016/08/335411.shtml (accessed 22 Aug 2016).

42. BP. 2016. Statistical Review of World Energy 2016. Available at http://www.bp.com/

43. Lu, X., M. B. McElroy, W. Peng, S. Liu, C. P. Nielsen, and H. Wang. 2016. Challenges faced by China compared with the US in developing wind power. Nat. Energy 1:16061.

44. NEA. 2016. Administrative tasks for the guaranteed full purchase of renewable electric power.

45. NEA. 2016. Measures for the guaranteed full purchase of renewable electric power.

46. NEA. 2016. Establishment of monitoring and early warning mechanisms to promote the sustained and healthy development of the wind power industry.

47. Central Committee of the Communist Party of China and the State Council. 2015. Opinions on further deepening of China's power sector reform.

48. Zhang, S., Y. Jiao, and W. Chen. 2017. Demand-side management (DSM) in the context of China's on-going power sector reform. Energy Pol. 100:1–8.

49. NDRC. 2015. Circular on the issuance of a set of supporting documents for the reform of the electric power system.

50. NDRC. 2016. Circular on pilot projects for reform of transmission and distribution tariffs.

51. NDRC. 2016. Circular on implementation of the two-part electricity price system and base electricity tariffs.

52. Wang, J., C. N. Bloyd, Z. Hu, and Z. Tan. 2010. Demand response in China. Energy 35:1592–1597.

53. NDRC & NEA. 2016. Approval for a comprehensive electric power system reform pilot in Hubei, Sichuan, Liaoning, Shaanxi and Anhui.

54. NDRC & NEA. 2016. Approval on plans to establish electric power trading centers in Beijing & Guangzhou.

55. Lo, A. Y. 2013. Carbon trading in a socialist market economy: can China make a difference? Ecol. Econ. 87:72–74.

56. Swartz, J. 2016. China's National Emissions Trading System: Implications for Carbon Markets and Trade; ICTSD.

57. De Boer, D., R. Roldao, and H. Slater. 2015. The 2015 China Carbon Pricing Survey, August 2015. China Carbon Forum.

58. State Council. 2016. 13th FYP period work program for controlling greenhouse gas emissions.

59. CEC. 2016. Regulations on the management of carbon emissions trading (draft).

60. NDRC. 2016. Key tasks for launching the national carbon emissions trading market.

61. IEA. 2015. Medium-Term Market Report 2015.

62. State Council. 2013. 12th FYP development plan for energy.

63. CEC. 2017. Detailed electricity statistics by the China Electricity Council; several years used.

64. NDRC. 2008. 11th Five Year Plan for Renewable energy development, (NDRC Energy 2008, nr. 610) (Original in Chinese). Available at http://www.sdpc.gov.cn/zcfb/zcfbghwb/200803/t20080318_579693.html.

65. NEA. 2016. 2015 wind power installations and production by province.

66. SGCC. 2016. Overview of UHV projects. Available at http://www.sgcc.com.cn/xwzx/gsyw/gtgz/tgysdgc_index.shtml.

67. NEA. 2016. 2015 National renewable power development monitoring and evaluation report.

68. NEA. 2016. 2015 PV Statistics.

Appendix

Table A1. Original language names and links for key Chinese documents.

Source	Title	Original language title	Link to (partially) translated version
CEC, 2016	Regulations on the management of carbon emissions trading (draft)	《碳排放权交易管理条例》（送审稿）	
CEC, 2017	Detailed electricity statistics by the China Electricity Council	2016年电力统计基本数据一览表	http://chinaenergyportal.org/en/2016-detailed-electricity-statistics/
CCCPC and the State Council, 2015	Opinions on further deepening of China's power sector reform	中共中央 国务院关于进一步深化电力体制改革的若干意见(中发〔2015〕9号)	http://chinaenergyportal.org/en/opinions-of-the-cpc-central-committee-and-the-state-council-on-further-deepening-the-reform-of-the-electric-power-system-zhongfa-2015-no-9/
NDRC, 2012	12th FYP development plan for renewable energy	可再生能源发展"十二五"规划	
NDRC, 2015	Circular on the issuance of a set of supporting documents for the reform of the electric power system	关于印发电力体制改革配套文件的通知(发改经体[2015]2752号)	
NDRC, 2016	13th FYP development plan for renewable energy	可再生能源发展"十三五"规划	http://chinaenergyportal.org/en/13th-fyp-development-plan-renewable-energy/
NDRC, 2016	Circular on implementation of the two-part electricity price system and base electricity tariffs	关于完善两部制电价用户基本电价执行方式的通知(发改办价格[2016]1583号)	
NDRC, 2016	Circular on pilot projects for reform of transmission and distribution tariffs	关于全面推进输配电价改革试点有关事项的通知(发改价格[2016]2018号)	
NDRC, 2016	Key tasks for launching the national carbon emissions trading market	关于切实做好全国碳排放权交易市场启动重点工作的通知 发改办气候[2016]57号	
NDRC, 2016	Notice on adjustments to feed-in-tariffs for onshore wind and PV power	关于调整光伏发电陆上风电标杆上网电价的通知(发改价格[2016]2729号)	http://chinaenergyportal.org/en/notice-on-adjustments-to-feed-in-tariffs-for-onshore-wind-and-pv-power/
NDRC, 2016	Notice on solar thermal power generation benchmark tariff policy	关于太阳能热发电标杆上网电价政策的通知(发改价格[2016]1881号)	http://chinaenergyportal.org/en/notice-solar-thermal-power-generation-benchmark-tariff-policy/
NDRC & NEA, 2016	13th FYP for the electricity sector	电力发展"十三五"规划(2016–2020年)	http://chinaenergyportal.org/en/13th-fyp-for-the-electricity-sector-full-text/
NDRC & NEA, 2016	Approval for a comprehensive electric power system reform pilot in Hubei, Sichuan, Liaoning, Shaanxi and Anhui	关于同意湖北等5省开展电力体制改革综合试点的复函(发改经体[2016]1900号)	
NDRC & NEA, 2016	Approval on plans to establish electric power trading centers in Beijing & Guangzhou	关于北京、广州电力交易中心组建方案的复函(发改经体[2016]414号)	
NDRC & NEA, 2016	Circular on promoting the orderly development of coal-fired power sector	关于促进我国煤电有序发展的通知(发改能源[2016]565号)	
NDRC & NEA, 2016	Provisionary measures for priority dispatch of renewable peaking power generation units	关于印发《可再生能源调峰机组优先发电试行办法》的通知(发改运行[2016]1558号)	http://chinaenergyportal.org/en/provisionary-measures-for-priority-dispatch-of-renewable-peaking-power-generation-units/

Table A1. (Continued)

Source	Title	Original language title	Link to (partially) translated version
NDRC MIIT & NEA, 2016	Made in China 2025 – Plan of action for energy equipment	中国制造2025—能源装备实施方案 (发改能源[2016]1274号)	
NEA, 2015	Tasks for using wind power for clean city-heating	国家能源局综合司关于开展风电清洁供暖工作的通知 国能综新能[2015]306号	
NEA, 2016	13th FYP development plan for biomass energy	生物质能发展"十三五"规划	http://chinaenergyportal.org/en/13th-fyp-development-plan-for-biomass-energy/
NEA, 2016	13th FYP development plan for hydro power	水电发展"十三五"规划	http://chinaenergyportal.org/en/13th-fyp-hydro-power/
NEA, 2016	13th FYP development plan for wind power	风电发展"十三五"规划	http://chinaenergyportal.org/en/13th-fyp-development-plan-for-wind-power/
NEA, 2016	13th FYP for the electricity sector — Press conference transcript	《电力发展"十三五"规划》新闻发布会	http://www.nea.gov.cn/xwfb/20161107zb1/index.htm
NEA, 2016	2015 National renewable power development monitoring and evaluation report	国家能源局关于2015年度全国可再生能源电力发展监测评价的通报 国能新能[2016]214号	
NEA, 2016	2015 PV Statistics	2015年光伏发电相关统计数据	http://chinaenergyportal.org/en/2015-pv-installations-utility-and-distributed-by-province/
NEA, 2016	2015 wind power installations and production by province	2015年风电产业发展统计数据	http://chinaenergyportal.org/en/2015-wind-installations-and-production-by-province/
NEA, 2016	Administrative tasks for the guaranteed full purchase of renewable electric power	关于做好风电、光伏发电全额保障性收购管理工作的通知(发改能源[2016]1150号)	
NEA, 2016	Circular on further regulating the planning and construction of coal fired power plants	关于进一步调控煤电规划建设的通知 国能电力[2016]275号	
NEA, 2016	Circular on the construction of solar thermal power generation demonstration projects	国家能源局关于建设太阳能热发电示范项目的通知 国能新能[2016]223号	
NEA, 2016	Establishment of monitoring and early warning mechanisms to promote the sustained and healthy development of the wind power industry	国家能源局关于建立监测预警机制促进风电产业持续健康发展的通知 国能新能[2016]196号	http://chinaenergyportal.org/establishment-of-monitoring-and-early-warning-mechanisms-to-promote-the-sustained-and-healthy-development-of-the-wind-power-industry/
NEA, 2016	Measures for the guaranteed full purchase of renewable electric power	关于印发《可再生能源发电全额保障性收购管理办法》的通知(发改能源[2016]625号)	
State Council, 2013	12th FYP development plan for energy	能源发展"十二五"规划	
State Council, 2016	13th FYP period work program for controlling greenhouse gas emissions	"十三五"控制温室气体排放工作方案	

PERMISSIONS

All chapters in this book were first published in ESE, by John Wiley & Sons Ltd.; hereby published with permission under the Creative Commons Attribution License or equivalent. Every chapter published in this book has been scrutinized by our experts. Their significance has been extensively debated. The topics covered herein carry significant findings which will fuel the growth of the discipline. They may even be implemented as practical applications or may be referred to as a beginning point for another development.

The contributors of this book come from diverse backgrounds, making this book a truly international effort. This book will bring forth new frontiers with its revolutionizing research information and detailed analysis of the nascent developments around the world.

We would like to thank all the contributing authors for lending their expertise to make the book truly unique. They have played a crucial role in the development of this book. Without their invaluable contributions this book wouldn't have been possible. They have made vital efforts to compile up to date information on the varied aspects of this subject to make this book a valuable addition to the collection of many professionals and students.

This book was conceptualized with the vision of imparting up-to-date information and advanced data in this field. To ensure the same, a matchless editorial board was set up. Every individual on the board went through rigorous rounds of assessment to prove their worth. After which they invested a large part of their time researching and compiling the most relevant data for our readers.

The editorial board has been involved in producing this book since its inception. They have spent rigorous hours researching and exploring the diverse topics which have resulted in the successful publishing of this book. They have passed on their knowledge of decades through this book. To expedite this challenging task, the publisher supported the team at every step. A small team of assistant editors was also appointed to further simplify the editing procedure and attain best results for the readers.

Apart from the editorial board, the designing team has also invested a significant amount of their time in understanding the subject and creating the most relevant covers. They scrutinized every image to scout for the most suitable representation of the subject and create an appropriate cover for the book.

The publishing team has been an ardent support to the editorial, designing and production team. Their endless efforts to recruit the best for this project, has resulted in the accomplishment of this book. They are a veteran in the field of academics and their pool of knowledge is as vast as their experience in printing. Their expertise and guidance has proved useful at every step. Their uncompromising quality standards have made this book an exceptional effort. Their encouragement from time to time has been an inspiration for everyone.

The publisher and the editorial board hope that this book will prove to be a valuable piece of knowledge for researchers, students, practitioners and scholars across the globe.

LIST OF CONTRIBUTORS

Anton Eberhard
Graduate School of Business, University of Cape Town, Private Bag X3, Rondebosch 7701, South Africa

Tomas Kåberger
Energy and Environment, Chalmers University of Technology, SE-412 96 Gothenburg, Sweden

Radhakumari Muktham
Chemical Engineering Division, CSIR – Indian Institute of Chemical Technology, Hyderabad 500007, India
School of Applied Sciences, Royal Melbourne Institute of Technology, Melbourne 3083, Australia

Andrew S. Ball and Suresh K. Bhargava
School of Applied Sciences, Royal Melbourne Institute of Technology, Melbourne 3083, Australia

Satyavathi Bankupalli
Chemical Engineering Division, CSIR – Indian Institute of Chemical Technology, Hyderabad 500007, India

Sonil Nanda and Janusz A. Kozinski
Department of Earth and Space Science and Engineering, Lassonde School of Engineering, York University, Toronto, Ontario, Canada

Sivamohan N. Reddy
Department of Chemical Engineering, Indian Institute of Technology Roorkee, Roorkee, Uttarakhand, India

Sushanta K. Mitra
Department of Mechanical Engineering, Lassonde School of Engineering, York University, Toronto, Ontario, Canada

Thomas Parker
WA3RM AB, Lund, Sweden

Anders Kiessling
Swedish University of Agricultural Sciences, Uppsala, Sweden

Nies Reininghaus, Clemens Feser, Benedikt Hanke, Martin Vehse and Carsten Agert
NEXT ENERGY EWE Research Centre for Energy Technology, University of Oldenburg, Carl-von-Ossietzky-Str. 15, 26129 Oldenburg, Germany

Yanwei Sun and Jialin Li
School of Architectural Civil Engineering and Environment, Ningbo University, Ningbo 315211, China

Run Wang
School of Resources and Environment, Hubei University, Wuhan 430062, China

Jian Liu
Zhejiang Academy of Social Science, Hangzhou 310007, China

Swati Takiyar and K. G. Upadhyay
M.M.M. University of Technology, Gorakhpur, Uttar Pradesh, India

Vivek Singh
National Hydroelectric Power Corporation Ltd., Faridabad, Haryana, India

Yales Vivadinar, Widodo W. Purwanto and Asep H. Saputra
Department of Chemical Engineering, Universitas Indonesia, Depok, Indonesia

Jiekang Wu, Zhishan Wu and Xiaoming Mao
School of Automation, Guangdong University of Technology, Guangzhou, Guangdong 510006, China

Fan Wu
Guangxi Bo Yang Electric Power Survey and Design Co., Ltd., Guangxi Power Grid Co., Ltd., Nanning, Guangxi 530023, China

Weiming Xiong, Yuanzhe Yang, Yu Wang and Xiliang Zhang
Institute of Energy, Environment and Economy, Tsinghua University, 100084 Beijing, China

Yanqun Yu
College of Engineering, Ocean University of China, Qingdao, Shandong Province 266100, China
College of Mechanical and Electronic Engineering, China University of Petroleum, Qingdao, Shandong Province 266580, China

Zongyu Chang
College of Engineering, Ocean University of China, Qingdao, Shandong Province 266100, China

Yaoguang Qi and Xin Xue
College of Mechanical and Electronic Engineering,
China University of Petroleum, Qingdao, Shandong
Province 266580, China

Jiannan Zhao
School of Petroleum Engineering, China University
of Petroleum, Qingdao, Shandong Province 266580,
China

Francesco Meneguzzo and Lorenzo Albanese
Institute of Biometeorology, CNR, via Caproni 8,
50145 Firenze, Italy

Rosaria Ciriminna and Mario Pagliaro
Institute for the Study of Nanostructured Materials,
CNR, via U. La Malfa 153, 90146 Palermo, Italy

Zhou Luyao
National Thermal Power Engineering and
Technology Research Center, North China Electric
Power University, Beijing 102206, China
Jiangsu Maritime Institute, Nanjing, 210000, China

Xu Cheng, Xu Gang, Bai Pu and Yang Yongping
National Thermal Power Engineering and
Technology Research Center, North China Electric
Power University, Beijing 102206, China

Neha Gupta
Centre for Energy Studies, Indian Institute of
Technology Delhi Hauz Khas, New Delhi 110016,
India

Gopal N. Tiwari
Bag Energy Research Society (BERS), 11B, Gyan
Khand IV, Indirapuram, Ghaziabad 201010, Uttar
Pradesh, India

Tobias Kohler and Karsten Müller
Institute of Separation Science and Technology,
Friedrich-Alexander-Universität Erlangen-Nürnb-
erg, Egerlandstr. 3, 91058 Erlangen, Germany

Jorrit Gosens
Department of Technology Management and
Economics, Division of Environmental Systems
Analysis, Chalmers University of Technology, Vera
Sandbergs Allé 8, SE-412 96 Göteborg, Sweden

Tomas Kåberger
Department of Space, Earth and Environment,
Division of Physical Resource Theory, Chalmers
University of Technology, Rännvägen 6B, SE-412
96 Göteborg, Sweden

Yufei Wang
Development Research Center of the State Council,
Zhonghua Dizhi Dasha, Hepingjie 13 Jia 20, 100013
Chaoyang, Beijing, China

Index

www.ingramcontent.com/pod-product-compliance
Lightning Source LLC
Chambersburg PA
CBHW082033190326
41458CB00010B/3351